# DUSTY OBJECTS
# IN THE UNIVERSE

# ASTROPHYSICS AND SPACE SCIENCE LIBRARY

A SERIES OF BOOKS ON THE RECENT DEVELOPMENTS
OF SPACE SCIENCE AND OF GENERAL GEOPHYSICS AND ASTROPHYSICS
PUBLISHED IN CONNECTION WITH THE JOURNAL
SPACE SCIENCE REVIEWS

PROCEEDINGS
VOLUME 165

# DUSTY OBJECTS IN THE UNIVERSE

PROCEEDINGS OF THE FOURTH INTERNATIONAL WORKSHOP
OF THE ASTRONOMICAL OBSERVATORY OF CAPODIMONTE
(OAC 4), HELD AT CAPRI, ITALY, SEPTEMBER 8–13, 1989

Edited by

## E. BUSSOLETTI

*Istituto Universitario Navale, Napoli, Italy*

and

## A. A. VITTONE

*Osservatorio Astronomico di Capodimonte, Napoli, Italy*

## KLUWER ACADEMIC PUBLISHERS

DORDRECHT / BOSTON / LONDON

Library of Congress Cataloging in Publication Data

```
Osservatorio astronomico di Capodimonte.  International Workshop (4th
  : 1989 : Capri, Italy)
   Dusty objects in the universe / edited by Ezio Bussoletti and
Alberto Angelo Vittone.
      p.   cm. -- (Astrophysics and space science library ; 165.)
   "Proceedings of the Fourth International Workshop of the
Astronomical Observatory of Capodimonte (OAC4), held at Capri,
Italy, September 8-13, 1989."
   Includes index.

   1. Cosmic dust--Congresses.  2. Astrophysics--Congresses.
3. Cosmochemistry--Congresses.   I. Bussoletti, Ezio, 1947-
II. Vittone, Alberto.  III, Title.  IV. Series: Astrophysics and
space science library ; v. 165.
QB791.085  1989
523.1'125--dc20                                            90-40285
```
ISBN-13: 978-94-010-6782-9        e-ISBN-13: 978-94-009-0661-7
DOI: 10.1007/978-94-009-0661-7

Published by Kluwer Academic Publishers,
P.O. Box 17, 3300 AA Dordrecht, The Netherlands.

Kluwer Academic Publishers incorporates
the publishing programmes of
D. Reidel, Martinus Nijhoff, Dr W. Junk and MTP Press.

Sold and distributed in the U.S.A. and Canada
by Kluwer Academic Publishers,
101 Philip Drive, Norwell, MA 02061, U.S.A.

In all other countries, sold and distributed
by Kluwer Academic Publishers Group,
P.O. Box 322, 3300 AH Dordrecht, The Netherlands.

*Printed on acid-free paper*

# TABLE OF CONTENTS

## POSTERS

# EDITORS' FOREWORD

Solid matter in space is crucial in accounting for many processes. In these last years a great improvement of the general knowledge of the problem has been possible due to the increase, in number and quality, of observations and of the laboratory efforts to simulate "cosmic" dust. Theoreticians have also given their contribution in solving some questions and in posing others.

Continuing an effort started in 1987, the Astrophysical Groups operating in Naples have considered it useful to organize a Workshop in Capri from September $8^{th}$ to $13^{th}$ 1989 in order to assess the state of the art in such an interesting field as that of "Dusty Objects in the Universe".

The subject is, obviously, too wide to be discussed in detail. However, the presence of many experts in different areas provided an updated general picture. Laboratory work has been reviewed, as well as recent observations and theoretical interpretations which concern processes occurring in our Galaxy and in external galaxies.

The Workshop gave to participants the unique opportunity of exchanging ideas and of discussing on different themes. In addition, the meeting has been enriched by the presentation of the major technological projects related to ongoing airborne and space missions.

Invited papers were prepared with the aim of giving the state of the art about specific subjects; contributed papers and some selected poster papers presented, on the contrary, very recent results in the various fields.

In addition to the scientific efforts, the Capodimonte Observatory and the Istituto Universitario Navale sought to give a warm welcome to the participants. Thanks to several sponsors, the LOC could organize some excursions and shows to entertain people during their spare time.

In particular, the LOC wishes to thank for their contributions, the Italian Ministry of University and Scientific Research, the Regione Campania, Aeritalia Space Systems Groups and Selenia Spazio.

A special thank goes to Mrs G. Iaccarino, Mr G. Cuccaro, Mr R. Trentarose of the OAC and to Miss E. Pignatiello, Mr P. Adamo of the Istituto Universitario Navale who have contributed to solve any logistical problem favouring the success of the Workshop.

A very special thank is due to Mrs A. D'Orsi and E. Acampa for their valuable help in preparing these proceedings.

## Scientific Organizing Committee

F. Bertola – Italy
E. Bussoletti – Italy
J.M. Greenberg – The Netherlands
M. Jura – USA
D. Lemke – FRG
J.S. Mathis – USA
P.G. Mezger – FRG
J.L. Puget – France
M. Rowan-Robinson – UK
K. Sellgren – USA
A.A. Vittone – Italy


## Local Organizing Committee

E. Covino
D. de Martino
C. Fusco
D. Mancini
A.A. Vittone (Chairman)

# LIST OF PARTICIPANTS

BAR-NUN A., University Tel Aviv, Israel
BARSONY M., University Berkeley, USA
BERTOLA F., University of Padova, Italy
BLANCO A. University of Lecce, Italy
BOHME D.K., York University, Canada
BUSARELLO G., Astronomical Observatory, Naples, Italy
CAMERON M., Max Plank Institut für Extraterrestrische Physik, FRG
CASACCI C., Gruppo Sistemi Spaziali, Torino, Italy
COLANGELI L., University of Cassino, Italy
CORCIULO G.G., Astronomical Observatory of Naples, Italy
COVINO E., Astronomical Observatory of Naples, Italy
DE MARTINO D., Astronomical Observatory of Naples, Italy
DE SIMONE V., University of Naples, Italy
EFSTATHIOU A., Queen Mary College, University of London, UK
EVANS N., University of Keele, UK
EVANS Rh., University of Wales College, Cardiff, UK
FERNANDEZ M., Observatorio Astronomico de Madrid, Spain
FERRARA G., Istituto Universitario Navale, Naples, Italy
FONTI S., Istituto Universitario Navale, Naples, Italy
FULLE M., Astronomical Observatory of Trieste, Italy
FUSCO C., Istituto Universitario Navale, Naples, Italy
GIARD M., Institut d'Astrophisique Spat., LPSP/IAS, Toulouse, France
GIURICIN G., University of Trieste, Italy
GOMEZ M.T., Astronomical Observatory, Naples, Italy
GREENBERG J.M., University of Leiden, Holland
HASHIMOTO O., University of Tokyo, Japan
JURA M., University of California, Los Angeles, USA
KESSLER M., Astronomy Division, ESA/ESTEC, Noordwijk, The Netherlands
KRÄTSCHMER W., Max Plank Institut für Kernphysik, Heidelberg, FRG
LAGAGE P.O., Service d'Astrophysique, CEN Saclay, France
LEMKE D., Max Planck Institut für Astronomie, Heidelberg, FRG
MANDOLESI R., CNR TESRE, Bologna, Italy
MENNELLA V., Astronomical Observatory, Naples, Italy
MEZGER P.G., Max Plank Institut für Radioastronomie, Bonn, FRG
NAPIER B., Royal Observatory Edinburg, Scotland
NAPOLITANO L., University of Naples, Italy
OROFINO V., University of Naples, Italy
PIRRONELLO V., University of Calabria, Italy
RIFATTO A., University of Padova, Italy
RIGUTTI M., Astronomical Observatory, Naples, Italy
ROWAN-ROBINSON M., Queen Mary College University of London, UK
SAKATA A., University of Electro-Communications, Tokyo, Japan
SALES N., CESR/CNRS, Toulouse, France

SELLGREN K., University of Hawaii, USA
SERRA G., CESR-CNRS/UPS, Toulouse, France
STRAZZULLA G., University of Catania, Italy
SVESTKA J., Prague Observatory, Czechoslovakia
VITTONE A.A., Astronomical Observatory, Naples, Italy
VULTAGGIO M., Istituto Universitario Navale, Naples, Italy
WDOWIAK T.J., University of Alabama at Birmingham (UAB), USA
WICKRAMASINGHE N.C., University of Wales College of Cardiff, UK
ZEILINGER W., University of Padova, Italy

# CARBONACEOUS SOLID MATERIALS: FROM LABORATORY TO SPACE

A. Blanco[1], E. Bussoletti[2,3], L. Colangeli[4], S. Fonti[2], C. Fusco[2],
V. Mennella[3], V. Orofino[1], A. Vittone[3]

[1]*Physics Department, University of Lecce, Lecce, Italy*
[2]*Istituto Universitario Navale, Naples, Italy*
[3]*Osservatorio Astronomico di Capodimonte, Naples, Italy*
[4]*Engineering Faculty, University of Cassino, Cassino, Italy*

ABSTRACT. We present here an updated review of laboratory results obtained from different carbonaceous materials in grain and molecular forms. Temperature behaviour of bands falling between 3 $\mu$m and 13 $\mu$m is reported for some PAHs. FIR spectra are measured for ten different solids. Finally, we have also identified several IR space sources which can be studied by means of ISO instruments to contribute a possible clarification on UIBs.

## 1 Unidentified IR bands in space and in laboratory

The family of the so-called "unidentified infrared bands" (UIBs) has been observed in emission from a wide variety of galactic and extragalactic sources as well as from comets Halley, Wilson and Bradfield, and in absorption towards strong galactic IR sources as GC-IRS 7. Most of these bands are considered as the "fingerprints" of C-C and C-H resonances in carbon-based species (see Sellgren, these proceedings). Various kinds of carbonaceous materials have been proposed in the past to interpret UIBs: 1) small carbon grains with functional groups attached to their surface (Duley and Williams, 1981); 2) hydrogenated amorphous carbon (HAC) particles (Borghesi et al., 1987; Goebel, 1987); 3) collections of single complex polycyclic aromatic hydrocarbon molecules –PAHs– (Léger and Puget, 1984; Allamandola et al., 1985); 4) quenched carbonaceous composite (QCC) (Sakata and Wada, 1989); 5) coal tar (Wdowiak, 1987); vitrinites (Papoular et al., 1989). Although different in physical and chemical properties, these materials have in common the presence of carbon and hydrogen as major elemental components so that they can produce features similar to the UIBs observed in space. However, the optical properties of each of these compounds change according to some specific parameters such as the dimension and the internal structure. In a growing dimension scale PAHs occupy the lowest stage as they are free molecules, while coal tar, QCC, vitrinites and HAC submi-cron grains are solids, although their physical and optical propeties can be considered in some way "intermediate" between molecules and bulk solids. In terms of internal

1

*E. Bussoletti and A. A. Vittone (eds.), Dusty Objects in the Universe, 1–8.*

structure PAHs are, in principle, simple graphite-like layers, while coal tar can be considered a collection of various PAH molecules. QCC, HAC grains and vitrinites are characterized by more or less "disordered" structure, although on a short scale ($\simeq$10 Å) "islands" of C atoms arranged in $sp^2$ (graphite-like) or $sp^3$ (diamond-like) coordination can be present, as also suggested by both the interpretation of extinction measurements (Colangeli et al., 1989a) and by Raman spectroscopy (Fonti et al., 1989). In space it appears reasonable that carbonaceous grains and PAHs probably are formed and cohexist together in the outflows of carbon sources and a continnum may occur from PAHs to disordered carbon grains so that UIBs can be probably produced by a mixture of both these two components. These hypotheses seem also supported from laboratory results. A comparative analysis of the wavelengths of occurrence for C-C and C-H resonances in absorption spectra of various types of PAHs, PAH mixtures and HAC grains (see table 1) suggests that a combination of various components is required to match the astronomical spectra (Blanco et al., 1988a). HAC grains lack some of the UIBs, while coal tar and PAH collections show also features not detected in space; on the contrary, mixtures of HAC grains and PAHs resemble the observations much better than the two separate components. Nevertheless, many questions remain still unsolved, due also to the present limitations of laboratory data available on the candidate materials. In particular we have to stress that: a) it is not clear to which extent "absorption" data can be adapted to simulate features detected in "emission" in space conditions; b) the actual mechanism(s) of emission are still not well defined; c) it is unclear how physical parameters such as, for example, the temperature can affect the optical properties of carbonaceous materials. One relevant problem of the attribution of UIBs to a specific class of carbonaceous compounds concerns the 3.3–3.4 $\mu$m bands. In fact, according to table 1 a variety of aromatic ($sp^2$ configuration) and aliphatic ($sp^3$ configuration) resonances fall in this wavelength range. Laboratory data show that the relative intensity, R=I(3.3 $\mu$m)/I(3.4 $\mu$m), between the 3.3 $\mu$m (aromatic) and the 3.4 $\mu$m (aliphatic) bands, can vary from R < 1 to R > 1, both when the "rank" of vitrinites increases and when HAC and QCC are annealed at temperatures higher than about 500° C (Colangeli et al., 1989b; Sakata and Wada, 1989). This behaviour might provide a "key" to solve the puzzling identification of the 3.3–3.4 $\mu$m bands in space. In fact, astronomical observations towards high temperature environments, such as HII Regions and Planetary Nebulae (T $\simeq$ 1000 K), show that: 1) a dominant 3.3 $\mu$m band and a shoulder at 3.4 $\mu$m are often observed; 2) however, R can vary in a wide range, from source to source; 3) in four Planetary Nebulae R decreases from 1.7 to 1.3 with the age of the sources (Magazzú and Strazzulla, 1989); this change can be attributed to modifications in the nature of the carriers, possibly induced by ion irradiation; 4) the Nova Cen 1986 (V842 CEN, T $\simeq$ 800 K) shows R $\simeq$ 1.7 (Hyland and McGregor, 1989). On the contrary, in cooler environments, such as the galactic center source GC-IRS7 (Butchart et al., 1986) and comets Halley, Wilson and Bradfield (Brooke et al., 1989) a band at 3.4 $\mu$m appears well dominant over the 3.3 $\mu$m feature. The comparison of laboratory results with astronomical data could suggest that "different states" of the same basic material allow to explain the evolution of the 3.3–3.4 $\mu$m bands observed in various environments: the hotter the environment, the higher the aliphatic degree of the carbonaceous materials. Of course, this approach does not account for fine changes in the band profiles, which have to be understood better in terms of local evolutionary changes (Colangeli et al., 1989b).

# 2   Temperature behaviour of PAHs absorption bands

PAH molecules are considered among the most likely candidates to explain the UIBs. They are believed to exist in space in the ionized state (Allamandola et al., 1985), partially hydrogenated (Cohen et al., 1985; 1989) and widespread in a large variety of astrophysical conditions. Up to now only room temperature absorption properties of some neutral PAH species have been used to simulate the astronomical spectra and to infer the physical and chemical properties of the interstellar PAH species (Léger and Puget, 1984; Allamandola et al., 1985; de Muizon et al., 1987; Léger and d'Hendecourt, 1987; Léger et al., 1989). This was done under the implicit assumption that optical properties of PAHs do not depend on temperature. To test this hypothesis Blanco et al. (1988b) measured the transmittance of three different representative PAH molecules such as coronene ($C_{24}H_{12}$), chrysene ($C_{18}H_{12}$) and 1-methylcoronene ($C_{25}H_{14}$), in the temperature range 300÷520 K. We report in figure 1 an updated set of results obtained in subsequent measurements made in order to confirm our previous findings. Similar measurements on coronene have also been made by Bernard et al. (1989) who do not find substantial changes in any band intensity.

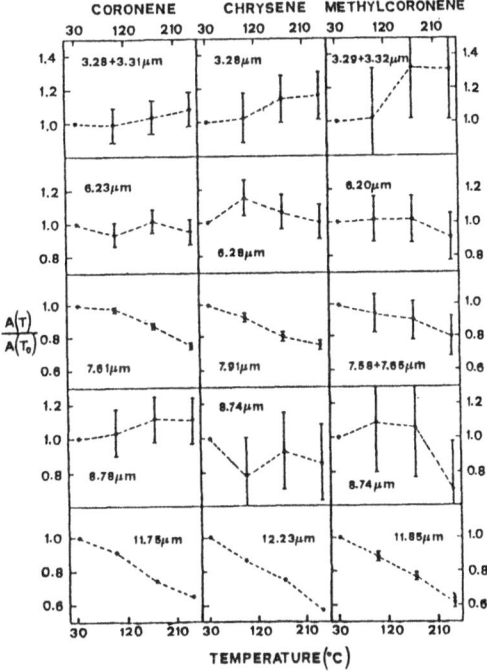

**Figure 1:** Relative integrated absorbance at different temperatures of some bands falling at UIBs wavelenghts and present in the IR spectrum of the PAH molecules measured in the laboratory. The matrix is KBr.

On the other hand, very recent results by Wdowiak (this volume) are in basic agreement with our data. From table 2 we note that, apart from the feature at 7.7 $\mu$m, whose assignment is doubtful and discussed in detail elsewhere (Blanco et al., 1990a), all the other bands showing a consistent change in intensity are commonly assigned to out-of-plane bending vibrations. We tentatively interpret this behaviour as due to interactions between out-of-plane groups and surrounding molecular environment, i.e. to solid state effects. Since the present measurements have been performed on PAHs embedded in KBr and CsI matrices, our results cannot be extrapolated to situations where PAHs are in the gas phase. However, taking into account that in space PAHs may exist in the solid phase too, attached to and/or embedded in solid carbonaceous grains, our data can be used to clarify some problems concerning UIBs.

## 3  UIBs and space observations with ISO

Future satellite observations with high spectral and spatial resolution capabilities will be of great help to try to clarify some open problems about UIBs carriers. Hints for such problems come both from laboratory and from astronomical observations:

a) changes in the ratio between the integrated intensities of the 11.3 and 3.3 $\mu$m features can be investigated at different positions in the same source and in various sources at different physical conditions. This to check the hypothesis of the temperature dependence of such ratio;

b) similar investigations can be carried out in the $3.3 \div 3.6$ $\mu$m region, where laboratory spectra of carbonaceous materials show a family of structures whose intensities change according to the ambient conditions;

c) a new feature at 5.2 $\mu$m has been recently observed in a planetary nebula (Cohen et al., 1989). Since many PAH spectra show also this feature, its detection in other sources can be important to strengthen the hypothesis that PAHs are the carriers for the UIBs;

d) the position and shape of the 7.7 $\mu$m band can be thoroughly studied in order to correlate this feature with the physical conditions existing in the sources (Cohen et al., 1989);

e) a careful analysis of the $11 \div 13$ $\mu$m spectral region, where several minor features seem to be present, can help in clarifying the problem of hydrogenation of UIBs carriers (Tielens et al., 1987).

The Infrared Satellite Observatory (ISO), will be an ideal tool to study also the above mentioned problems. The on board instrumentation (Lemke and Kessler, this volume) provides the required high spectral and spatial resolution, together with a quite good signal to noise ratio. In this framework possible selected targets observable with ISO are reported in table 3, together with information on their sizes and fluxes at 3, 7, and 11 $\mu$m. The most suitable instruments to perform the observations are also listed for each source in the last column.

# 4  FIR spectral trends of various carbonaceous materials

In order to increase the knowledge of optical properties of carbon based materials which can be formed in the atmosphere of C-rich objects, we have performed FIR transmission measurements on several kinds of astrophysically interesting particles, by using a SPECAC Polarizing Michelson Interferometer [spectral range = 40 $\mu$m ÷ 1000 $\mu$m]. Similar measurements were done in the past by Tanabé et al. (1983), Borghesi et al. (1986) and Bussoletti et al. (1987) up to 300 $\mu$m. The materials examined are:

a) two types of $\beta$-SiC (Silicon Carbide) and three types of $\alpha$-SiC (SiC N, SiC 1200, SiC 600);

b) three types of amorphous carbon obtained by burning benzene in air (BE), by arc discharge between amorphous carbon electrodes (AC) and graphite electrodes (GR-AR). Particle size and morphological properties as well as details of the production methods are reported elsewhere (Bussoletti et al., 1987, Fonti et al., 1989);

c) two types of commercial graphite powder.

The results concerning the FIR spectral indices are summarized in table 4, while the complete set of data, together with a detailed discussion of the measurements and astrophysical implications, are reported elsewhere (Blanco et al., 1990b). The results presented here are basically in agreement with those already reported by other authors. We want to note, however, that the spectral index of a given material can change considerably from one sample to the other. These variations are probably due both to the purity degree of the sample and to the nature and relative abundances of the impurities. These findings can be related to the wide variation in the slopes of observed FIR fluxes of astrophysical sources (Sopka et al., 1985, Jura, 1986). Full advantages of the present laboratory measurements will be possible, however, when a large number of sources will be observed, with higher spectral resolution, in a range extended towards longer wavelength.

**Table 1:** Identification of IR bands in space and in laboratory.

| Observations | | | | Laboratory | | | Assignment |
|---|---|---|---|---|---|---|---|
| UIBs | IRS7 | P/Halley | HAC | CHAR | COAL TAR | CHAR +HAC | |
| | | | | $\lambda$ ($\mu$m) | | | |
| 3.28 | 3.29 | 3.28 | | 3.27 | 3.29 | | 3.28: sp$^2$CH aromatic |
| | | | | | | | 3.31: sp$^2$CH$_2$ olefinic |
| | | | | | | | 3.33: sp$^2$CH olefinic |
| | | | | 3.38 | | 3.38 | 3.38: sp$^3$CH$_3$ asym. |
| | | | | 3.39 | | | 3.39: sp$^2$CH$_2$ olefinic |
| 3.4 | 3.4 | 3.4 | | 3.42 | | 3.42 | 3.42: sp$^3$CH$_2$ asym. |
| | | | | 3.44 | | 3.44 | 3.43: sp$^3$CH |
| | 3.48 | | | 3.48 | | | 3.48: sp$^3$CH$_3$ sym. |
| 3.51 | | 3.51 | 3.50 | | 3.50 | 3.5 | 3.51: sp$^3$CH$_2$ sym. |
| 5.62 | | | 5.78 | 5.88 | | 5.86 | CO stretching in -CHO |
| 6.29 | | | 6.29 | 6.19 | 6.25 | 6.25 | skeletal C=C in plane |
| 6.9 | | | 6.85 | 6.94 | 6.94 | 6.94 | asym. deform. in -CH$_3$ |
| 7.27 | | | 7.27 | 7.27 | 7.25 | 7.27 | CH roking in -CHO sym. deform. in -CH$_3$ |
| 7.7 | | | | 7.6 | 7.94 | 7.6 | skeletal C=C |
| 8.6 | | | | 8.62 | | | in plane aromatic CH bending |
| | | | | 9.01 | | | |
| | | | | 9.43 | | | |
| | | | | 9.66 | 9.66 | | |
| 11.3 | | | 11.3 | 11.4 | 11.4 | 11.4 | out-of-plane arom. CH bending (1 adjacent H) |
| | | | | | 11.9 | | "    (2  "  H) |
| | | | | | 12.1 | | "    (2  "  H) |
| | | | | | 12.8 | 13.2 | "    (3  "  H) |

Note: CHAR is an aliphatic PAHs mixture.

6

**Table 2:** PAHs absorption bands and their assignments.

| $\nu(cm^{-1})$ | $\lambda(\mu m)^{(x)}$ | Assignment |
|---|---|---|
| 3030 | 3.3 | aromatic C-H stretch |
| 1615 | 6.2 | aromatic C-C stretch |
| 1490 | 6.7(*) | C-C vibrational modes in non compact PAHs (chrysene) |
| 1300 | 7.7(*) | aromatic C-C stretch |
| 1150 | 8.7 | aromatic C-H in plane bend |
| 950-740 | 10.5-13.5 (*) | aromatic C-H out-of-plane bending modes |
| 545 | 18.3 (*) | skeleted C-C-C out-of-plane bending |

(x) Wavelengths are purely indicative since they can slightly vary from one molecule to the other.
(*) Bands with a definite variation with temperature

**Table 3:** Objects observable with ISO.

| NAME | | Angular Size | $F_\nu$(Jy) $3\mu m$ | $7\mu m$ | $11\mu m$ | Instrument |
|---|---|---|---|---|---|---|
| **A) PLANETARY NEBULAE** | | | | | | |
| A30 | | 130" | 0.3 | 1 | 3 | Isophot-S |
| IC418 | | 11" | 0.5 | 2 | 10 | Isophot-S+SWS |
| J900 | | 8" | ... | ... | 1 | Isophot-S |
| M1-11 | | ... | 0.5 | ... | 7 | Isophot-S+SWS |
| NGC 6572 | | 14" | 0.3 | ... | 2 | Isophot-S |
| BD+30°3639 | | 6" | ... | 20 | 50 | SWS |
| NGC 7027 | | 12" | 2 | 70 | 200 | SWS |
| IC 5117 | | < 1" | 0.3 | ... | 10 | Isophot-S+SWS |
| CPD-56°8032 | | ... | 5 | 70 | 100 | SWS |
| He 2-113 | | ... | ... | 30 | ... | SWS |
| NGC 6302 | | 54" | ... | 0.1 | ... | Isophot-S |
| NGC 6790 | | ... | ... | 2 | 5 | Isophot-S+SWS |
| **B) REFLECTION NEBULAE** | | | | | | |
| NGC 7023 | | 4'x2' | 0.02 | ... | 1 | Isophot-S |
| NGC 2023 | | 6'x6' | 0.02 | ... | 0.6 | Isophot-S |
| NGC 2068 | | 6'x5' | 0.02 | ... | 0.3 | Isophot-S |
| HD 44179 | | 40' | 60 | ... | 300 | SWS |
| P18 | | ... | ... | ... | ... | Isophot-S+SWS |
| IRAS 05044-0325 | | ... | ... | 2 | 5 | Isophot-S+SWS |
| IRAS 08513-4201 | | ... | ... | 6 | 3 | Isophot-S+SWS |
| **C) HII REGIONS** | | | | | | |
| NGC 1976 | | 5'x5' | > 2 | ... | >10 | SWS+Isophot-S |
| NGC 6618 | | 4' | 2 | ... | ... | SWS+Isophot-S |
| GL437 | | 18' | 0.2 | ... | 10 | Isophot-S+SWS |
| M1-78 | | ... | 0.3 | ... | 30 | Isophot-S+SWS |
| G 333.6-0.2 | | ... | ... | 100 | 700 | SWS |
| RCW 108 | | ... | ... | 10 | ... | SWS |
| MWC 922 | | ... | ... | 100 | ... | SWS |
| **D) OTHER SOURCES** | | | | | | |
| M82 | G | 7'x2' | 0.4 | ... | 4 | Isophot-S+SWS |
| R22 | PPN | ... | ... | 30 | ... | SWS |
| HR 4049 | PPN | ... | ... | 50 | 20 | SWS |
| He 2-77 | HII/NP | ... | ... | 3 | ... | Isophot-S+SWS |
| V 842 Cen | NOVA | ... | 0.6 | ... | ... | Isophot-S+SWS |

NOTES: G = Galaxy; PPN = Proto Planetary Nebula; HII/PN= HII region or Planetary Nebula.

**Table 4:** FIR spectral indices of the samples analysed in this work see text) at different wavelength intervals (in $\mu$m).

| Sample | $\lambda_{max}$ | $40-\lambda_{max}$ | 40-300 | Other works |
|---|---|---|---|---|
| $\beta$-SiC (1) | 800 | 0.9 | 0.8 | 0.8[a], 1.10[b] |
| $\beta$-SiC (2) | 800 | 1.2 | 1.3 | 1.37[c] |
| $\alpha$-SiC N (1) | 900 | 1.4 | 1.4 | |
| $\alpha$-SiC 600 (1) | 900 | 1.3 | 1.3 | 1.4[b] |
| $\alpha$-SiC 1200 (1) | 900 | 0.8 | 0.8 | 0.8[a], 1.0[b] |
| BE | 700 | 0.9 | 0.8 | 0.7[d] |
| AC | 1000 | 1.0 | 1.0 | 0.6[c], 0.9[d] |
| GR-He | 900 | 1.2 | 1.2 | |
| Graphite (3) | 800 | 2.1 | 2.0 | 2.18[c] |
| Graphite (4) | 700 | 2.1 | 2.1 | |

(1) From Elektroschmeltzwerk GmBH; (2) From Showa Denko; (3) From Ultra Carbon; (4) From Ventron.

(a) Borghesi et al. (1986) in the 40-100 $\mu$m range; (b) Borghesi et al. (1986) in the 100-300 $\mu$m range; (c) Tanabè et al. (1983) in the 40-250 $\mu$m range; (d) Bussoletti et al. (1987) in the 30-300 $\mu$m range.

# Acknowledgments

This work is finantially supported partially by grants of MPI and of A.S.I. (Italian Space Agency) under the contracts PSN-88-RB-020 and ASI-88-RB-060.

# References

Allamandola, L.J., Tielens, A.G.G.M., Barker, J.R.: 1985, Astrophys. J. Lett., **290**, L25.

Bernard, J.P., D'Hendecourt, L.B., Léger, A.: 1989, Astron. Astrophys., **220**, 245.

Blanco, A., Bussoletti, E., Colangeli, L.: 1988a, Astrophys. J., **334**, 875.

Blanco, A., Borghesi, A., Fonti, S., Orofino, V., Bussoletti, E., Colangeli, L.: 1988b, in "Dust in the Universe" (M.E. Bailey and D.A. Williams, eds.), p. 287, Cambridge University Press.

Blanco, A., Fonti, S., Orofino, V.: 1990a, Astrophys. J., submitted.

Blanco, A., Fonti, S., Fusco, C., Rizzo, F.: 1990b, Infrared Phys., in press.

Borghesi, A., Bussoletti, E., Colangeli, L.: 1987, Astrophys. J. **314**, 422.

Borghesi, A., Bussoletti E., Colangeli, L., Orofino, V., Guido, M., Nunziante-Cesero, S.: 1986, Infrared Phys., **26**, 37.

Brooke, T.Y., Knacke, R.F., Owen, T.C., Tokunaga, A.T.: 1989, NASA CP 3036 (L.J. Allamandola and A.G.G.M. Tielens, eds.), in press.

Bussoletti, E., Colangeli, L., Borghesi, A., Orofino, V.: 1987, Astron. Astrophys. Suppl., **70**, 257.

Butchart, I., McFadzean, A.D., Whittet, D.C.B., Geballe, T.R., Greenberg, J.M.: 1986, Astron. Astrophys., **154**, L5.

Cohen, M., Allamandola, L.J., Tielens, A.G.G.M.: 1985, Astrophys. J. Lett., **299**, L93.

Cohen, M., Tielens, A.G.G.M., Bregman, J., Witteborn, F.C., Rank, D.M., Allamandola, L.J., Wooden, D.H., DeMuizon, M.: 1989, Astrophys. J., **341**, 246.

Colangeli, L., Schwehm, G., Bussoletti, E., Fonti, S., Blanco, A., Orofino, V.: 1989a, Astrophys. J., in press.

Colangeli, L., Bussoletti, E., Papoular, R., Mennella, V.: 1989b, Icarus, in press.

8

de Muizon, M., d'Hendecourt, L.B., Geballe, T.R.: 1987, In Polycyclic aromatic hydrocarbons and astrophysics (A. Léger, L. d'Hendecourt, and N. Boccara, eds.), p. 287. D. Reidel Publ. Comp., Dordrecht.

Duley, W.W., Williams, D.A.: 1981, M.N.R.A.S., 196, 269.

Fonti, S., Blanco, A., Bussoletti, E., Colangeli, L., Lugarà M., Mennella, V., Orofino, V., Scamarcio, G.: 1989, Infrared Phys., in press.

Goebel, J.H.: 1987, in "Polycyclic Aromatic Hydrocarbons and Astrophysics" (A. Léger, L. d'Hendecourt and N. Boccara, eds.), p. 329. D. Reidel Publ. Comp., Dordrecht.

Hyland, A.R., McGregor, P.J.: 1989, NASA CP 3036 (L.J. Allamandola and A.G.G.M. Tielens, eds.).

Jura, M.: 1986, Astrophys. J., 303, 327. Kessler, M.F. 1990; this volume.

Léger, A., Puget, J.L.: 1984, Astron. Astrophys. 137, L5.

Léger, A., d'Hendecourt, L.B.: 1987, in "Polycyclic Aromatic Hydrocarbons and Astrophysics" (A. Léger, L. d'Hendecourt, and N. Boccara, eds.) p. 223, D. Reidel Publ. Comp., Dordrecht.

Léger, A., d'Hendecourt, L.B., Dwfourneau, D.: 1989, Astron. Astrophys., in press.

Lemke, D.: 1990, this volume.

Magassú, A., Strassulla G.: 1989, Astrophys. J. Lett., in press.

Papoular, R., Conard, J., Giuliano, M., Kister, J., Mille, G.: 1989, Astron. Astrophys., in press.

Sakata, A., Wada, S.: 1989, in "Interstellar Dust" (L.J. Allamandola and A.G.G.M. Tielens, Eds.) p. 191. Kluwer Acad. Publ., Dordrecht.

Sellgren, K.: 1990, this volume.

Sopka, R.J., Hildebrand, R., Jaffe, D.T., Gatley, I., Roelling, T., Werner, M., Jura, M., Zuckerman, B.: 1985, Astrophys. J., 294, 242.

Tanabé, T., Nakada, Y., Kamijo, F., Sakata, A.: 1983, P.A.S.J., 35, 397.

Tielens, A.G.G.M., Allamandola, L.J., Barker J.R., Cohen, M.: 1987, in "Polycyclic Aromatic Hydrocarbons and Astrophysics" (A. Léger, L. d'Hendecourt, and N. Boccara, eds.) p. 273, D. Reidel Publ. Comp., Dordrecht.

Wdowiak, T.J.: 1987, in "Polycyclic Aromatic Hydrocarbons and Astrophysics" (A. Léger, L. d'Hendecourt and N. Boccara, eds.), p. 327. D. Reidel Publ. Comp., Dordrecht.

Wdowiak, T.J.: 1990, this volume.

# CHEMICAL ALTERATIONS IN ION IRRADIATED FROZEN HYDROCARBONS

G.A. BARATTA [1], G. STRAZZULLA [2]
[1]Osservatorio Astrofisico, Citta' Universitaria I-95125 Catania, Italy
[2]Istituto di Astronomia, Citta' Universitaria I-95125 Catania, Italy

## 1.  Introduction

For the last 10 years many experimental results have been obtained on the chemical and physical changes induced by ion and electron irradiation of organic materials. Those results have been reviewed in recent literature with a view to their astrophysical relevance (e.g. Strazzulla and Johnson 1989, Andronico et al. 1987, Lanzerotti et al. 1987, Foti et al. 1984, Johnson et al. 1984).

We have to outline that the presence of even complex organic materials in the general interstellar medium, in circumstellar envelopes, and in many objects of the Solar System is now widely accepted.

The organics on/in the various astronomical objects are continuously modified by external radiation, mainly UV photons and energetic ions determining both their formation starting from simple carbon-containing species and their further evolution.

In particular, some physico-chemical effects induced by keV-MeV ions colliding on frozen gases and organic materials have been studied in some laboratories, among which in Catania.

In this paper we discuss on some recent experiments performed by bombarding frozen hydrocarbons with keV ions. In particular we have studied the induced chemical effects by using "in situ" IR spectroscopy.

## 2.  In "Situ" IR Spectra of Ion-Irradiated Frozen Hydrocarbons

In figure 1 it is shown a very schematic view of the experimental apparatus we used to irradiate frozen gases and obtain "in situ" IR spectra. A scattering chamber was faced trough KBr windows, to the FTIR Perkin-Elmer (mod. 1710) spectrophotometer (4400-400 $cm^{-1}$ =2.27-25 $\mu$m) of the Catania (Italy) Institute of Astronomy. Vacuum was better than $10^{-7}$ mbar. Frosts were accreted onto a silicon substratum put in contact with a cold finger (77-300 K) by admitting gases into the

9

*E. Bussoletti and A. A. Vittone (eds.), Dusty Objects in the Universe, 9–16.*
© 1990 *Kluwer Academic Publishers.*

chamber, trough a needle valve. The accreted ices were bombarded by 3 keV Ar or He ions having a 2x2 cm$^2$ spot on the target (greater than the spot of the IR beam).

Analogous experiments have been previously done (Moore 1982, Moore et al. 1983) by irradiating various mixtures of $H_2O$, $NH_3$, $CH_4$, $N_2$, $C_3H_8$, $CO_2$ and CO with ≃ MeV protons and, from the analysis of the IR spectra, the synthesis of $C_2H_6$, $N_2O$, NO, $C_2H_4$, and $C_3H_8$ was reported.

Here we present results on the irradiation of benzene ($C_6H_6$) and butane ($C_4H_{10}$). The use of (relatively) simple hydrocarbons is justified by the fact that simpler targets can better get insight into the details of the processes induced by incoming ions.

In figure 2 they are shown the spectra obtained after 4000 sec of deposition of benzene (P≅ 5 10$^{-6}$ mbar) on a cold silicon substratum (77 K) with (upper spectrum) and without (lower spectrum) contemporary irradiation with 3 keV He ions at an energy fluence of ~100 eV/mol. From figure 2 we can see that in addition to their own, many new bands appear in the spectrum of irradiated benzene. This testifies for the complex chemistry induced by ions, already evidenced by other techniques (Foti et al. 1984). In table I the observed bands are reported together with their identification.

The new synthesized species contain triple carbon bonds (C≡C). All the three IR-active bands of acetylene (H-C≡C-H) (Khanna et al. 1988) are present in the spectrum. The intense band at ~2110 cm$^{-1}$ is attributable to monosobstituted acetylenes H-C≡C-R. The substituent group could be e.g. $CH_3$ (H-C≡C-$CH_3$) or phenyl (H-C≡C-$C_6H_5$). The intensity of the band could be indicative of conjugation with a carbonyl group (Bellamy 1975) that would testify for the extreme reactivity of irradiated samples, trapping oxygen from the residual gas in the scattering chamber.

In figure 3 they are shown the spectra obtained after 1000 sec of deposition of butane (P≅ 5 10$^{-6}$ mbar) on the cold substratum (77 K) with (upper spectrum) and without (lower spectrum) contemporary irradiation with 3 keV He ions at an energy fluence of ~10 eV/mol. From figure 3 we can see that also in this case, many new bands appear in the IR spectrum. In table II the observed bands are reported together with their identification.

The new synthesized species contain double and triple carbon bonds. Acetylene and monosubstituted acetylene are identified by the bands at, respectively, 3300 and 3228 cm$^{-1}$. The band at ~2110 cm$^{-1}$ is however lacking, indicating the absence of conjugation. Most of the new detected bands seems however attributable to vinyl-type (-CH=CH$_2$) alkenes. It is also evident from figure 3 that the intensity ratio of the 2956 cm$^{-1}$ band to the 2925 cm$^{-1}$ one, decreases indicating a greater number of CH$_2$ groups into respect to the CH$_3$ ones. This is consistent with the observed dehydrogenation of irradiated hydrocarbons. For methane, H$_2$ has been recognized, by mass spectroscopy, as the most relevant emitted specie (Brown et al. 1987), although at the beginning

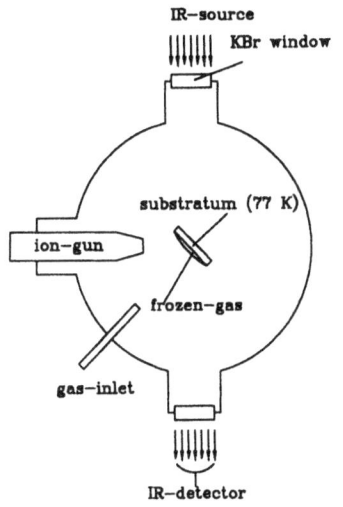

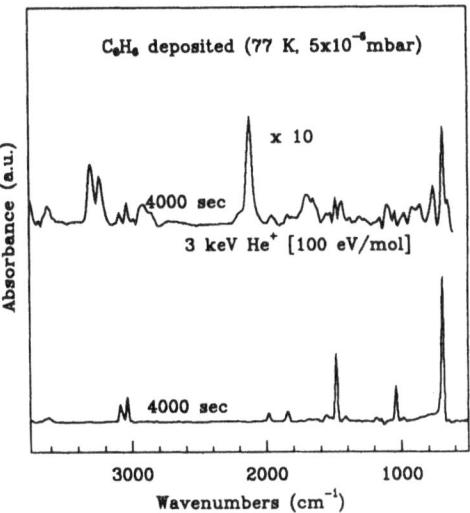

Fig.1 Schematic of the experimantal apparatus used to obtain IR spectra of irradiated frozen gases

Fig.2 IR spectra of frozen benzene as deposited without (bottom) and with contemporary irradiation (~100 eV/mol).

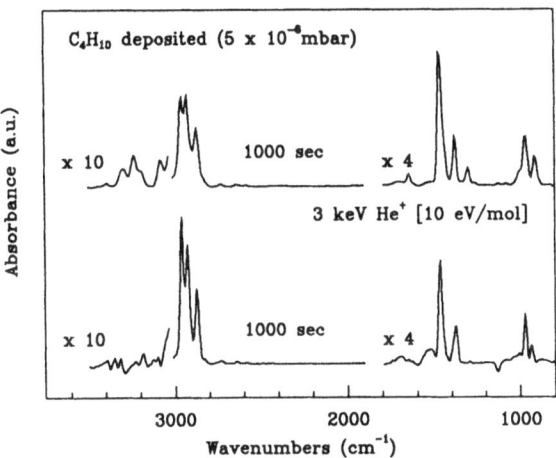

Fig.3 IR spectra of frozen butane as deposited without (bottom) and with contemporary irradiation (~10 eV/mol).

TABLE I. - *IR bands observed from ion-irradiated benzene (3 keV He ions, 100 eV/$C_6H_6$ mol).*

| Band cm$^{-1}$ ($\mu$m) | Assignment | Specie |
|---|---|---|
| 3300 (3.03) | ≡C-H stretching | monosubst.acetylenes |
| 3235 (3.09) | C-H asymm.stretch. | acetylene |
| 3088 (3.24) | C-H aromatic | benzene |
| 3036 (3.29) | C-H aromatic | benzene |
| *2960 (3.38) | -$CH_3$ asymm.stretch. | long chain, cross linked |
| *2925 (3.42) | -$CH_2$ asymm.stretch. | organic material |
| *2860 (3.50) | -$CH_3$ symm.stretch. | |
| 2116 (4.73) | C≡C stretching | monosubst.acetylenes |
| 1984 (5.04) | comb.bands | benzene |
| 1840 (5.43) | comb.bands | benzene |
| *1680 (5.95) | C=O stretching | long chain, cross linked |
| *1640 (6.10) | C=C stretching | organic material, Oxygen contamination from vacuum |
| 1480 (6.76) | C=C stretching | benzene |
| *1450 (6.89) | C-H asymm. def. | organic material |
| 1440 (6.94) | comb.mode | acetylene |
| *1380 (7.25) | C-H symm.def. | organic material |
| *1100 (9.1) | C-H rocking | organic material |
| 1040 (9.62) | C-H in plane bending | benzene |
| *~885 (11.30) | C-H out of plane bend. | aromatic compounds |
| 760 (13.16) | C-H out of plane bend. | acetylene |
| 688 (14.53) | C-H out of plane bend. | benzene |

* These bands are also observed in the refractory organic residue left over by ion irradiation.

**TABLE II** - *IR bands observed from ion-irradiated butane (3 keV He ions, ~10 eV/$C_4H_{10}$mol).*

| Band $cm^{-1}$ ($\mu$m) | Assignment | Specie |
|---|---|---|
| 3300 (3.03) | $\equiv$C-H stretching | monosubst.acetylenes |
| 3228 (3.10) | C-H asymm.stretch. | acetylene |
| 3080 (3.25) | C-H stretch. | alkenes (vinyl) |
| 3020 (3.31) | C-H stretch. | alkenes (vinyl) |
| 2956 (3.38) | -$CH_3$ asymm.stretch. | butane |
| 2925 (3.42) | -$CH_2$ asymm.stretch. | butane |
| 2868 (3.49) | -$CH_3$ symm.stretch. | butane |
| 1644 (6.08) | C=C stretch. | alkenes |
| 1468 (6.81) | -$CH_2$- deform. | butane |
| 1380 (7.25) | C-$CH_3$ symm.def. | butane |
| 1300 (7.69) | C-H in plane deform. | alkenes (vinyl) |
| ~1000 (10) | C-H out of plane def. | alkenes (vinyl) |
| 964 (10.37) | ring breathing | cyclobutane |
| 916 (10.92) | $CH_2$ out of plane def. | alkenes (vinyl) |

of irradiation emission of $CH_4$ and $C_nH_m$ has been observed (about 100 mol/1.5 MeV He ion) (Foti et al. 1987, Lanzerotti et al. 1987). After a fluence equivalent to about $10^{14}$ ions/$cm^2$ (1.5 MeV He) this yield becomes almost zero: as reported many times in recent literature, with a view to the astrophysical implications (Andronico et al. 1987, Calcagno et al. 1985, Johnson et al. 1984, Strazzulla and Johnson 1989), a refractory organic residue is left over after extraction of the irradiated target from the scattering chamber. The IR bands marked by an asterisk in table I are those observed also in such a residue. Further bombardment of this organic material causes its evolution towards what we will call in the next IPHAC (Ion Produced Hydrogenated Amorphous Carbon). IPHAC can be produced by irradiating most of the carbon containing materials: in a sense this new material "forgets" what it was before irradiation. Also IPHAC, as all the other organic materials, exhibits a dehydrogenation, evidenced by changes in its IR spectrum as shown in figure 4.

## 3.    Structural Properties of IPHAC

The vibrational spectra provide valuable evidence to the structure of carbonaceous compounds. To exemplify this, Raman spectra of some relevant carbon compounds have been obtained in our laboratory and exhibited in figure 5.

Diamond presents a first order feature at 1340 $cm^{-1}$ and a broad, structured band extending from 2200 to 2700 $cm^{-1}$. Highly oriented pyrolytic graphite (HOPG) shows the $E_{2g}$ mode band at 1580 $cm^{-1}$ and a second order structured band at 2700-2800 $cm^{-1}$. Powder graphite exhibits an additional band at 1355 $cm^{-1}$. Some authors (Tuinistra and Koening 1970) suggested a linear relationship between the intensity ratio of the 1355 $cm^{-1}$ to the 1580 $cm^{-1}$ bands and the reciprocal of crystalline diameter $L_a$. That result has been questioned (Nakamizo et al. 1974) although, undubitably, greater $L_a$ are accompanied by a less intense 1355 $cm^{-1}$ band. In the case of figure 5 the $L_a$ values would result of about 60 A.

The IPHAC, whose spectrum is shown in figure 5 has been obtained by irradiating frozen benzene with 3 keV Ar ions (~ 50 $eV/C_6H_6$ mol). Its spectrum is quite similar to those of amorphous carbons a-C or hydrogenated amorphous carbons a-C:H produced e.g. by sputtering, evaporation (Robertson 1986) or even arc-discharge (Blanco et al. 1988). The spectra of all of these amorphous carbons are dominated by a large peak at ~ 1550 $cm^{-1}$ with a shoulder at 1350-1400 $cm^{-1}$. The 1550 $cm^{-1}$ band has been interpreted as providing strong evidence in favor of graphite ($sp^2$) bonding. The intensity of the shoulder relative to the 1550 $cm^{-1}$ peak could be indicative of the diameter of graphite microcrystals (Robertson 1986) or of the relative amount of $sp^3$ (diamond-like) ligands Yoshikawa et al. 1988). Work is in progress in our laboratory to distinguish between these two possibilities and get better insight into the structural properties of IPHAC.

Even if the pristine material contains large-order crystalline domains, ion irradiation causes the destruction of such an order.

In figure 6 the Raman spectra of HOPG, pristine and after irradiation with 3 keV He ions (~ $1.5 \times 10^{16}$ ions/$cm^2$), are shown. It is evident the ion-induced destruction of the crystallinity.

## 4.    Astrophysical Considerations

The astrophysical applications of the results from laboratory vibrational spectroscopy of pristine and ion or UV irradiated materials have been mainly based on the comparison of the laboratory spectra with those obtained from astronomical observations. The comparison, although being a necessary condition, cannot be considered enough to indicate a

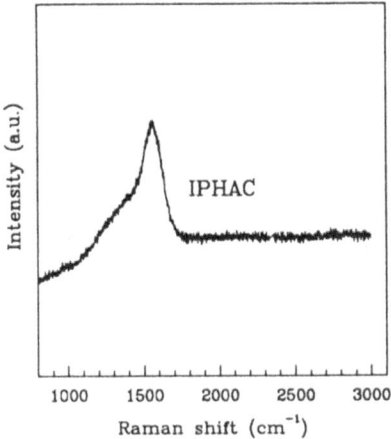

Fig.4 IR spectra of organic residue from ion irradiated frozen hydrocarbon (IPHAC)

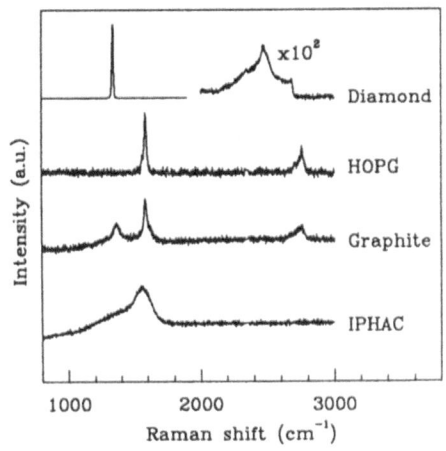

Fig.5 Raman spectra of carbonaceous materials, from top to bottom: Diamond, High Oriented Pyrolitic Graphite, microcrystalline ($L_a \sim 100$ Å) Graphite, IPHAC (from benzene irradiated by 3 keV Ar$^+$, ~50 eV/mol).

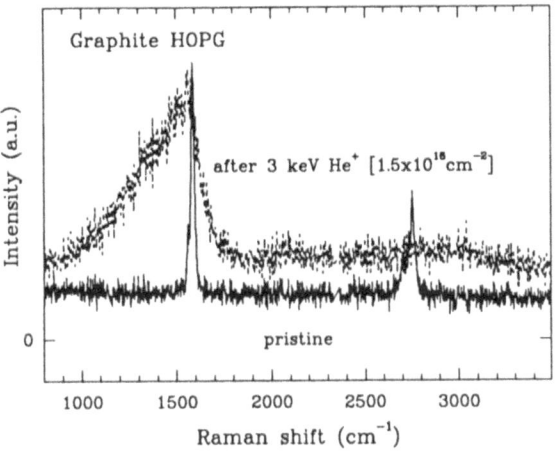

Fig.6 Raman spectra of HOPG pristine and after ion–irradiation

given material as responsible for a given astronomical feature. This both because, very often, different materials can as well (or badly) mimic the astronomical spectra and because the different physical conditions in the laboratory and in space prevent, often, to do a reliable comparison. Thus, in our opinion an acceptable criterion to establish the carriers of a given feature is not only to reproduce the spectrum, but rather to explain the formation and the evolution of that carrier under the influence of laboratory-tested processes of the kind discussed in the present paper.

## Acknowledgements

This research has been partially supported by the Italian Ministero della Pubblica Istruzione (Fondi 40%) and by Italian Space Agency (ASI).

## REFERENCES

Andronico, G., Baratta, G.A., Spinella F., and Strazzulla G.: 1987, Astron. Astrophys. 184, 333

Bellamy, L.J.: 1975 The infra-red spectra of complex molecules (Chapman and Hall, London)

Blanco, A., Bussoletti, E., Colangeli, L., Fonti, S., and Orofino V.: 1988, Infrared Phys. 28, 383

Brown, W.L., Foti, G., Lanzerotti, L.J., Bower J.E., and Johnson, R.E.: 1987, Nucl. Inst. Methods B19/20, 899

Calcagno, L., Foti, G., Torrisi, L., and Strazzulla, G.: 1985 Icarus 63, 31

Foti, G., Calcagno, L., Sheng, K.L., and Strazzulla, G.: 1984, Nature (London) 310, 126

Foti, G., Calcagno, L., Zhu, F.Z., and Strazzulla, G.: 1987, Nucl. Instr. Methods in Phys. Res. B24-25, 522

Khanna, R., Ospina, M.J., and Zhao, G.: 1988 Icarus 73, 527

Johnson, R.E., Lanzerotti, L.J., and Brown, W.L.: 1984, Adv.Space Res. 4 n9, 41

Lanzerotti, L.J., Brown, W.L., and Marcantonio, K.J.: 1987, Astrophys.J. 313, 910

Moore,M.H., and Donn, B:1982, Astrophys.J.Lett. 257, L47-L50

Moore,M.H., Donn, B., Khanna,R., and A' Hearn, M.F.: 1983 Icarus 54, 388

Nakamizo, M., Kammereck, R., and Walker Jr, P.L.: 1974, Carbon 12, 259

Robertson, J.: 1986, Adv. in Phys. 35, 317

Strazzulla, G., and Johnson, R.E.: 1989, in Comets in the Post-Halley Era, edited by R. Newburn, M. Neugebauer and J. Rahe (Kluwer publ. co.), in press

Tuinistra, F., and Koenig, J.L.: 1970, J. Chem. Phys. 53, 1126

Yoshikawa, M., Katagiri, G., Ishida, H., Ishitani, A., and Akamatsu, T.: 1988, in XI Int. Conf. on Raman Spectr., ed. by R.J.H. Clark and D.A. Long (J. Wiley and Sons), p. 339

# INVESTIGATION OF COSMIC DUSTS USING SYNCHROTRON RADIATION

A. Evans
*Department of Physics, University of Keele, U.K.*

ABSTRACT. We are investigating a variety of cosmic materials (meteorites, Brownlee particles, 'synthetic' dusts etc.) using synchrotron radiation. Some preliminary results using the *EXAFS* and X-ray diffraction techniques are described.

## 1    Introduction

In 1987 we initiated a study of cosmic solids using the Science and Engineering Research Council Synchrotron Radiation Source (SRS) at Daresbury Laboratory. Our original aims were to use the techniques of *EXAFS* and X-ray powder diffraction to investigate the local structural environment of various cosmic solids and to determine their degree of crystallinity; these factors are of course strongly affected by the thermal history of, and the processing suffered by, the material, particularly during the early history of the solar system. We also intended using the same techniques to investigate 'synthetic' materials, produced in the laboratory to simulate the optical properties of cosmic dusts; these materials have generally been produced under known conditions and their histories and compositions are well-known and so they provide a 'control' for the investigation of genuine cosmic dusts.

Here we shall review the work we have done to date, which consists mainly of *EXAFS* and X-ray diffraction.

## 2    The Daresbury Synchrotron Radiation Source

The Daresbury SRS operates at electron energy 2 GeV in a magnetic field of 1.2 T, with an electron current of typically 300 mA. The electrons circulate in bunches of width 170 psec, successive bunches being separated by 2 nsec. The resulting photon flux at a wavelength of 1 Å is typically $\sim 10^{12}$ photons $s^{-1}$ $mA^{-1}$ $mrad^{-1}$ for a 2 GeV, 1 A beam [see e.g. Catlow, Greaves (1986) for details of the Daresbury SRS]. The spectral distribution of the Daresbury SRS is shown in figure 1. Synchrotron radiation provides an intense, broadband beam of electromagnetic radiation which allows the investigation of condensed matter to an extent that is not possible with many of the conventional methods. Furthermore the techniques that we have used allow the measurement of the

*E. Bussoletti and A. A. Vittone (eds.), Dusty Objects in the Universe, 17–25.*
© 1990 *Kluwer Academic Publishers.*

properties of the *bulk* solid, rather than simply the properties of an exposed surface. This contrasts with the case of, for example, analytical SEM, which analyses only the exposed surface of a meteoritic sample.

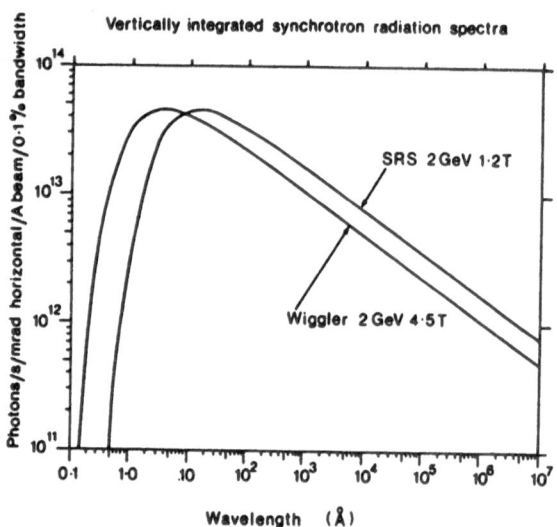

**Figure 1:** Spectral distribution of the Daresbury SRS [after Catlow and Greaves (1986)].

## 3  Extended X-ray absorption fine structure

Absorption edges—caused by the ionization of atomic or ionic species—are of course widely observed in the ultraviolet, optical and infrared spectra of astronomical objects. Since the availability of satellite-borne X-ray spectrometers, such as that on the *Einstein* observatory (Giacconi et al., 1978), equivalent absorption edges have also been observed at X-ray wavelengths (e.g. Schattenburg and Canizares, 1986). Indeed it was pointed out over twenty years ago [Overbeck (1965); Hayakawa (1970)] that these X-ray absorption edges could potentially tell us something about the chemical composition of interstellar grains.

In the case of absorption edges at X-ray wavelengths, of course, the ionization occurs as the result of the removal of an electron from an inner (K– or L–) shell rather than an outer electron. If the absorbing atom is free the electron de Broglie wave propagates outward spherically symmetrically. However if, as would be the case for an interstellar grain, the atom is locked up in the solid state (or even in a molecule like CO) the electron de Broglie wave is scattered by neighbouring atoms and we get constructive and destructive interference between the outgoing and scattered waves. As a result the spectrum immediately shortward of the absorption edge is not (as it would be in the case of an isolated absorbing atom) smooth: there is a modulation on the high energy side of the absorption edge which is referred to as Extended X-ray Absorption Fine Structure—*EXAFS*.

*EXAFS* is a characteristic of the solid state, and provides direct information about the immediate environment of the absorbing atom. Indeed the absorption spectrum carries a signature that is characteristic of the identity and disposition of the scattering atoms around the absorbing atom [see e.g. Stern (1978) for a review]. The possibility of observing *EXAFS* in interstellar grains in situ has been discussed by Martin (1970) and more recently by Evans (1986).

It is customary to present *EXAFS* data in the form $\chi = (\mu - \mu_0)/\mu_0$, where $\mu$ is the absorption coefficient defined in the usual way $(I = I_0 \exp[-\mu x])$ and $\mu_0$ is the absorption coefficient appropriate for free atoms (i.e. not in the solid state). It may be shown that the Fourier transform of $\chi$ gives directly the co-ordination number, distance and identity of scattering atoms (Sayers et al., 1971). The various stages in the extraction and interpretation of the *EXAFS* data are shown in figure 2; the final step is the Fourier

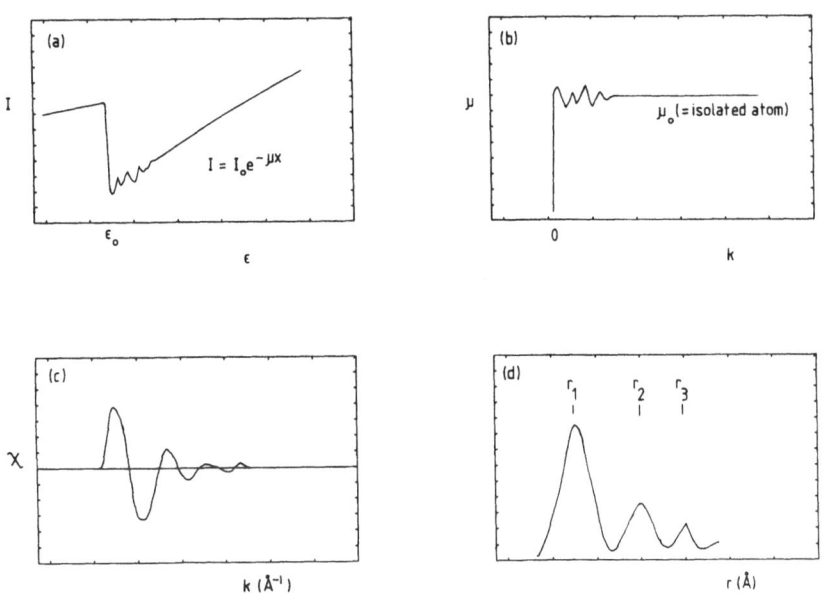

**Figure 2:** From absorption edge to atomic environment: steps in the extraction of information from *EXAFS*. (a) Absorption edge; (b) absorption coefficient $\mu$; (c) normalized $\mu$; (d) Fourier transform of (c). Axes arbitrary.

transform of the normalized absorption $\chi$ and the fitting of the transform with a model based on the nature and disposition of scattering atoms around the absorbing atom.

We have investigated the *EXAFS* at the K–edges of nickel (8.33 keV) and iron (7.15 keV) in iron-nickel meteorites (Cressey et al. 1989a). A typical meteoritic absorption edge (the iron K–edge of the Butler meteorite), including *EXAFS*, is shown in figure 3; the Fourier transform is shown in figure 4 (cf. figure 2).

The metallic phases of these objects consist of iron-nickel alloys containing some 4 to 40% Ni by weight (Reuter et al. 1988). The meteoritic alloy normally occurs as two co-existing phases, namely kamacite, which has body-centred cubic (BCC) structure and taenite, which is faced-centred cubic (FCC). The iron-nickel equilibrium phase diagram (Romig and Goldstein, 1980) indicates that the kamacite phase is stable to $\sim 7 - 8\%$ wt nickel content but as the nickel content increases, kamacite is precipitated in a matrix of FCC taenite. The relative amounts of the phases present are determined not only by the characteristics of the primary body from which they originated, but also by the fragmentation of the primary and the subsequent thermal history of the individual fragments.

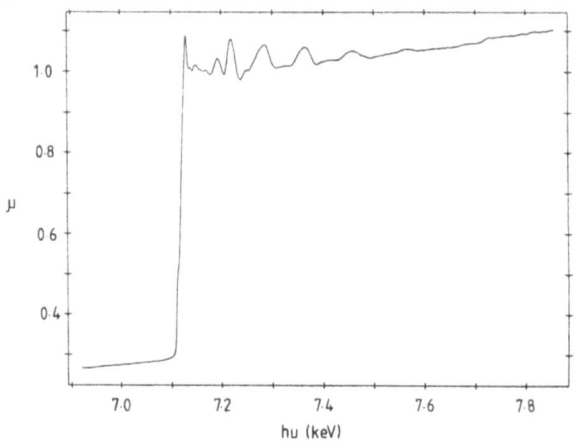

**Figure 3:** *EXAFS* at the iron K–edge of the Butler meteorite; vertical scale arbitrary.

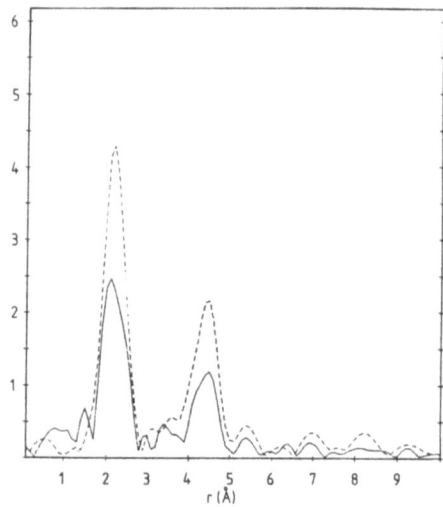

**Figure 4:** Fourier transform of the normalized *EXAFS* for the Butler meteorite; vertical scale arbitrary.

Detailed analysis of the *EXAFS* data allows the determination of the relative proportions of kamacite and taenite in BCC and FCC phases. While most iron-nickel meteorites conform with expectation, one meteorite (Santa Catharina) shows clear, hitherto unsuspected, differences, suggesting that this object has suffered a thermal history that is substantially different from the other meteorites we have investigated. This result emphasizes the benefits of using the *EXAFS* technique: previous determinations have used conventional (optical micrograph and SEM) methods, both of which give kamacite-taenite proportions in terms of the *relative surface area* occupation of the section under consideration and do not probe the bulk of the meteorite.

## 4  X-ray diffraction

As is well-known a crystal can act like a three-dimensional diffraction grating and X-ray diffraction is widely exploited to investigate crystal structure. A necessary (though not sufficient) condition that atomic planes (separated by distance $d_{hkl}$) scatter radiation of wavelength $\lambda$ in phase is the Bragg condition $n\lambda = 2d_{hkl}\sin\theta$, where $n$ is an integer and $\theta$ is essentially the angle of scattering. The indices $hkl$ are integers (the usual Miller indices) which identify specific planes in the crystal.

Experimentally the Bragg condition is achieved either by illuminating a single crystal with 'white' X-rays or by illuminating a powdered sample (which has crystal planes orientated randomly) with monochromatic X-rays.

### 4.1  Powder diffraction

We have looked at various phases of a number of carbonaceous chondrite meteorites using this technique. The X-ray diffraction pattern (X-ray counts vs. $2\theta$) for a fine ($\lesssim 37\,\mu$m) sieve fraction of the Allende meteorite is illustrated in figure 5.

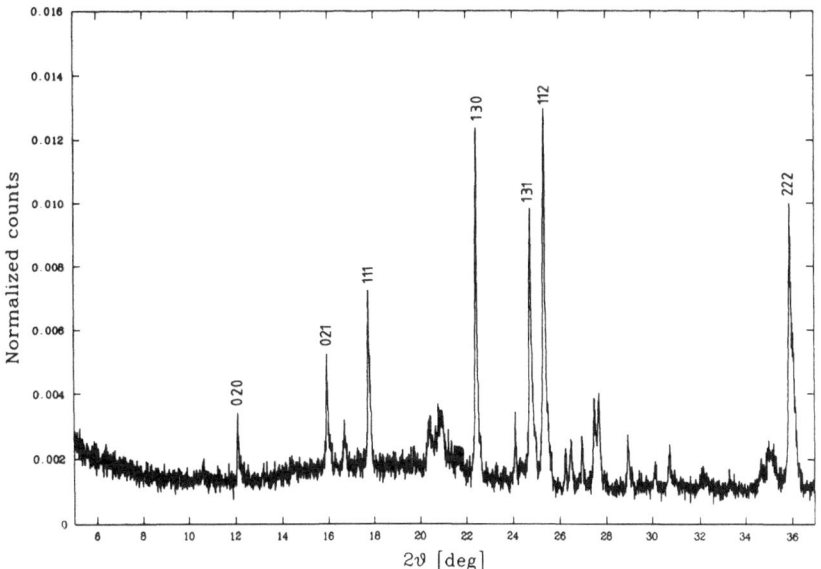

**Figure 5:** X-ray powder diffraction pattern for the Allende meteorite; $\lambda = 1.087$Å.

The more prominent peaks (most of which are due to olivine) are identified by their Miller indices [hkl]. The structure (bond lengths, angles etc.) of the solid under investigation can generally be refined using a technique due to Rietveld (1969).

For a purely crystalline sample we would expect to see features only at values of $2\theta$ corresponding to the Bragg condition, with essentially zero intensity at all other values of $2\theta$. We see in figure 5, however, that there is a substantial 'continuum' which results from scattering by an amorphous component. This arises partly in the glass capillary used to present the meteoritic sample to the X-ray beam but we strongly suspect, from comparison with the X-ray diffraction pattern produced by amorphous silicates made available by the Toronto group, that there is a significant amorphous component in the meteoritic silicate.

We can now exploit the high intensity of the SRS, and the consequent high angular resolution, by looking at the structure in a single peak, specifically the peak labelled [130] in figure 5. This peak is of special significance because the $d$–spacing of the [130] planes in silicates depend in a simple way on the relative proportions of Mg and Fe in the sample (note again how this technique gets at the *bulk* properties). For example, if we have a silicate $Mg_xFe_{1-x}SiO_3$, then

$$d_{130} = \alpha - \beta x, \tag{1}$$

where $\alpha$ and $\beta$ are known constants (note that the spacing *increases* as the relative proportion of iron to magnesium increases).

In figure 6 we compare different fractions of the Allende and Murchison meteorites. The solid line is the [130] peak for the Allende fine sieve fraction shown in figure 5; the broken line is the same feature as seen in the Allende sieve residue (i.e. particles of dimension $\gtrsim 37\,\mu m$), while the dash-dot line is the olivine [130] profile for Murchison sieve residue ($\gtrsim 37\,\mu m$).

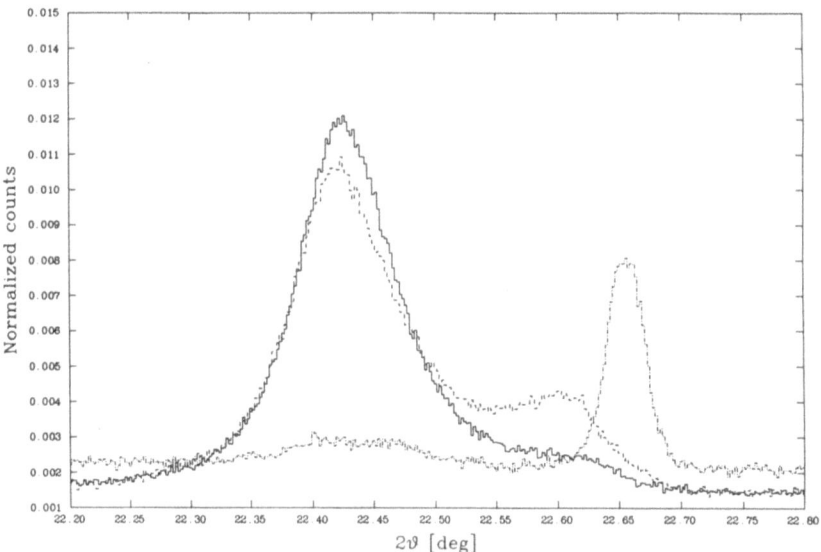

**Figure 6:** The [130] peak for the Allende and Murchison meteorites; see text for details.

In the case of the Allende data the peak at $\simeq 22.4°$ represents an iron-rich ($x \sim 0.5$) olivine but the peak is resolved and its width suggests the presence of a range of olivines in solid solution. In the larger fraction there is a separate peak (at $\simeq 22.6°$) which is due to a magnesium-rich ($x \simeq 0.9$) component. These results suggest that, in the larger grains, the iron-rich and magnesium-rich components are quite distinct and this is borne out by SEM data, which clearly show particles having iron-rich mantles which become increasingly magnesium-rich towards the core. In the Murchison sample however the [130] peak is due almost entirely to the magnesium-rich ($x \simeq 0.97$) component. Full details may be found in Cressey et al. (1989b).

## 4.2 Single crystal diffraction

We have also done some preliminary work on diffraction from single olivine crystals from carbonaceous chondrite meteorites. Figure 7 shows a single crystal of olivine from Allende (dimensions $\sim 20$ $\mu$m) mounted on 10 $\mu$m glass fibre ready for presentation to the (focussed) X-ray beam.

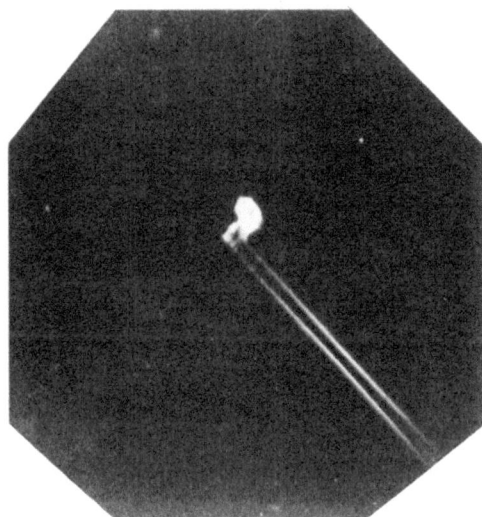

**Figure 7:** Single 20 $\mu$m crystal of silicate from Allende *(Photograph kindly supplied by G. Cressey).*

The X-ray diffraction from a 40 $\mu$m fragment of a white chondrule from the Allende matrix is shown in figure 8. Each diffraction spot is double, and corresponds to the doubling of the [130] peak seen in figure 6.

Although the work described in this sub-section is still at a preliminary stage it does demonstrate the awesome potential of the SRS for investigating in detail single crystals having dimensions $\lesssim 100$ $\mu$m.

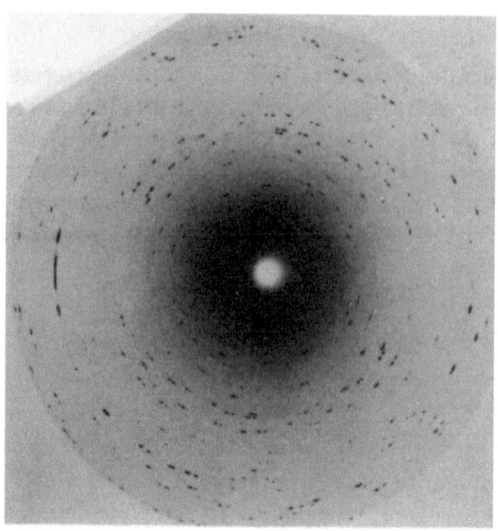

**Figure 8:** X-ray diffraction pattern from a 40 μm fragment of a white chondrule from Allende.

### 4.3 Future plans

We have plans to exploit the entire bandwidth of the Daresbury SRS in the future and our plans for the coming year (1990) are as follows:

- Photoacoustic spectroscopy [see Patel and Tam, (1981) for a review] of meteoritic and 'synthetic' dusts, to measure the absorption–as opposed to the extinction–coefficient, over the wavelength range 10–3000 Å. By chopping the SRS and making time-resolved measurements we hope to be able to investigate in detail the kinetics of photon absorption in fine dusts.

- Extending our preliminary work on the X-ray diffraction by single particles/crystals by investigating Brownlee particles, IDPs etc.

- Soft X-ray Absorption Fine Structure *(SOXAFS)* of meteoritic and synthetic silicates, to investigate the local environments of the light elements, such as C, Mg, Si. This work will give us bond lengths (e.g. Si–O) which can be correlated with emissivity properties in the infrared.

In the longer term we intend to extend the photoacoustic spectroscopy to cover the entire bandwidth allowed by the SRS (see figure 1). This will provide the *absorption* efficiency of cosmic materials, at high wavelength resolution (∼ 1 Å), from X-ray to millimeter wavelengths. A further benefit will be that these measurements will provide as homogeneous a set of data as is possible.

The high intensity of the SRS allows the measurement of time-dependent processes that may occur too rapidly (in the laboratory) for investigation using other techniques. For example, diffraction patterns such as that shown in figure 8 can be obtained on a timescale of seconds using an area detector and we expect that structural changes which accompany the thermal and radiative annealing of silicates and carbon (such as that suggested to occur in inter- and circum-stellar space) can be seen directly using this technique.

## Acknowledgements

This work is carried out in collaboration with J.S. Albinson, G. Cressey, G.N. Greaves, C.M.B. Henderson, R. Hutchison, S. Lunt, I. Munro, S.P. Thompson and R. Zhu. Meteoritic samples are provided by the Mineralogy Department of the British Museum (Natural History), and samples of 'synthetic' dusts have been provided by workers at the Universities of Toronto and Lecce. We thank the United Kingdom Science and Engineering Research Council for support and (through its Beam-time Allocation Panel) for the generous allocation of beam time on the SRS.

## References

C.R.A., Greaves, G.N.: 1986, Chemistry in Britain, **22**, 806.

Cressey, G., Dent, A.J., Dobson, B., Evans, A., Greaves, G.N., Henderson, C.M.B., Hutchison, R., Thompson, S.P., Zhu, R.: 1989a, Geochim. Cosmochim. Acta, submitted.

Cressey, G., Cressey, B.A., Evans, A., Thompson, S.P., Hutchison, R., Greaves, G.N., Henderson, C.M.B.: 1989b, Geochim. Cosmochim. Acta, submitted.

Evans, A.: 1986, M.N.R.A.S., **223**, 219.

Giacconi, R. et al.: 1978, Astrophys. J., **230**, 540.

Hayakawa, S.: 1970, Prog. Theor. Phys., **43**, 1224.

Martin, P.G.: 1970, M.N.R.A.S., **149**, 221.

Overbeck: 1965, Astrophys. J., **141**, 864.

Patel, C.K.N., Tam, A.C.: 1981, Rev. Mod. Phys., **53**, 517.

Reuter, K.B., Williams, D.B., Goldstein, J.I.: 1988, Geochim. Cosmochim. Acta, **52**, 617–626.

Rietveld, H.M.: 1969, J. Appl. Cryst., **65**, 65.

Romig, A.D., Goldstein, J.I.: 1980, Metall. Trans., **11A**, 1151–1159.

Sakumar, V., Goldstein, J.I.: 1988. Geochim. Cosmochim. Acta, **52**, 715–726.

Sayers, D.E., Stern, E.A., Lytle, F.W.: 1971. Phys. Rev. Lett., **52**, 1204–1207.

Schattenburg, M.L., Canizares, C.R.: 1986. Astrophys. J., **301**, 759.

Stern, E.A.: 1978, Contemp. Phys., **19**, 289–310.

# THE PARTICLES AROUND MASS-LOSING CARBON STARS

M. JURA
Department of Astronomy
University of California
Los Angeles CA 90024 USA

ABSTRACT. Mass losing carbon stars are responsible for about half of the mass lost by intermediate mass stars in the solar neighborhood. These stars typically have dust to gas ratios by mass near $5\ 10^{-3}$; most of the matter heavier than helium is apparently contained within solid grains. The relative amounts of large grains ($>10^{-6}$ cm) and small particles in these outflows is not yet reliably known and may vary from star to star.

## 1. INTRODUCTION

Most intermediate mass stars (those with initial main sequence masses between 1 and $\sim 5\ M_{\odot}$) become white dwarfs of approximately 0.7 $M_{\odot}$ (Iben and Renzini 1983). Most of this mass loss occurs during the second ascent of the red giant branch, the Asymptotic Giant Branch (AGB). It seems that most of this mass loss occurs during a short interval (perhaps $10^5$ years) of high mass loss ($\sim 10^{-5}\ M_{\odot}$ yr$^{-1}$) (Knapp and Morris 1985, Jura and Kleinmann 1989). The typical luminosity, L, and effective temperature of a mass losing AGB star are $10^4\ L_{\odot}$ and 3000 K, respectively.

The mass-losing red giants typically are surrounded by envelopes of solid dust grains and molecules. These envelopes have been extensively studied with radio and infrared techniques (see, for example, Morris and Zuckerman 1985, Kwok and Pottasch 1987, Johnson and Zuckerman 1989) and a large number of species have been identified (Olofsson 1989, Jura 1989). The characteristic outflow velocity, v, is measured from the molecular line profiles and is typically 15 km s$^{-1}$. The mass loss rate, dM/dt, displays a great range from values as low as perhaps $10^{-8}\ M_{\odot}$ yr$^{-1}$ to a maximum of about $10^{-4}\ M_{\odot}$ yr$^{-1}$. It seems that for most of these stars the mass loss is driven by radiation pressure on dust grains that form in the circumstellar envelope (Jura 1986, Lucy, Robertson and Sharp 1986) and we therefore have the relationship, $vdM/dt \leq L/c$ The inequality holds when the mass loss rate is less than its saturation value. Since most AGB stars have a luminosity of $10^4\ L_{\odot}$ and outflow velocities of 15 km s$^{-1}$, the characteristic maximum mass loss rate is about $1.3\ 10^{-5}\ M_{\odot}$ yr$^{-1}$.

Stars derive most of their luminosity from nuclear reactions that occur in their interiors. In the AGB phase, material from the interior is mixed with that at the surface and the elemental abundances of the photosphere of a red giant can be very different from the star's

E. Bussoletti and A. A. Vittone (eds.), Dusty Objects in the Universe, 27–34.
© 1990 Kluwer Academic Publishers.

initial composition. In most cases, it appears that in the atmospheres of the red giants, the major elements are still hydrogen and helium as measured from the spectra of planetary nebulae (Zuckerman and Aller 1986). However, the next most common elements, carbon, nitrogen and oxygen, may have very nonstandard compositions, and this can greatly affect the chemistry and the nature of the solid particles that form. In particular, the [C]/[O] ratio is critical. In the usual thermodynamic description for material in cool (T $\sim$ 3000 K) stellar atmospheres, we expect the CO molecule, because of its very high binding enregy, to contain as much carbon and oxygen as possible (Salpeter 1977). If [C]/[O] is less than 1, there is excess oxygen and the material is oxygen-rich while if [C]/[O] is larger than 1, there is excess carbon and the material is carbon-rich. S-type stars have [C] $\sim$ [O] to within a factor of about 1.05 (Scalo and Ross 1976).

In the Yale Bright Star Catalog (Hoffleit and Jaschek 1982) of the $\sim$10,000 optically brightest stars in the sky, there are only about 20 carbon-rich stars, and it would seem that these objects are quite rare. However, mass-losing stars are bright in the infrared, and are not conspicuous at optical wavelengths. The most famous example is IRC+10216 which is the brightest object in the sky at 10 $\mu$m outside the solar sytem, yet it is barely detectable optically. About half of the mass loss from AGB stars results from carbon-rich stars (Knapp and Morris 1985, Zuckerman and Aller 1986, Jura and Kleinmann 1989), and these stars are the subject of this paper.

After the AGB, the star in most cases becomes an optically visible planetary nebula (Iben and Renzini 1983, Kwok 1982). Studies of planetary nebulae may also provide insight into the nature of carbon particle production by mass-losing stars.

## 2. PHYSICAL CONDITIONS IN THE OUTFLOWS

In the simplest picture, the outflow from a mass-losing cool carbon-rich star can be divided into three zones (see, for example, Lafont, Lucas and Omont 1982, Glassgold *et al.* 1987). There is an inner zone (r less than $10^{14}$ or $10^{15}$ cm, depending upon the mass loss rate), dominated by three-body reactions, where grains and complex particles are synthesized. In this region, matter is accelerated to its terminal outflow velocity which characteristically is 15 km s$^{-1}$, and the evolving chemical composition tends to "freeze-out" (McCabe, Smith and Clegg 1979) as it flows outwards and the pressure drops. Further from the star ($10^{14}$ to $10^{17}$ cm), there is a zone of lower density where two body chemical reactions dominate the processing of the material. Finally, far from the star (greater than $10^{16}$ or $10^{17}$ cm), the collision time between gas particles is long compared to the dynamic outflow time. In this outermost region, photoprocessing and photodissociation by the ambient interstellar ultraviolet radiation field are the dominant processes. Particularly abundant molecules, such as CO and H$_2$, can be self-shielding and need not be protected by dust (Morris and Jura 1983, Mamon, Glassgold and Huggins 1988).

Certain limited aspects of the grain formation and processing in the outflows from red giants can be modelled in a satisfactory fashion. For example, there is good agreement between observations and models for those oxygen-rich stars where dust particles accrete solid water ice mantles onto their surfaces (Jura and Morris 1985). However, the region where the grains are created and grow is still not fully understood. Here, we adopt a semi-empirical approach to the study of the nature and composition of the circumstellar grains.

Theoretical models for the growth and evolution of solid grains in circumstellar envelopes are given by Frenklach and Feigelson (1989), Gail and Sedlmayr (1984, 1986, 1987, 1988), Keller (1987) and Salpeter (1974)

## 3. BROAD BAND OBSERVATIONS OF THE GRAINS

One of the most strking results of observations of mass outflows from late-type giants is that there is a strong correlation between the gas loss rate and the amount of grains measured by their infrared emission (see, for example, Jura 1986b). From the observed infrared emission, it is possible to infer the dust loss rate, and this value can be compared to the gas loss rate derived from radio CO measurements which have now been obtained for a large number of carbon stars (Knapp and Morris 1985, Nguyen-Q-Rieu 1987, Olofsson, Eriksson and Gustafsson 1987, 1988, Zuckerman and Dyck 1986a,b, Zuckerman, Dyck and Claussen 1986, Zuckerman and Dyck 1989). While there may be fluctuations from star to star, the dust to gas ratio by mass appears to have an average value of about $4.5 \ 10^{-3}$ (Jura 1986). This result assumes that the opacity at 60 $\mu$m of the grains is 150 cm$^2$ gm$^{-1}$ (Borghesi, Bussoletti and Colangeli 1985). If, for example, the infrared opacity is larger than this value, then less mass is present in the grains to account for the observed infrared emission.

The inferred dust to gas ratios implies for solar abundances that most of the refractory material that conceivably could be condensed into solids actually does so. Indirect observational evidence of the gas also supports this conclusion. For example, only about 0.01 of the silicon is contained within gas phase SiO, consistent with the view that most of the silicon is contained within grains (Morris et al. 1979).

Not only do the grains contain a large fraction of the material other than hydrogen and helium, it seems that the circumstellar grains also have sizes not too different from those of interstellar grains. With considerable uncertainty, there is evidence that many of the grains have sizes within an order of magnitude of 0.1 $\mu$m. (i) Optical observations of the intensity of H$\alpha$/H$\beta$ emission lines in some carbon stars indicate considerable circumstellar reddening (Cohen 1979). This implies grains sizes not much larger than 0.1 $\mu$m. (ii) Molecules such as HCN are photodissociated by ambient interstellar ultraviolet photons as they flow out of the star (see, for example, Jura and Kroto 1990). Interpretation of the spatial distribution of these molecules indicates that the extinction probably continues to rise in the ultraviolet from the visible. Therefore, many of the particles in circumstellar envelopes are probably not much larger than 0.1 $\mu$m. (iii) Polarization of the optical (Shawl 1974) and near infrared radiation (Tamura et al. 1988 indicates that many of the grains cannot be much smaller than 0.1 $\mu$m; otherwise they would not be effective scatterers. (iv) In the outer circumstellar envelope (the region with r $>$ $10^{15}$ cm), the gas is heated by the grains steeaming supersonically through the gas (Goldreich and Scoville 1976). This heating rate is sensitive to the grain size, and at least in the case of IRC+10216, the best studied circumstellar envelope around a carbon star, it appears that the mean grain size is about 0.04 $\mu$m (Kwan and Hill 1977). However, it some stars, the gas temperature profile may be very different from that in IRC+10216 (Jura, Kahane and Omont 1988).

The overall infrared energy distribution of the light from circumstellar dust shells seems to be well underrstood (Rowan-Robinson and Harris 1983a,b, Rowan-Robinson et al. 1986;

30

Sopka *et al.* 1985, Jura 1986b, Le Bertre 1987, 1988, Martin and Rogers 1987). Models of the observations are most consistent with a grain emisivity that is modelled by a power law which varies more like $\lambda^{-1.2}$ rather than $\lambda^{-2.0}$. In carbon stars, this suggests that the grains are more likely to be composed of amorphous carbon rather than graphite (Borghesi, Bussoletti and Colangeli 1985) even though the emissivity as a function of frequency in the infrared of graphite is not clearly well represented by a simple power law (Martin and Rogers 1987).

## 4. SPECTROSCOPY OF CIRCUMSTELLAR DUST

Although the broad band measurements can provide useful constraints on the nature of the dust, they do not provide nearly as much information on the detailed structure of the grains as does spectroscopy. Merrill (1979) and Kleinmann, Gillett and Joyce (1981) have note that many carbon-rich stars display the SiC feature in emission at 11.3 $\mu$m. Although detailed studies of the *IRAS* data show that not all the emission features in the carbon stars are identical with each other (Baron *et al.* 1987, Papoular 1988), it seems that $\alpha$-SiC is the most likely form of the carrier of the 11.3 $\mu$m feature (Baron *et al.* 1987, Pegourie 1988). An additional feature which is seen in carbon stars losing a large amount of mass is a broad emission near 30 $\mu$m (Forrest, Houck and McCarthy 1981). This feature has been attributed to solid MgS (Goebel and Moseley 1985). Laboratory spectra of candidate solids are discussed by Nuth *et al.* (1985).

Draine (1984) has predicted the presence of a feature at 11.52 $\mu$m in the spectra of carbon-rich stars if the grains are composed of graphite. The absence of this feature in the spectrum of IRC+10216 is consistent with the view that the material is more likely to be amorphous carbon rather than graphite.

In carbon-rich planetary nebulae such as NGC 7027, much of the radiation is carried in discrete bands rather than in a continuum (Russell, Soifer and Willner 1977). The feature at 3.3 $\mu$m is almost certainly produced by C-H stretch (Duley and Williams 1981), and a very promising identification of these species is PAH's (polycyclic aromatic hydrocarbons, Leger and Puget 1984, Leger, d'Hendecourt and Boccara 1987, Omont 1986). However, there are a number of other possible carriers (see Sellgren 1989). At the moment, the physical situation is very unclear; there does seem to be good evidence for a large population of very small grains or large molecules of some sort.

## 5. CIRCUMSTELLAR $HC_7N$: A TEST OF GRAIN EVOLUTION?

The sizes and kinds of many of the carbon particles that are formed in the outflows from these stars are not known. Such information is critical for our understanding of the origin of PAH-like particles and other complex species such as $C_{60}$ and related carbon-cage molecules (Kroto 1988). Here we note that the unusual spatial distribution of $HC_7N$ in the circumstellar envelope around the carbon star AFGL 2688 (Nguyen-Q-Rieu, Winnberg and Bujarrabal 1986) may serve as a valuable tool of the carbon particle chemistry.

There have been a number of models for the gas phase synthesis of the polyynes such as $HC_7N$ in interstellarr and circumstellar environments (Freeman and Millar 1983, Kroto et al. 1987, Leung, Herbst and Heubner 1984, Mitchell, Huntress and Prasad 1979, Nejad and Millar 1987, Schiff and Bohme 1979, Stahler 1984). These models may be valid for the formation of the simpler polyynes such as $HC_3N$ around IRC $+10216$ (Bieging and Nguyen-Q-Rieu 1988). However, these calculations seem unlikely to be able to explain the spatially extended $HC_7N$ around AFGL 2688. Instead, grain-grain collisions (Duley and Williams 1984) seem most promising (Jura and Kroto 1990).

AFGL 2688 displays a highly bipolar geometry for its mass loss (Ney et al. 1975). We expect along the poles that the grains can stream very rapidly and collide with each other. Jura and Kroto (1990) have hypothesized that the grain-grain collisions that occur will result in fragmentation into carbon chains such as $HC_7N$. While this argument is quite uncertain, it does at least have the qualitative ability to reproduce the observed spatial distribution of the circumstellar molecules. We take Figures 1 and 2 from Jura and Kroto (1990). Figure 1 shows the comparison between the observed (solid lines) and predicted (dashed lines) for the spatial variation of the $NH_3$. The observations are taken from Nguyen-Q-Rieu et al. (1986).

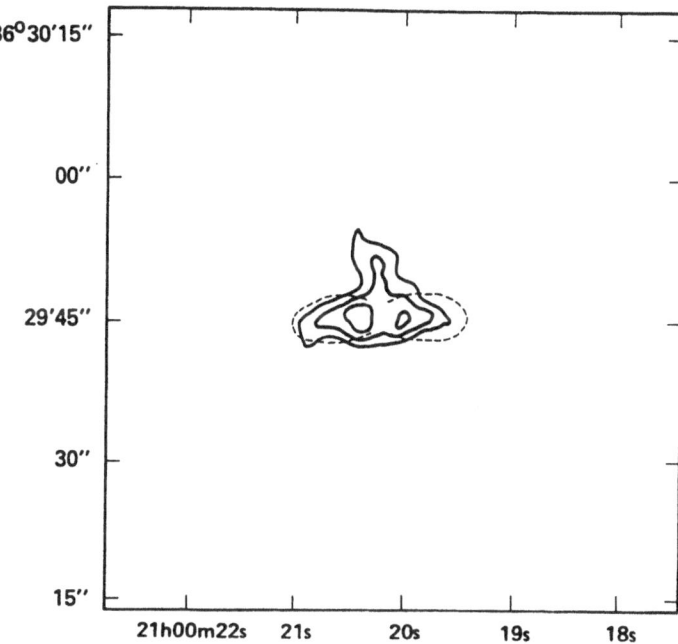

*Fig. 1.* From Jura and Kroto (1990), the observed and computed spatial distribution of $NH_3$ around AFGL 2688.

The basic idea in this theoretical calculation is that $NH_3$ is synthesized in the inner region of the outflow and then is chemically inert ("frozen-out") until it is photodissociated by

32

the ambient interstellar ultraviolet radiation field. The spatial distribution of the NH$_3$ has a flattened appearance because the outflow is mostly confined to a disk. In Figure 2 we display the comparison between theory (dashed line) and observations (solid line) for the spatial distribution of the HC$_7$N molecule. The basic idea of this model is that the HC$_7$N is synthesized by grain-grain collisions at sufficiently high speed which occurs in the lobes of the bipolar outflow.

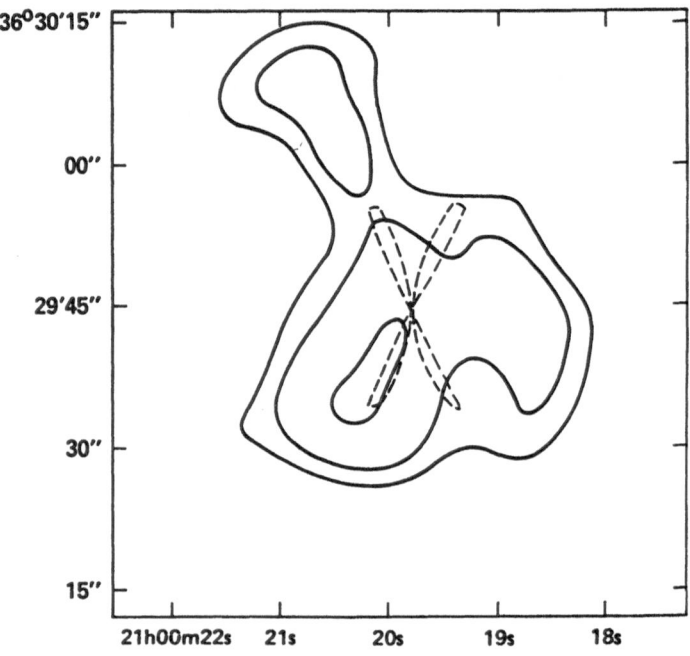

*Fig. 2.* From Jura and Kroto (1990), the observed and computed spatial distribution of HC$_7$N around AFGL 2688.

We conclude that the spatial distribution of the HC$_7$N is consistent with the hypothesis that this molecule is produced in the lobes of the bipolar outflow. It may be important to include grain destruction as well as grain synthesis in the outflows from carbon stars in order to understand fully the origin and evolution of carbon particles.

This work has been partly supported by NASA. I thank S. Kleinmann, H. Kroto, M. Morris and B. Zuckerman for their comments.

6. REFERENCES

Baron, Y., de Muizon, M., Papoular, R., and Pegourie, B. 1987, *Astr. Ap.*, 186, 271.

Bieging, J. H., and Nguyen-Q-Rieu 1988, *Ap. J. (Letters)*, 329, L107.

Borghesi, A., Bussoletti, E., and Colangeli, L. 1985, *Astr. Ap.*, 142, 225.

Cohen, M. 1979, *M.N.R.A.S.*, 186, 837.

Draine, B. T. 1984, *Ap. J. (Letters)*, 277, L71.

Duley, W. W., and Williams, D. A. 1981, *M.N.R.A.S.*, 196, 269.

Duley, W. W., and Williams, D. A. 1984, *M.N.R.A.S.*, 211, 97.

Forrest, W. J., Houck, J. R., and McCarthy, J. F. 1981, *Ap. J.*, 248, 195.

Freeman, A., and Millar, T. J. 1983, *Nature*, 301, 402.

Frenklach, M., and Feigelson, E. D. 1989, *Ap. J.*, 341, 372.

Gail, H.-P., and Sedlmayr, E. 1984, *Astr. Ap.*, 133, 320.

Gail, H.-P., and Sedlmayr, E. 1985, *Astr. Ap.*, 148, 183.

Gail, H.-P., and Sedlmayr, E. 1987, *Astr. Ap.*, 171, 197.

Gail, H.-P., and Sedlmayr, E. 1988, *Astr. Ap.*, 206, 153.

Goebel, J. H., and Moseley, S. H. 1985, *Ap. J. (Letters)*, 290, L35.

Glassgold, A. E., Mamon, G. A., Omont, A., and Lucas, R. 1987, *Astr. Ap.*, 180, 183.

Goldreich, P., and Scoville, N. 1976, *Ap. J.*, 205, 144.

Hoffleit, D., and Jaschek, C. 1982, *The Bright Star Catalogue* (4th ed.; New Haven: Yale University Observatory).

Iben, I., and Renzini, A. 1983, *Ann. Rev. Astr. Ap.*, 21, 271.

Johnson, H., and Zuckerman, B. 1989, in *IAU Colloquium No. 106: Evolution of Peculiar Red Giant Stars* (Cambridge: Cambridge University Press).

Jura, M. 1986a, *Irish Astr. J.*, 17, 322.

Jura, M. 1986b, *Ap. J.*, 303, 327.

Jura, M. 1989, in press.

Jura, M., Kahane, C., and Omont, A. 1988, *Astr. Ap.*, 201, 80.

Jura, M., and Kleinmann, S. 1989, *Ap. J.*, 341, 359.

Jura, M., and Kroto, H. 1990, *Ap. J.*, in press.

Jura, M., and Morris, M. 1985, *Ap. J.*, 292, 487.

Keller, R. 1987, in *Polycyclic Aromatic Hydrocarbons and Astrophysics*, ed. A. Leger, L. d'Hendecourt and N. Boccara (Dordrecht: Reidel), p. 387.

Kleinmann, S. G., Gillett, F. C., and Joyce, R. R. 1981, *Ann. Rev. Astr. Ap.*, 19, 411.

Knapp, G. R., and Morris, M. 1985, *Ap. J.*, 292, 640.

Kroto, H. 1988, *Science*, 242, 1139.

Kroto, H. W., Heath, J. R., O'Brien, S. C., Curl, R. F., and Smalley, R. E. 1987, *Ap. J.*, 314, 352.

Kwan, J., and Hill, F. 1977, *Ap. J.*, 215, 781.

Kwok, S. 1982, *Ap. J.*, 258, 280.

Kwok, S., and Pottasch, S. R. 1987, in *Late Stages of Stellar Evolution (Dordrecht: Reidel)*.

Lafont, S., Lucas, R., and Omont, A. 1982, *Astr. Ap.*, 106, 201.

Le Bertre, T. 1987, *Astr. Ap.*, 176, 107.

Le Bertre, T. 1988, *Astr. Ap.*, 203, 85.

Leger, A., and Puget, J.-L. 1984, *Astr. Ap.*, 137, L5.

Leger, A., d'Hendecourt, L., and Boccara, N. 1987, *Polycyclic Aromatic Hydrocarbons and Astrophysics* (Dordrecht: Reidel).

Leung, C. M., Herbst, E., and Huebner, W. F. 1984, *Ap. J. Suppl.*, 56, 231.

34

Lucy, L. B., Robertson, J. A., and Sharp, C. M. 1986, *Astr. Ap.*, 154, 267.
Mamon, G. A., Glassgold, A. E., and Huggins, P. J. 1988, *Ap. J.*, 328, 797.
Martin, P. G., and Rogers, C. 1987, *Ap. J.*, 322, 374.
McCabe, E. M., Smith, R. C., and Clegg, R. E. S. 1979, *Nature*, 281, 263.
Merrill, K. M. 1979, *Ap. Space Sci.*, 65, 199.
Mitchell, G. F., Huntress, W. T., and Prasad, S. S. 1979, *Ap. J.*, 233, 102.
Morris, M., and Jura, M. 1983, *Ap. J.*, 264, 546.
Morris, M., Redman, R., Reid, M. J., and Dickinson, D. F. 1979, *Ap. J.*, 229, 257.
Morris, M., and Zuckerman, B. 1985 *Mass Loss from Red Giants* (Dordrecht: Reidel).
Nejad, L. A. M., and Millar, T. J. 1987, *Astr. Ap.*, 183, 279.
Ney, E. P., Merrill, K. M., Becklin, E. E., Neugebauer, G., and Wynn-Williams, C. G. 1975, *Ap. J. (Letters)*, 198, L129.
Nguyen-Q-Rieu, Epchtein, N., Truong-Bach, and Cohen, M. 1987, *Astr. Ap.*, 180, 117.
Nguyen-Q-Rieu, Winnberg, A., and Bujarrabal, V. 1986, *Astr. Ap.*, 165, 204.
Nuth, J. A., Moseley, S. H., Silverberg, R. F., Goebel, J. H., and Moore, W. J. 1985, *Ap. J. (Letters)*, 290, L41.
Olofsson, H. 1989, in *Evolution of Peculiar Red Giants*, eds. H. Johnson and B. Zuckerman (Cambridge: Cambridge University Press), p. 321..
Olofsson, H., Eriksson, K., and Gustafsson, B. 1987, *Astr. Ap.*, 183, L13.
Olofsson, H., Eriksson, K., and Gustafsson, B. 1988, *Astr. Ap.*, 196, L1.
Omont, A. 1986, *Astr. Ap.*, 164, 159.
Papoular, R. 1988, *Astr. Ap.*, 204, 138.
Pegourie, B. 1988, *Astr. Ap.*, 194, 335.
Rowan-Robinson, M., and Harris, S. 1983a, *M.N.R.A.S.*, 202, 767.
Rowan-Robinson, M., and Harris, S. 1983b, *M.N.R.A.S.*, 202, 797.
Rowan-Robinson, M., Lock, T. D., Walker, D. W., and Harris, S. 1986, *M.N.R.A.S.*, 222, 273.
Russell, R. W., Soifer, B. T., and Willner, S. P. 1977, *Ap. J. (Letters)*, 217, L149.
Salpeter, E. E. 1974, *Ap. J.*, 193, 585.
Salpeter, E. E. 1977, *Ann. Rev. Astr. Ap.*, 15, 267.
Scalo, J. M., and Ross, J. E. 1976, *Astr. Ap.*, 48, 219.
Schiff, H. I., and Bohme, D. K. 1979, *Ap. J.*, 232, 740.
Sellgren, K. 1989, in press, these proceedings.
Shawl, S. 1974, in *Planets Stars and Nebulae Studied with Photopolarimetry*, ed. T. Gehrels (Tucson: University of Arizona), p. 821.
Sopka, R. J., Hildebrand, R., Jaffe, D. T., Gatley, I., Roellig, T., Werner, M., Jura, M., and Zuckerman, B. 1985, *Ap. J.*, 294, 242.
Stahler, S. 1984, *Ap. J.*, 281, 209.
Tamura, M., Hasegawa, T., Ukita, N., Gatley, I., McLean, I. S., Burton, M. G., Tayner, J. T., McCaughrean, M. J. 1988, *Ap. J. (Letters)*, 326, L17.
Zuckerman, B., and Aller, L. H. 1986, *Ap. J.*, 301, 772.
Zuckerman, B., and Dyck, H. M. 1986a, *Ap. J.*, 304, 394.
Zuckerman, B., and Dyck, H. M. 1986b, *Ap. J.*, 311, 345.
Zuckerman, B., Dyck, H. M., and Claussen, M. 1986, *Ap. J.*, 304, 401.
Zuckerman, B., and Dyck, H. M. 1989, *Astr. Ap.*, 209, 119.

# THE INFRARED INTERSTELLAR EMISSION FEATURES

K. Sellgren

*Institute for Astronomy, University of Hawaii, U.S.A.*

ABSTRACT. The infrared emission features, at 3.3, 6.2, 7.7, 8.6, and 11.3 $\mu$m, are observed throughout the interstellar medium of our own and other galaxies. This review presents recent observational results on spectroscopy of the features, with an emphasis on the more complex picture of the features resulting from the higher spectral resolution and higher sensitivity currently available. Laboratory work on identification of the feature emitting material is also reviewed, and different candidate materials for the features are compared.

## 1 Introduction

The infrared interstellar emission features have provided a challenge for the last fifteen years for observational astronomers and laboratory astrophysicists. Five strong emission features, at 3.3, 6.2, 7.7, 8.6, and 11.3 $\mu$m, were first discovered in the planetary nebula NGC 7027 (Gillett, Forrest, and Merrill 1973; Merrill, Soifer, and Russell 1975; Russell, Soifer, and Willner 1977), and have since been observed in other planetary nebulae, H II regions, bipolar nebulae, reflection nebulae, Herbig Ae stars, Wolf-Rayet stars, novae, the plane of the Milky Way Galaxy, and starburst galaxies (Giard et al., 1988; Hyland and McGregor 1989; and references in Aitken 1981, Allamandola 1984, and Willner 1984). These features are broad and smooth when observed at high spectral resolution, consistent with an origin in a solid state material or in large molecules but ruling out atomic line or molecular band emission. The features are invariably observed together, and always appear in emission. They have not been to date detected in absorption, even toward heavily reddened sources. The material from which the emission features arise was completely unidentified until five years ago, when groundbreaking papers by Léger and Puget (1984) and Allamandola, Tielens, and Barker (1985) independently identified the emission features with a class of organic molecules called polycyclic aromatic hydrocarbons (PAHs). Since then, observational astronomers and laboratory astrophysicists have devoted tremendous efforts to observational tests of this proposed identification and to laboratory work on both PAHs and alternate materials aimed at improving the fit to observations.

This review has two goals: first, to present the most current observational results on the emission features, and second, to present an even-handed and unbiased look at all the laboratory work on the identification of the features. A new generation of infrared spectrometers for astronomy has vastly improved the spectral resolution and sensitivity

*E. Bussoletti and A. A. Vittone (eds.), Dusty Objects in the Universe, 35–47.*

with which the features can be studied in space, and has revealed a wealth of detail in the feature emission that was not previously suspected. The quality and complexity of current observations certainly challenge laboratory astrophysicists in their work on candidate materials, but also provides them with an opportunity to study in detail this ubiquitous component of the interstellar medium.

There have been several recent reviews in the literature on the infrared emission features. Observational overviews have been presented by Bregman (1989) and Allamandola (1989) while more theoretical approaches based on the PAH model have been presented by Allamandola, Tielens, and Barker (1989) and Puget and Léger (1989).

## 2    General characteristics of the features

The infrared emission features are observed virtually everywhere there is a supply of interstellar dust and ultraviolet radiation. In planetary nebulae where dust is being formed, however, the fraction of the total infrared luminosity radiated in the 7.7 $\mu$m feature is proportional to the C/O ratio in the planetary (Cohen et al., 1986, 1989), strongly suggesting that the emitting material includes carbon. Allamandola and Norman (1978) and Duley and Williams (1981) first pointed out that the wavelengths of the emission features correspond to vibrational transitions in organic molecules. While the infrared emission features thus are associated with a carbonaceous component of the interstellar medium, they apparently do not arise from the same carbonaceous material thought to be responsible for the 2200 Å absorption feature. Recent observations of dust forming near HR 4049 show the infrared features in emission but not the 2200 Å feature in absorption (Buss, Lamers, and Snow 1989; Geballe et al., 1989a; Cohen et al., 1989).

The relative strengths of different emission features are observed to vary, but the ratios of features close together in wavelength, such as the 6.2 $\mu$m/7.7 $\mu$m ratio, vary by the same amount as the ratios of features far apart in wavelength, such as the 3.3 $\mu$m/11.3 $\mu$m ratio (Cohen et al., 1986). This argues against excitation as an explanation for the variation in feature ratios. The observations of the feature ratio variations from source to source, and spatially within a single source (Bregman et al., 1989), are most simply explained if the features arise from a mixture. This could be either a mixture of compositions, a mixture of grain sizes, or both.

The features are observed not only in carbon-rich planetary nebulae, where the emitting material is formed, but also in H II regions and reflection nebulae, where the dust is a mixture of older carbon-rich and oxygen-rich dust which has survived for a long time in the general interstellar medium. Recent observations of the plane of the Milky Way galaxy have also shown the 3.3 $\mu$m feature in emission (Giard et al., 1988). The features are thus ubiquitous in the interstellar medium, which demands that the feature material either is difficult to destroy or is easily reformed. If the feature material is frequently destroyed and reformed, then a similar mixture of composition and/or size must be created each time to account for the relative constancy of the feature ratios. Observations suggest that the features are indeed destroyed inside H II regions (Aitken et al., 1979; Sellgren 1981; Bregman et al., 1989; Roche, Aitken, and Smith 1989; Sellgren, Tokunaga, and Nakada 1990) and in active galactic nuclei (Aitken and Roche 1985; Roche and Aitken 1985).

The emission mechanism for the features is almost certainly an non-equilibrium mechanism. Two kinds of models have been proposed: non-equilibrium thermal emission from very small grains, and UV-excited vibrational fluorescence from very large molecules. In the first model, very small grains are briefly heated to very high temperatures by the absorption of single UV photons (Sellgren 1984). In the second model, UV photons absorbed by isolated large molecules, such as PAHs, share their energy among different vibrational modes, which then radiate the infrared emission features via fluorescence (Allamandola, Tielens, and Barker 1985). Both models have in common a very small size for the emitting particle, 5–15 Å in radius. This size is in the uncertain physical regime where it is not clear whether a molecular or solid state description of the emission mechanism is more appropriate; both treatments are used even for the PAH molecule (Allamandola, Tielens, and Barker 1989; Léger, d'Hendecourt, and Défourneau 1989). In a real astrophysical situation a size distribution is likely to exist, perhaps including both very large molecules and very small grains, so it is possible that both mechanisms are important.

## 3  Spectroscopy of the features

The first observations of the infrared emission features, obtained with a spectral resolution $R = \lambda/\Delta\lambda = 50$–100, discovered the five main features shown in table 1, plus an additional weak feature at 3.4 $\mu$m. Current observations made with higher resolution ($R = 400$–1500) and sensitivity now show a large variety of new features. These new features include both weaker features and broad structure in the spectrum; table 1 lists the new features currently known. In the following the current observations of each of these features is discussed.

Table 1: Wavelengths of infrared interstellar emission features.

| Main features | Weaker features | Broad features |
|---|---|---|
| $\lambda$ ($\mu$m) | $\lambda$ ($\mu$m) | $\lambda$ ($\mu$m) |
| 3.3 | 3.40, 3.46, 3.52, 3.57 | 3.2–3.6 Plateau |
| 6.2, 7.7, 8.6 | 5.2, 5.7, 6.9 | 6–9 Broad Bump |
| 11.2 | 11.9, 12.7 | 11–13 Plateau |

### 3.1  The main features

The 3.3 $\mu$m feature is observed, at $R = 1500$, to show two distinctly different feature profiles. A broad 3.3 $\mu$m feature is observed in planetary nebulae, H II regions, and galaxies, and is very reproducible from source to source (Nagata et al., 1988; Tokunaga et al., 1989; Sellgren, Tokunaga, and Nakada 1990). The other profile is twice as narrow, and is so far only observed in two sources, in the central star of the bipolar nebula HD 44179 and in the Herbig Ae star Elias 1 (Tokunaga et al., 1988, 1989). Both the broad and narrow 3.3 $\mu$m feature profiles, however, have central wavelengths which are identical within the errors. The width of the 3.3 $\mu$m feature varies spatially in HD 44179, changing from a narrow profile near the star to a broad profile far from the star (Tokunaga et al., 1988; Geballe et al., 1989*b*).

The 7.7 μm feature, in contrast to the 3.3 μm feature, shows definite variations in the central wavelength of the feature. The peak wavelength is near 7.6 μm for reflection nebulae and H II regions, and near 7.8–7.9 μm for planetary nebulae (Cohen et al., 1989). The central wavelength thus appears to depend on some physical condition in the source, such as the degree of ionization, temperature of the exciting star, or age of the material. There is some evidence that the 7.7 μm feature may in fact be two separate features, accounting for the shift in central wavelength (Cohen et al., 1989).

High resolution observations of the 11.3 μm feature also show variations in the feature profile (Witteborn et al., 1989). Observations of two planetary nebulae, a Wolf-Rayet star, and a bipolar nebula show that all of these sources have similar basic profiles, but that some sources appear to have additional features superposed on the 11.3 μm feature profile.

## 3.2  The weaker features

The 3.4 μm feature, because of its strength, is often included with the main features. However, even early low resolution observations questioned whether the 3.4 μm feature always appeared with the five main features (Aitken 1981). Higher resolution observations by Geballe et al. (1985) showed that the 3.4 μm feature in fact consisted of two independent components, a narrow feature at 3.40 μm and a plateau of emission from 3.2 to 3.6 μm. Geballe et al. (1985) found the strengths of the 3.40 μm feature and the plateau, relative to the main 3.3 μm feature, show strong variations which are independent of each other. Recent high resolution observations ($R = 400$–$1500$) have shown that there is actually a series of narrow features, at 3.40, 3.46, 3.52, and 3.57 μm, as well as the broad and smooth plateau feature (de Muizon et al., 1986; Nagata et al., 1988; Geballe et al., 1989*b*). Geballe et al. (1989*b*) have observed spatial variations in the strength of these features relative to the main 3.3 μm feature in both HD 44179 and the Orion Bar. In Orion they find that the 3.40 μm/3.3 μm and 3.52 μm/3.3 μm ratios increase with distance from the exciting star, while the 3.46 μm/3.3 μm ratio decreases. Magazzù and Strazzulla (1989) find that the 3.40 μm/3.3 μm ratio increases as the diameter of a planetary nebula increases, and interpret this as an evolution in the material with the age of the planetary.

There are several weak features in the 5–7 μm region. Two features, at 5.7 and 6.9 μm, were first discovered by Bregman et al. (1983). A weak feature at 5.2 μm has been predicted by proponents of the PAH identification for the main features (Léger, d'Hendecourt, and Défourneau 1989). This 5.2 μm feature was recently searched for and detected by Allamandola et al. (1989).

Higher resolution spectroscopy has been critical to the discovery of another feature at 12.7 μm. At lower resolution ($R = 50$–$60$), this feature is indistinguishable from [Ne II] emission at 12.8 μm, which is prominent in H II regions and planetary nebulae. Roche, Aitken, and Smith (1989) and Witteborn et al. (1989), however, observed the Orion Bar with $R = 140$–$400$ and clearly showed the presence of a resolved feature at 12.7 μm on which the unresolved 12.8 μm [Ne II] emission was superposed. Roche, Aitken, and Smith (1989) find the spatial distribution of 12.7 μm emission in the Orion Bar follows that of the 11.3 $\mu$m main feature. This is in contrast to the strong spatial variations in the 3.40 $\mu$m/3.3 $\mu$m ratio observed in the Orion Bar (Geballe et al., 1989$b$). Witteborn et al. (1989) have also found a weak feature at 11.9 $\mu$m in several sources.

### 3.3  The broad structure

Spectra of sources showing the infrared emission features show that there is broad structure in the 3–14 $\mu$m region which underlies the main and weak emission features. The 3.2–3.6 $\mu$m emission plateau on the red side of the main 3.3 $\mu$m feature, discovered by Geballe et al. (1985), was discussed in the previous section. A similar plateau of emission on the red side of the main 11.3 $\mu$m feature, covering 11–13 $\mu$m, has also been reported by Cohen, Tielens, and Allamandola (1985), de Muizon et al. (1986), Bregman et al. (1989), Roche, Aitken, and Smith (1989), and Witteborn et al. (1989). The strongest emission from broad structure is from the broad bump of emission at 6–9 $\mu$m, which lies underneath the main features at 6.2 and 7.7 $\mu$m (Cohen et al., 1986; Bregman et al., 1989). Some sources with strong emission from the main features, such as HD 44179, show little or no evidence for broad structure, while in other sources such as NGC 7027 the broad structure dominates the appearance of the spectrum. Bregman et al. (1989) and Roche, Aitken, and Smith (1989) also find that the spatial distributions of the main features in the Orion Bar are different from the spatial distribution of the broad structure. This suggests the main emission features and the broad structure may arise from different components.

### 3.4  The continuum

The continuum emission of most sources showing the infrared emission features, such as H II regions and planetary nebulae, is due to thermal emission from warm dust in radiative equilibrium with the exciting star or is due to free-free emission. In reflection nebulae, however, a near infrared continuum from 1 to 13 $\mu$m is observed which is not due to ordinary thermal emission from dust, free-free emission, or reflected starlight (Sellgren, Werner, and Dinerstein 1983; Sellgren 1984; Sellgren et al., 1985). These reflection nebulae also show the infrared emission features, and the near infrared continuum is spatially associated with the features (Sellgren, Werner, and Dinerstein 1983). Sellgren (1984) has interpreted this near infrared continuum emission as due to thermal emission from very small grains (10 Å radius) briefly heated to high temperatures (1000 K) by the absorption of a single UV photon.

## 4  Laboratory results

Table 2 gives the wavelengths of the main interstellar emission features, and their identifications with specific vibrational transitions for each proposed laboratory model of the features. While the identification of the 3.3 and 11.3 $\mu$m features is generally agreed upon, no agreement among models is seen for the vibrational transitions responsible for the 6.2, 7.7, and 8.6 $\mu$m features.

40

Table 2: Laboratory identifications for the infrared interstellar emission features.

| MAIN FEATURES | | |
| --- | --- | --- |
| $\lambda$ ($\mu$m) | Identifications | Model |
| 3.3 | Aromatic C-H stretch | AC, Coal, PAH, QCC |
| 6.2 | Aromatic C=C in plane stretch | AC, Coal, HAC, Orgueil, PAH, QCC |
| | Aromatic amine -NH$_2$ deformation | AC |
| | Conjugated carbonyl C=O | Coal |
| | Nitrogen-substituted graphitic carbon | NAC |
| | Cross-conjugated ketone C=O stretch | QCC |
| 7.7 | Aromatic amine CN stretch | AC |
| | Aromatic and aliphatic C-O stretch | Coal |
| | Nitrogen-substituted graphitic carbon | NAC |
| | Blend, aromatic C=C stretch | PAH |
| | Cross-conjugated ketone | QCC |
| 8.6 | Terminal ether or alcohol groups | HAC |
| | Aromatic C-H in-plane bend | PAH |
| | Cross-conjugated ketone | QCC |
| 11.3 | Aromatic C-H out-of-plane bend (solo H) | AC, Coal, HAC, Orgueil, PAH, QCC |
| | Aromatic C=C vibration | HAC |

References for table: – AC: Duley and Williams, 1981; Borghesi, Bussoletti and Colangeli, 1987. Coal (Vitrinite): Papoular et al., 1989. HAC: Blanco, Bussoletti and Colangeli, 1988; Ogmen and Duley, 1989. NAC (Nitrogenated Amorphous Carbon): Sapertein, Metin and Kaufman, 1989. Orgueil (carbonaceous residue from meteorite): Wdowiak, Flickinger and Cronin, 1988. PAH: Allamandola, Tielens and Barker, 1989; Léger, d'Hendecourt and Défourneau, 1989. QCC: Sakata et al., 1987.

This disagreement reflects the difficulty of finding a laboratory substance which matches all of the interstellar emission features in wavelength and intensity without predicting additional strong features not observed in the interstellar medium. The problem is most severe for the 7.7 $\mu$m feature; its identification is most problematical, yet it is the strongest by far of the observed interstellar features, and therefore is the most important to identify correctly. In the following only the main features are compared to laboratory substances, since any model must at a minimum be able to reproduce the five main features to be considered a correct identification.

## 4.1 Polycyclic aromatic hydrocarbons

The suggestion that 10 Å radius grains heated to 1000 K were required to explain the near infrared continuum emission of reflection nebulae (Sellgren 1984) led Léger and Puget (1984) to consider the composition of such hot tiny grains. Léger and Puget argued that polycyclic aromatic hydrocarbons (PAHs) were the only particles of 10 Å radius that would be able to survive heating to the high temperatures required. Léger and Puget (1984), and independently Allamandola, Tielens, and Barker (1985), realized that PAHs also had absorption spectra with strong similarities in wavelength to the infrared interstellar emission features. Léger and Puget (1984) modeled the features with a calculation of the emission of coronene ($C_{24}H_{12}$), using a thermal (blackbody) approximation for the emission, while Allamandola, Tielens, and Barker (1985) instead calculated the UV-excited infrared vibrational fluorescence of PAH molecules, with chrysene ($C_{18}H_{12}$) as their example. Both groups derived PAH abundances requiring a few percent of the cosmic carbon abundance. Both groups also argued that PAHs in space must be highly dehydrogenated, to account for the dominance of the 11.3 $\mu$m feature, attributed to a CH bend of an H atom on an aromatic ring with no neighboring H atoms. Léger, d'Hendecourt, and Défourneau (1989) find that compact PAHs provide a better match to the interstellar features than do non-compact PAHs; compact PAHs are also preferred in terms of survival under interstellar conditions. Léger and Puget (1984), ironically enough, were unable to explain the near infrared continuum in reflection nebulae with PAH emission, despite that having been the motivation for their examination of small particles capable of surviving high temperatures. Allamandola, Tielens, and Barker (1985) attributed the near infrared continuum to electronic fluorescence or phosphorescence of PAH molecules, or to emission from overtone and combination bands of PAHs.

The PAH model was a breakthrough in our understanding of the emission features, because it simultaneously accounted for the feature wavelengths, the small particle size, and the emission mechanism for the features. It has gained widespread acceptance for these reasons. Its proponents have considered the astrophysics of PAHs in detail, and have also made many predictions which can be tested by observers. However, these strengths are balanced by a number of problems in the PAH model. Léger and Puget (1984) pointed out serious wavelength mismatches between coronene and the interstellar features: coronene has features at 8.85 and 11.9 $\mu$m, while the features are at 8.6 and 11.3 $\mu$m in space. These discrepancies are well outside observational error. The strong 5.2 $\mu$m feature seen in coronene and other PAHs also is only seen very weakly in the interstellar medium. Wdowiak (1989) has predicted that partially dehydrogenated coronene will have an absorption at 11.2–11.3 $\mu$m instead of 11.9 $\mu$m, which might resolve one problem.

Higher resolution spectroscopy also poses problems for PAHs. Wdowiak (1986) and Sakata et al. (1990) have pointed out that the absorption profiles of the PAH molecules coronene, chrysene, and hexabenzocoronene are a very bad fit to the observed emission profiles of the 3.3 and 11.3 $\mu$m interstellar features. These PAHs, furthermore, are the PAHs whose absorption wavelengths are closest to the interstellar feature wavelengths. Sakata et al. (1990) have shown that the 3.3 $\mu$m absorption wavelengths of the overwhelming majority of PAHs are nowhere near the wavelength of the interstellar 3.3 $\mu$m feature. Since the peak wavelength of the 3.3 $\mu$m feature is observed to be quite constant in space, they argue therefore that no mix of PAHs is likely to fit the interstellar feature.

One possible solution to these difficulties lies in the temperature of the material. Wdowiak (1989) has reported that at room temperature the 3.3 $\mu$m absorption profile of coronene is double peaked and a poor fit to the interstellar 3.3 $\mu$m feature, but that when coronene is reversibly heated to 713–788 K its 3.3 $\mu$m feature becomes single peaked and provides a good fit to the interstellar feature. Temperatures of 700–800 K are similar to those inferred for tiny particles in reflection nebulae (1000 K; Sellgren 1984). Blanco (1989) and Sakata (1989) also report alterations in the absorption spectra of several PAHs when heated. However, even at high temperature, the 7.7 $\mu$m feature in coronene is too narrow and weak, and the 5.2 $\mu$m feature too strong, compared to the interstellar emission (Wdowiak 1989).

Allamandola, Tielens, Barker (1985) have also pointed out that no individual PAH has a 7.7 $\mu$m feature at the same wavelength as the interstellar feature. They suggest a mixture of PAHs may be required. Léger, d'Hendecourt, and Défourneau (1989) also acknowledge the need for a mixture of PAHs to fit the interstellar features, especially the 7.7 $\mu$m feature. Allamandola, Tielens, Barker (1985) compared the interstellar 6.2 and 7.7 $\mu$m features to the Raman spectrum of auto soot, which they argued is a collection of PAHs. This approach has two difficulties: Raman spectra and infrared absorption spectra are not always identical, and auto soot may be a more amorphous and complex substance than a collection of individual PAH molecules.

## 4.2 Amorphous materials

The work described above on PAHs points to the need for mixtures of PAHs, and to the fact that the interstellar emission spectra are often better fitted by more amorphous substances such as auto soot (Allamandola, Tielens, and Barker 1985) than by individual PAHs (Léger and Puget 1984). Many laboratory workers have focussed on amorphous materials, including quenched carbonaceous composite (QCC), hydrogenated amorphous carbon (HAC), amorphous carbon (AC), nitrogenated amorphous carbon (NAC), and coal. Each of these is described in the following.

### 4.2.1 Quenched carbonaceous composite

Sakata et al. (1984) proposed that QCC, an amorphous carbonaceous material made by quenching radicals from a hydrocarbon plasma, could explain the infrared interstellar emission features. Sakata et al. (1987) improved their fit to the 7.7 and 8.6 $\mu$m interstellar features by allowing QCC to oxidize by exposure to air, resulting in oxidized filmy QCC. Sakata et al. (1987) argue that oxygen is required to create a cross-conjugated ketone structure which gives rise to the 7.7 and 8.6 $\mu$m features, features which the PAH model has trouble reproducing. However, even oxidized filmy QCC does not provide a perfect fit in the 7 $\mu$m region; the 6.9 $\mu$m feature is much stronger than the 7.7 $\mu$m feature in QCC, in contrast to the interstellar features; and the very strong 7.3 $\mu$m feature in QCC is not observed in the interstellar medium. Sakata et al. (1990) find that the 3.3 $\mu$m absorption profile of oxidized filmy QCC best fits the interstellar emission profile when heated; the central wavelength of heated QCC is identical to that of the interstellar 3.3 $\mu$m feature, although the width in absorption is still a little broader than even the broadest observed interstellar 3.3 $\mu$m emission feature.

### 4.2.2 Amorphous carbon and hydrogenated amorphous carbon

Duley and Williams (1981) were the first to suggest the interstellar emission features might be due to amorphous carbon (AC) grains. They envisioned large AC grains whose surface functional groups (CH, $CH_3$, and $NH_2$) produced the interstellar emission features. They also were the first to identify the 3.3 and 11.3 $\mu$m features with aromatic CH stretches and bends, an identification which is now commonly accepted (table 2). They were unable, however, to identify the 8.6 $\mu$m feature. Also, their identification of the 6.2 and 7.7 $\mu$m features with aromatic amines predicts a strong feature at 2.9–3.0 $\mu$m which is not observed in the interstellar emission features.

Borghesi, Bussoletti, and Colangeli (1987) have also proposed that amorphous carbon could explain the interstellar features. Their spectrum shows features at 6.3 and 11.3 $\mu$m which agree with the interstellar features, but also show other features which are either much stronger than is seen in the interstellar medium (3.4, 5.8, 6.9 $\mu$m) or which are not seen at all in the interstellar case (7.3 $\mu$m). Their substance also does not show the 3.3, 7.7, or 8.6 $\mu$m features. Heating their AC sample to 670 K improves the fit somewhat by weakening the 3.4 $\mu$m feature and strengthening the 6.2 $\mu$m feature. Ogmen and Duley (1989) have also compared hydrogenated amorphous carbon (HAC) to the interstellar emission features. Their spectrum of HAC actually fits the interstellar absorption spectrum better than the interstellar emission spectrum, but it does show features at 6.2, 8.7, and 11.6 $\mu$m which correspond to emission features. It however lacks features at 3.3 and 7.7 $\mu$m, and shows features at 2.9, 4.7, and 5.9 $\mu$m observed weakly or not at all in the interstellar emission features. They note that because oxygen reacts so strongly with HAC, even a small oxygen contamination in their sample add features due to oxygen bonds.

Blanco, Bussoletti, Colangeli 1988 have suggested that a mixture of HAC grains and PAH molecules is a better fit to the interstellar features, with some features coming from HAC grains and others from PAHs. They use a char created by vacuum pyrolysis of cellulose as their PAH mixture, and consider chars heated both to 330 C (mostly aliphatic hydrocarbons) and 460 C (mostly aromatic hydrocarbons). Each of these three components shows some of the interstellar features, but not all; each component also shows features which are either much stronger than observed in space (3.4, 5.9, 6.9 $\mu$m) or not seen in space at all (7.3 $\mu$m). No mixture measured by Blanco, Bussoletti, and Colangeli (1988) exactly reproduces the interstellar features, but it may be possible to do so if the correct mixture is chosen. Allamandola, Tielens, and Barker (1989) have also suggested that a mixture of PAH molecules, clusters of PAHs, and amorphous carbon grains is required to explain all the components of the interstellar emission features. They show a spectrum of char heated to 480 C which appears to be a good fit to the interstellar features.

Duley (1988) has argued that the structure of HAC grains is a collection of loosely connected clusters of HAC or AC. Each HAC cluster is composed of 5–8 PAH rings while each AC cluster is 20–40 PAH rings. Duley has made the novel suggestion that this disordered structure means the energy of an absorbed UV photon will be localized to a small (10–20 Å) region of the grain. Thus the UV photon energy absorbed by large (0.1 $\mu$m) grains is confined to small enough regions that these regions can radiate the interstellar emission features as though they were individual PAH molecules or tiny grains. One major problem with this idea, however, is the fact that the interstellar emission features are never observed in absorption, which is generally attributed to the small size of the grains. Léger, d'Hendecourt, and Défourneau (1989) have emphasized that the interstellar emission features and interstellar absorption features occur at completely non-overlapping wavelengths in the infrared. This seems to imply that the emission and absorption arises in two populations of grains. Colangeli et al. (1989) have suggested that perhaps the same hydrocarbons exist throughout the interstellar medium, but that whether emission (mostly aromatic) or absorption (mostly aliphatic) is observed depends on the temperature processing of the material. However, the observation of 3.4 $\mu$m aliphatic absorption in the Galactic plane (Butchart et al., 1986) contrasted with the detection of 3.3 $\mu$m aromatic emission from the Galactic plane (Giard et al., 1988) again argues instead for two distinct populations of hydrocarbons.

### 4.2.3 Coal tar and vitrinite

Several groups have investigated coal or coal tars as a laboratory analog to the interstellar emission features. Wdowiak (1986) found that coal tar, heated to 1400 K, provides a good fit to the 3.3 $\mu$m feature, but the remainder of the spectrum was not as good a fit to the interstellar spectrum. Papoular et al. (1989) have considered vitrinite, one component of coal. Vitrinite is a hydrogenated and partially ordered carbon but which has a higher oxygen content and is produced at higher temperature (500 C) than HAC or QCC. The 3.3, 6.2, 7.7, and 11.3 $\mu$m features in vitrinite are a good match to the interstellar feature; Papoular et al., argue that oxygen is essential to the good fit at 7.7 $\mu$m. However, vitrinite shows no 8.6 $\mu$m feature, its 6.9 $\mu$m feature is too strong, and it shows a 3.0 $\mu$m feature not observed in the interstellar emission features. Colangeli et al. (1989) have compared their previous work on HAC and chars to coal tar; again, coal tar provides a much better fit to the 3.3 $\mu$m feature than either HAC or chars, but also has problems with missing or extra features elsewhere in the spectrum.

### 4.2.4 Nitrogenated amorphous carbon

Several groups have found that the addition of oxygen is essential to achieving a good fit to the 7.7 $\mu$m feature (Sakata et al., 1987; Ogmen and Duley 1989; Papoular et al., 1989). This conclusion from laboratory work is in serious conflict with observations by Cohen et al. (1986, 1989) that the strength of the 7.7 $\mu$m feature increases as the C/O ratio increases in planetary nebulae where the features are formed. This discrepancy is only worsened by theoretical work indicating that when carbon is more abundant than oxygen, most of the oxygen is trapped in CO, a very stable molecule. However, Sakata et al. (1987) suggested that the addition of oxygen to QCC may in fact only make the 7.7 $\mu$m feature infrared active. This idea is borne out by work by Saperstein, Metin, and Kaufman (1989). They find that 6.3 and 7.4 $\mu$m features observed in the Raman spectrum of HAC are not observed in infrared absorption, but that adding small amounts of nitrogen to the HAC causes the features to be observed both in absorption and Raman spectra. They argue that the 6.3 and 7.4 $\mu$m absorption features, which are reasonable fits to the interstellar features, are not due to a nitrogen bond but that the addition of nitrogen in the HAC structure breaks the symmetry of HAC rings and makes the features seen in Raman spectra infrared active. Thus the problem of oxygen availability in carbon-rich planetary nebulae may be resolved by substitution of another element which could equally well cause a 7.7 $\mu$m feature to appear in an aromatic hydrocarbon.

## 4.3 Solar system analogs

Allamandola, Sandford, and Wopenka (1987) have compared the Raman spectrum of the interplanetary dust particle Essex to the infrared emission spectrum of the Orion Bar. This interplanetary dust particle was collected from the Earth's upper atmosphere, and thus is not a laboratory analog but an actual material from space to be compared to interstellar dust. They find a very good match between the 6.2 and 7.7 $\mu$m features in Essex and in Orion. However, one must always take care in comparing Raman spectra and infrared absorption or emission spectra. Wdowiak, Flickinger, and Cronin (1988) have studied the carbonaceous residue of the Orgueil meteorite, which is a carbonaceous chondrite type meteorite. Again, this is material from space rather than a laboratory substance. They find a reasonable fit to the 6.2, 7.7, and 11.3 $\mu$m features, especially after heating the residue to 500 C. No. 3.3 $\mu$m feature is seen, however.

# 5 Future work

The laboratory identification work to date shows that many different substances containing aromatic hydrocarbons provide a rough match to the infrared interstellar emission features. All of the proposed substances, however, have problems with predicting all of the observed interstellar features without predicting additional strong features not seen in the interstellar medium; with matching the observed wavelengths precisely; and with matching the observed feature intensity ratios. In general more amorphous materials fit the feature profiles better in detail than do individual PAHs.

Many questions remain. Will amorphous materials still provide a better fit to the features than PAHs if the amorphous material is in 10 Å radius grains? Can the idea that a 10 Å size region of a larger grain responds like an isolated grain or molecule to UV photons be verified in the laboratory? Can a laboratory substance, or mixture of substances, be found that reproduces all the wavelengths and relative strengths of the features, even at high spectral resolution, without producing extra features not observed in space? Can the near infrared continuum also be reproduced? One major problem is that laboratory conditions do not reproduce interstellar conditions. The feature material in space is small (5–15 Å), isolated, and partially dehydrogenated; has a range of sizes and/or compositions; and is observed in emission at high temperature (1000 K) after excitation by a UV photon. If PAHs are present, they are probably ionized. Laboratory materials are usually in much larger grains or clusters, or in a molecular solid; are fully hydrogenated and neutral; and are usually observed in absorption at 300 K. Cherchneff and Barker (1989) have observed infrared fluorescence from a PAH in the laboratory, excited by a UV laser. Their result verifying that the infrared emission and absorption spectra of a PAH are similar is crucial.

It is also important for laboratory and theoretical work to emphasize the astrophysics of a model. It is not enough for a material to match the wavelengths of the features, but the formation, destruction, abundance, emission mechanism, and size of the material must also match the observations. The most important aspect of any proposed model for the features, after accounting for all of the current observations, is to make new predictions that observers can test. It is only through the continued interaction of observations, laboratory work, and theory that this or any other astrophysical problem can be solved.

# References

Aitken, D.K., Roche, P.F., Spenser, P.M., Jones, B.: 1979, Astron. Astrophys., **76**, 60.

Aitken, D.K.: 1981, in *IAU Symposium 96, Infrared Astronomy*, eds. C.G. Wynn-Williams and D.P. Cruikshank (Dordrecht: Reidel), p. 207.

Aitken, D.K., Roche, P.F.: 1985, M.N.R.A.S., **213**, 777.

Allamandola, L.J., Norman, C.A.: 1978, Astron. Astrophys., **63**, L23.

Allamandola, L.J.: 1984, in *Galactic and Extragalactic Infrared Spectroscopy*, eds. M.F. Kessler and J.P. Phillips (Dordrecht: Reidel), p. 5.

Allamandola, L.J., Tielens, A.G.G.M., Barker, J.R.: 1985, Astrophys. J. Lett., **290**, L25.

Allamandola, L.J., Sandford, S.A., Wopenka, B.: 1987, Science, **237**, 56.

Allamandola, L.J., Bregman, J.D., Sandford, S.A., Tielens, A.G.G.M., Witteborn, F.C., Wooden, D.H.: 1989, Astrophys. J. Lett., **345**, L59.

Allamandola, L.J., Tielens, A.G.G.M., Barker, J.R.: 1989, Astrophys. J. Suppl., submitted.

Allamandola, L.J.: 1989, in *IAU Symposium 135, Interstellar Dust*, eds. L.J. Allamandola and A.G.G.M. Tielens (Dordrecht: Reidel), p. 129.

Blanco, A., Bussoletti, E., Colangeli, L.: 1988, Astrophys. J., **334**, 875.

Blanco, A.: 1989, this volume.

Borghesi, A., Bussoletti, E., Colangeli, L.: 1987, Astrophys. J., **314**, 422.

Bregman, J.D., Dinerstein, H.L., Goebel, J.H., Lester, D.E., Witteborn, F.C., Rank, D.M.: 1983, Astrophys. J., **274**, 666.

Bregman, J.D., Allamandola, L.J., Tielens, A.G.G.M., Geballe, T.R., Witteborn, F. C.: 1989, Astrophys. J., **344**, 791.

Bregman, J.D. 1989: in *IAU Symposium 135, Interstellar Dust*, eds. L.J. Allamandola and A.G.G.M. Tielens (Dordrecht: Reidel), p. 109.

Buss, R.H., Lamers, H.J.G.L.M., Snow, T.P.: 1989, preprint.

Butchart, I.B., McFadzean, A.D., Whittet, D.C.B., Geballe, T.R., Greenberg, J.M.: 1986, Astron.. Astrophys., **154**, L5.

Cherchneff, I., Barker, J.R.: 1989, Astrophys. J. Lett., **341**, L21.

Cohen, M., Tielens, A.G.G.M., Allamandola, L.J.: 1985, Astrophys. J. Lett., **299**, L93.

Cohen, M., Allamandola, L.J., Tielens, A.G.G.M., Bregman, J., Simpson, J.P., Witteborn, F.C., Wooden, D. Rank, D.: 1986, Astrophys. J., **302**, 737.

Cohen, M., Tielens, A.G.G.M., Bregman, J., Witteborn, F.C., Rank, D.M., Allamandola, L.J., Wooden, D.H., de Muizon, M.: 1989, Astrophys. J., **341**, 246.

Colangeli, L., Schwehm, G., Bussoletti, E., Fonti, S., Blanco, A., Orofino, V.: 1989, preprint.

de Muizon, M., Geballe, T.R., d'Hendecourt, L.B. Baas, F.: 1986, Astrophys. J. Lett., **306**, L105.

Duley, W.W., Williams, D.A. 1981, M.N.R.A.S., **196**, 269.

Duley, W.W.: 1988, M.N.R.A.S., **234**, 61P.

Geballe, T.R., Lacy, J.H., Persson, S.E., McGregor, P.J., Soifer, B.T.: 1985, Astrophys. J., **292**, 500.

Geballe, T.R., Noll, K.S., Whittet, D.C.B., Waters, L.B.F.M.: 1989a, Astrophys. J. Lett., **340**, L29.

Geballe, T.R., Tielens, A.G.G.M., Allamandola, L.J., Moorhouse, A., Brand, P.W.J.L.: 1989b, Astrophys. J., **341**, 278.

Giard, M., Pajot, F., Lamarre, J.M., Serra, G., Caux, E., Gispert, R., Léger, A., Rouan, D.: 1988, Astron. Astrophys., **201**, L1.

Gillett, F.C., Forrest, W.J., Merrill, K.M.: 1973, Astrophys. J., **183**, 87.

Hyland, A.R., McGregor, P.J.: 1989, abstract at *IAU Symposium 135, Interstellar Dust.*

Léger, A., Puget, J.L.: 1984, Astron. Astrophys., **137**, L5.

Léger, A., d'Hendecourt, L., Défourneau, D.: 1989, Astron. Astrophys., **216**, 148.

Magazzù, A., Strazzulla, G.: 1989, Astrophys. J. Lett., submitted.

Merrill, K.M., Soifer, B.T., Russell, R.W.: 1975, Astrophys. J. Lett., **200**, L37.

Nagata, T., Tokunaga, A.T., Sellgren, K., Smith, R.G., Onaka, T., Nakada, Y., Sakata, A.: 1988, Astrophys. J., **326**, 157.

Ogmen, M., Duley, W.W.: 1989, preprint.

Papoular, R., Conard, J., Guiliano, M., Kister, J., Mille, G.: 1989, Astron. Astrophys., **217**, 204.

Puget, J.L. , Léger, A. 1989, Ann. Rev. Astr. Astrophys., **27**, 161.

Roche, P. F., Aitken, D.K.: 1985, M.N.R.A.S., **213**, 789.

Roche, P.F., Aitken, D.K., Smith, C.H.: 1989, M.N.R.A.S., **236**, 485.

Russell, R.W., Soifer, B.T., Willner, S.P.: 1977, Astrophys. J. Lett., **217**, L149.

Sakata, A., Wada, S., Tanabe, T., Onaka, T.: 1984, Astrophys. J. Lett., **287**, L51.

Sakata, A., Wada, S., Onaka, T., Tokunaga, A.T.: 1987, Astrophys. J. Lett., **320**, L63.

Sakata, A., Wada, S., Onaka, T., Tokunaga, A.T.: 1990, Astrophys. J., in press.

Sakata, A.: 1989, this volume.

Saperstein, D.D., Metin, S.S., Kaufman, J.H.: 1989, Astrophys. J. Lett., **342**, L47.

Sellgren, K.: 1981, Astrophys. J., **245**, 138.

Sellgren, K., Werner, M.W., Dinerstein, H.L.: 1983, Astrophys. J. Lett., **271**, L13.

Sellgren, K.: 1984, Astrophys. J., **277**, 623.

Sellgren, K., Allamola, L.J., Bregman, J.D., Werner, M.W., Wooden, D.H.: 1985, Astrophys. J., **299**, 416.

Sellgren, K., Tokunaga, A.T., Nakada, Y.: 1990, Astrophys. J., in press.

Tokunaga, A.T., Nagata, T., Sellgren, K., Smith, R.G., Onaka, T., Nakada, Y., Sakata, A., Wada, S.: 1988, Astrophys. J., **328**, 709.

Tokunaga, A.T., Sellgren, K., Sakata, A., Wada, S., Onaka, T., Nakada, Y., Nagata, T.: 1989, abstract at *IAU Symposium 135, Interstellar Dust.*

Wdowiak, T.J.: 1986, B.A.A.S., **18**, 1030.

Wdowiak, T.J., Flickinger, G.C., Cronin, J.R. 1988, Astrophys. J. Lett., **328**, L75.

Wdowiak, T.J.: 1989, this volume.

Willner, S.P.: 1984, in *Galactic Extragalactic Infrared Spectroscopy*, eds. M.F. Kessler J.P. Phillips (Dordrecht: Reidel), p. 37.

Witteborn, F.C., Sford, S.A., Bregman, J.D., Allamola, L.J., Cohen, M., Wooden, D.H., Graps, A.L.: 1989, Astrophys. J., **341**, 270.

# QUENCED CARBONACEOUS COMPOSITE (QCC): THERMAL ALTERATION OF QCC AND PAHs

A. Sakata[1], S. Wada[2], T. Onaka[3], A.T. Tokunaga[4]
[1]*Dept. of Applied Phys. and Chem., Univ. of Electro-Comm., Tokyo, Japan*
[2]*Lab. of Chemistry, Univ. of Electro-Communications, Tokyo, Japan*
[3]*Dept. of Astronomy, Faculty of Science, Univ. of Tokyo, Japan*
[4]*Inst. for Astronomy, Univ. of Hawaii, Honolulu, U.S.A.*

# 1    Introduction

There is intense interest about the family of emission features at 3.3, 3.4, 6.2, 7.7, 8.6, and 11.3 $\mu$m. These features are observed in many diverse types of astronomical objects (reviewed by Allamandola, 1984; Willner, 1984; Léger and d'Hendecourt, 1987; Allamandola, Tielens, and Barker, 1987). It has been postulated that these features may arise from aromatic hydrocarbons (Duley and Williams, 1981; Léger and Puget, 1984; Allamandola, Tielens and Barker, 1985). It has also been proposed by Sakata et al. (1984, 1987) that the emission features may arise from "quenched carbonaceous composite" (QCC), obtained from hydrocarbon plasma.

Organic molecules and carbonaceous solid materials show specific absorbance peaks and band shapes that result from the stretching vibration of carbon-hydrogen bonds. High resolution observation of 3 micron feature were obtained by Nagata et al. (1988) and Tokunaga et al. (1988) to test the PAH and QCC hypotheses. They found that the shape of the 3.3 micron feature is a broad single-peaked feature and its peak position is located at 3.295±0.005 $\mu$m in all observed astronomical sources. These observations suggest that the material giving rise to the infrared emission features are of a very specific structure.

In order to understand what type of carbonaceous materials have an absorbance peak precisely at 3.295 micron, we have obtained: 1) measurements of the 3.3 micron features of PAH molecules such as coronene and crysene; 2) high resolution measurement of thermally-altered PAH molecules; 3) high resolution measurement of thermally-altered QCC. By comparing the observed data with the observed emission feature, we can draw some tentative conclusions about the structure of the carbonaceous material in space.

*E. Bussoletti and A. A. Vittone (eds.), Dusty Objects in the Universe, 49–55.*
© 1990 *Kluwer Academic Publishers.*

## 2  Experimental

The 3 micron spectra of our samples were measured with an infrared spectrophotometer (JASCO IR-810), with a resolving power of 0.003 micron.

The PAH samples at room temperature are solid except for benzene. The PAH molecules were commercial reagents and their spectra were obtained with the KBr pellet method. Thermal alteration of the PAH molecules was carried out as follows. A pyrex or quartz glass tube of 6 mm in diameter and 100 mm in length, containing a sample of about 20 mg, was evacuated. The tube containing the sample was cut off with fire flame and then heated in an electrical furnace at a fixed temperature for 30 minutes. The sample was cooled the room temperature and was measured with the infrared spectrometer.

QCC is a quenced carbonaceous material condensed from ejected gases out of a hydrocarbon plasma. Details of the experimental procedure have been given by Sakata et al. (1983, 1987). Filmy material, named as "filmy QCC", was condensed from the plasma onto the wall of the reaction chamber, and it was collected on a NaCl substrate set at a position 50 mm away from the orifice of the plasma chamber.

Thermal alteration of "filmy-QCC" was achieved as follows. The filmy QCC on the NaCl substrate was placed in a specially-made fused quartz container. The quartz container was then heated under vacuum in an electric furnace for 30 minutes at a fixed temperature. After heating, the sample was cooled to room temperature and was measured with the infrared spectrometer. The sample was placed in the quartz container again and the procedure was repeated at the next temperature.

## 3  Results and discussion

### 3.1  IR spectra of PAH molecules

To obtain IR spectra of high spectral resolution, we measured IR spectra of 20 PAHs. They are molecules composed of fused aromatic rings with the number of rings from 1 to 9 and a special 13 ring molecule, hexabenzocoronene.

15 out of 20 PAHs have peak absorbance at a shorter wavelength than the observed emission peak in astronomical sources at $3.295 \pm 0.005$ $\mu$m. Among the 20 PAHs only 3, (benzene, 1, 2, –5, 6, –dibenzanthracene, and benzo[a]pyrene) show a peak close to that the observed emission. The 3.3 micron band of PAHs has a complex structure with multiple components, each peak of which is sharper than the observed emission. But the whole spectral shape of PAHs around 3.3 $\mu$m is broader than the astronomical emission band. A summary of the PAH absorbance peaks will be given by Sakata et al. (1990).

In figure 1, we present IR spectra of three PAHs with a peak close to that of the observed emission and show IR spectra of crysene and coronene, which Allamandola, Tielens and Barker (1985) and Léger and d'Hendecourt (1987) have used as representative interstellar PAHs.

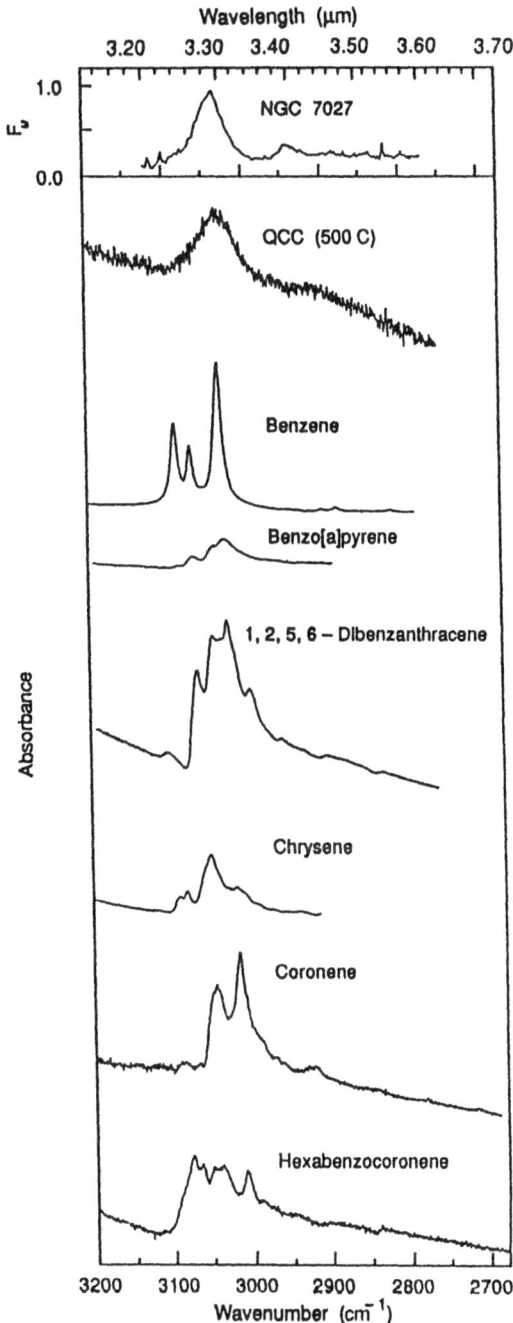

**Figure 1:** The spectra of various samples of PAHs, the film QCC after heating to 500 °C and NGC 7027 obtained By Nagata et al. (1988). The Pf-$\delta$ line at 3.297 $\mu$m has been removed from the spectrum of NGC 7027.

## 3.2 IR spectra of thermally altered PAHs

The structure of PAH molecules is classified to be of two types: one is peri-structure, which has a compact arrangement of aromatic rings such as pyrene, perylene, and coronene, and the other is acene-structure, which has an open, branched arrangement of aromatic molecules such as anthracene, crysene, and naphthacene.

In figure 2, we show the spectra of thermally altered PAHs.

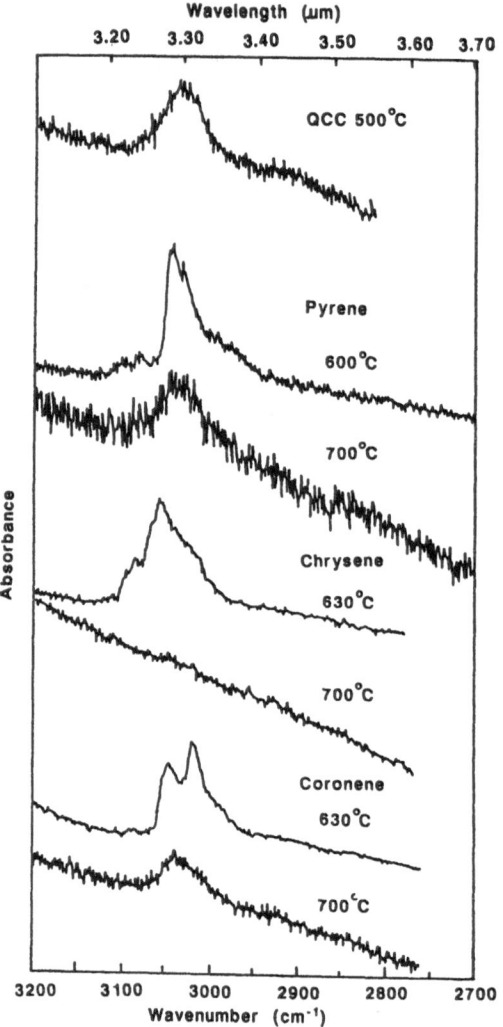

**Figure 2:** The spectra of thermally-altered PAHs and the thermally-altered film QCC.

The PAH molecules were not altered at temperatures less than 350 °C. At 400 °C, the PAH molecules were partially converted to dark carbonaceous matter, but IR spectra were same as those of unaltered PAH molecules. The change of IR spectra begins above

500 °C. At 700 °C, the PAHs were completely altered into black flake carbonaceous material.

The IR spectrum of crysene changed successively above 500 °C. At 630 °C, the IR spectrum of thermally-altered crysene changed into a single feature peaked at 3.279 $\mu$m, but at 700 °C this peak disappeared. Therefore at this temperature crysene was converted into a carbonaceous material which has very little hydrogen (i.e. little C-H bonds).

The IR spectra of pyrene and coronene changed abruptly at 700 °C. The 3.3 micron feature of thermally-altered pyrene and coronene was observed to be a single weak peak. Their peaks were located at 3.289±0.003 $\mu$m. Therefore the heated pyrene and coronene were thermally-altered into carbonaceous material which contain small amount of hydrogen also.

In summary, when PAHs were heated to 600~ 700°C, a single-peaked feature in their 3.3 micron IR spectrum was observed at a wavelength close to but at a slightly shorter wavelength than the emission feature observed in astrophysical sources.

## 3.3   IR spectra of thermally altered QCC

In figure 3, thermal alteration of f-QCC is shown.

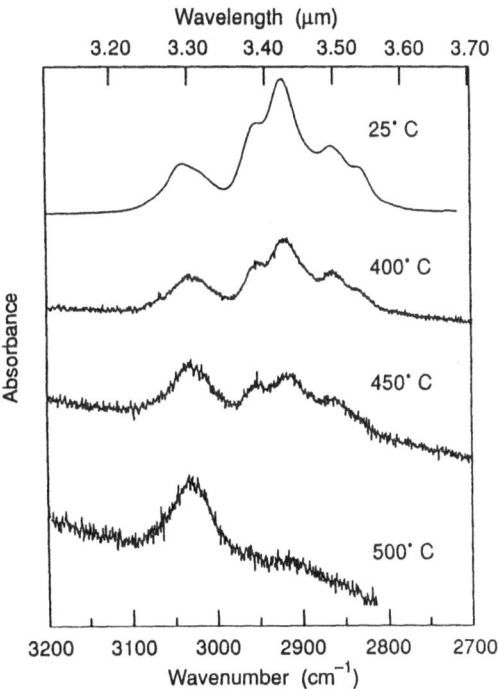

**Figure 3:** The spectra of filmy QCC at room temperature and after heating to variuos temperatures.

The brownish f-QCC at room temperature was converted into a dark black material at a temperature of about 300 °C. At 500 °C, a single-peaked feature survived. The peak is located at 3.295±0.003 μm. Thus the IR spectrum of the thermally-altered f-QCC matched well that of the observed emission feature as shown in figure 4. Further details are discussed by Sakata et al. (1990).

A slight difference of the peak wavelength among thermally-altered PAHs and QCC results from a difference of peripheral structure around the C-H bonds that survived heating to 600~ 700°C.

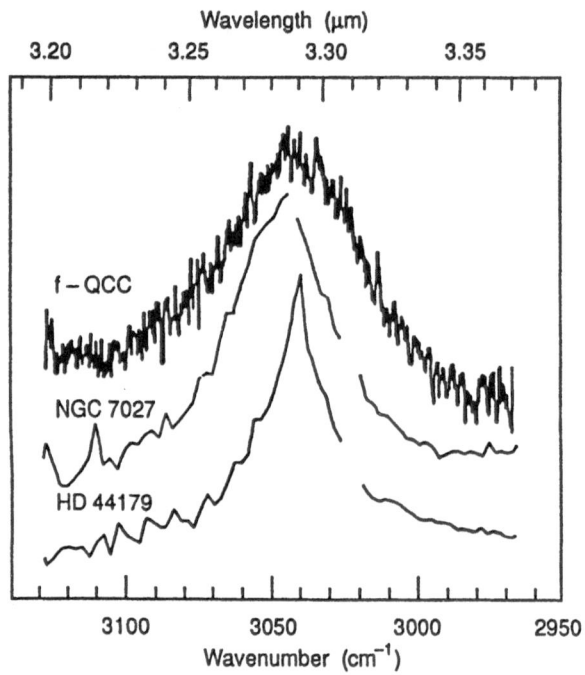

**Figure 4:** Comparison of the absorbance of thermally-altered filmy QCC to the emission spectra of NGC 7027 and HD 44179.

## 4  Conclusion

1. The 3.3 micron IR feature of PAH molecules did not match to those of the observed emission.

2. PAHs and f-QCC were converted into a dark carbonaceous material at an elevated temperature. These thermally-altered carbonaceous materials have a single-peaked feature around 3.29 μm. This peak is caused by a small amount of hydrogren attached to edges of the carbon skeletal structure of thermally-altered PAHs and QCC.

3. Among these thermally-altered materials, the 3.3 micron feature of thermally-altered film QCC at 3.295±0.003 μm matched best to that of the astrophysical sources at 3.295 ±0.005 μm.

# References

Allamandola, L.J.: 1984, "Galactic and extragalactic infrared astronomy", eds. K.F. Kessler, J.P. Phillips, Reidel, Dordrecht, p. 5.

Allamandola, L.J.: 1987, in "Physical processes in interstellar clouds", eds. G. Morfill, M. Scholer, Reidel, Dordrecht, p. 305.

Allamandola. L.J., Tielens, A.G.G.M., Barker, J.R.: 1985, Ap.J. (Letters), **290**, L25.

Duley, W.W., Williams, D.A.: 1981, M.N.R.A.S., **196**, 269.

Léger, A., d'Hendecourt, L.: 1987, in "Polycyclic aromatic hydrocarbons and astrophysics", eds. A. Léger, L. d'Hendecourt, N. Boccara, Reidel, Dordrecht, p. 223.

Léger, A., Puget, J.L.: 1984, Astr. Ap., **137**, L5

Nagata, Y., Tokunaga, A.T., Sellgren, K., Smith, R.G., Onaka, T., Nakada, Y., Sakata, A.: 1988, Ap. J., **326**, 157.

Sakata, A., Wada, S., Okutsu, Y., Shintani, H., Nakada, Y.: 1983, Nature, **301**, 493.

Sakata, A., Wada, S., Tanabé, T., Onaka, T.: 1984, Ap. J. (Letters), **287**, L51.

Sakata, A., Wada, S., Onaka, T., Tokunaga, A.T.: 1987, Ap. J. (Letters), **320**, L63.

Sakata, A., Wada, S., Onaka, T., Tokunaga, A.T.: 1990, Ap. J., in press.

Tokunaga, A.T., Nagata, Y., Sellgren, K., Smith, R.G., Onaka, T., Nakada, Y., Sakata, A., Wada, S.: 1988, Ap. J., **328**, 709.

Willner, S.P.: 1984, in "Galactic and extragalactic infrared astronomy", eds. K.F. Kessler, J.P. Phillips, Reidel, Dordrecht, p. 37.

# MINERAL GRAINS IN INTERSTELLAR SPACE

N.C. Wickramasinghe, F. Hoyle, S. Al-Mufti, T. Al-Jabory
*University of Wales College, Cardiff, U.K.*

ABSTRACT. Astronomical data relating to emission, absorption and polarisation properties of interstellar dust in the mid-infrared spectral region point to a composite mineral-organic grain model with overall mass proportions consistent with cosmic abundances.

## 1 Introduction

As early as 1974 one of the present authors had argued that the detailed shape of the 10m emission feature in the Trapezium nebula pointed to a significant contribution from condensed organic matter in a complex polymeric form (Wickramasinghe, 1974, 1975; Cooke and Wickramasinghe, 1977; Cooke, 1977). Silicate core-organic mantle grains were found to produce better agreement with the data than either organic material or silicate material alone (Cooke and Wickramasinghe, 1975). A similar conclusion has been reached long afterwards by Cohen et al. (1989).

Since the original discovery of the 10m feature in astronomical spectra (Woolf and Ney, 1969; Stein and Gillett, 1969) there has been a sustained effort by astronomers to attempt an explanation exclusively on the basis of mineral particles. Yet, significant discrepancies have persisted between calculated behaviour of silicate dust and the astronomical data, particularly in relation to emission from the Trapezium nebula. Figure 1 illustrates a problem that appears to be endemic for pure mineral grain models, where calculations of flux from amorphous and hydrated silicates heated to 175 K are compared with the normalised flux data for the Trapezium (Forrest et al., 1975a,b). Simple magnesium silicates are evidently lacking in absorption over the entire 8–9 $\mu$m wavelength interval.

A widely prevalent astronomical practice has been to ignore discrepancies of the type seen in figure 1 and to define a notional "astronomical silicate" that artificially accorded

57

*E. Bussoletti and A. A. Vittone (eds.), Dusty Objects in the Universe, 57–68.*
© 1990 *Kluwer Academic Publishers.*

58

with the Trapezium flux data (Draine and Lee, 1984).

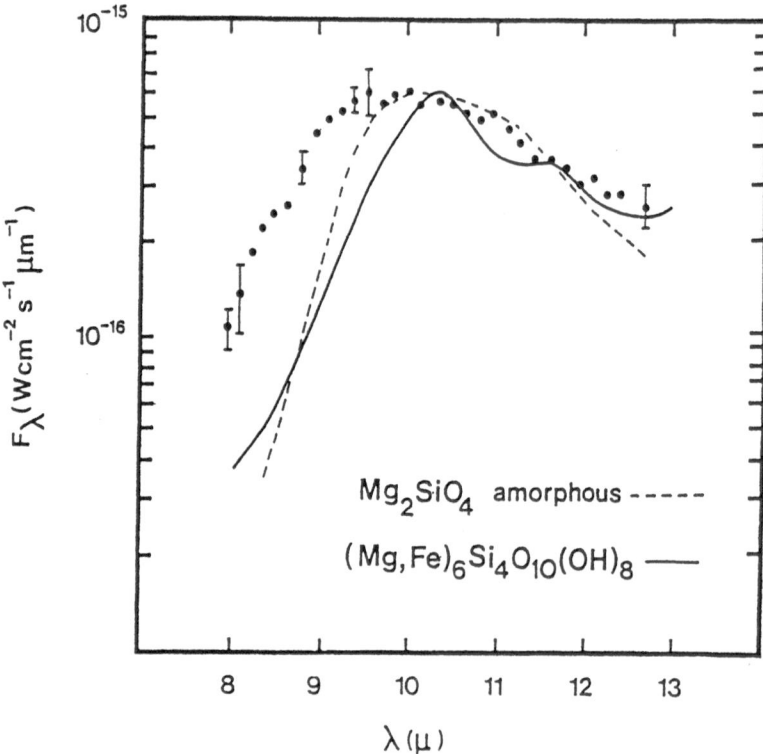

**Figure 1:** Flux data for the Trapezium nebula (dots) compared with the theoretical flux for models of magnesium silicate grains at a temperature of 175 K. Normalization is to F $= 6\times10^{-16}$ W cm$^{-2}$ s$^{-1}$ $\mu$m at the wavelength of maximum flux. The data is taken from Forrest et al. (1975a,b).

Such a practice is not one that has appealed to the present authors (Hoyle and Wickramasinghe, 1986). If the optical constants of the Draine-Lee "astronomical silicate" can be demonstrated to represent accurately the behaviour of a plausible mixture of cosmically abundant minerals, then this procedure might have been justified. If, on the other hand, plausible mineral mixtures cannot be found to match the tabulations of Draine (1985) the entire procedure should be viewed with caution. It is clear that organic polymers with C-N, C-O, C-C and C-O-C linkages must contribute to absorptions and emissions over the mid-infrared spectral region overlapping absorption bands arising from SiO$_4$ tetrahedra in silicate materials, particularly over the 8–12 $\mu$m and 15–25 $\mu$m wavebands. To neglect such contributions might have been justified in 1969, when there was no other evidence for the presence of interstellar organic grains, but it is not justified now with the evidence available for the widepsread occurrence of such material (Hoyle and Wickramasinghe, 1989).

The optical characteristics of minerals differ significantly from those of organic solids in the 8–30 $\mu$m waveband in several respects. Minerals have high values of k near the 10

$\mu$m and 20 $\mu$m peaks, k~1 to 2.5; whereas organic polymers (e.g. typified by bacteria) have k~0.1 at the corresponding peaks. Also, the refractive index n for silica and silicates has a strongly dispersive type of variation with wavelength near each of these absorption peaks, whereas that for organics is nearly constant. We shall show in section 4 that such differences in optical properties are relevant for interpreting observations of the polarization to extinction ratio in the 10 $\mu$m interstellar absorption band.

## 2 Mixtures of minerals

To decide whether plausible mixtures of minerals could produce an average 10 $\mu$m absorption band profile which is in agreement with the Trapezium data, we collected transmittance spectra of 89 minerals from standard atlases (Keller et al., 1952; Hunt et al., 1950). Each spectrum was electronically digitised and calibrated to give a $\tau(\lambda)$ curve that was normalised so that $\tau(9.5 \ \mu m) = 1$. The entire set of normalised spectra were then averaged to give a $\tau(\lambda)$ curve, so that each mineral contributed equally to the opacity at $\lambda = 9.5\mu$m. The list of minerals included in our analysis is set out in table 1.

For this mixture of minerals the best fit to the Trapezium flux data was obtained with a grain temperature T = 200 K. Figure 2 shows a normalised flux

$$F_\lambda \propto \bar{\tau}(\lambda)B_\lambda \qquad (200 \text{ K}) \qquad (1)$$

where $B_\lambda(T)$ denotes the Planck function. The closeness of the fit seen in figure 2 would have instantly vindicated the idea of a mineral mixture in an elegant way if the combination of atoms contained in these minerals was cosmically plausible.

A simple tally of the principal elements in the mixture obtained by adding the relevant suffixes in the molecular formulae, after standardisation to log A = 7.52 (where A is the abundance by number) gives the relative abundance distribution shown in the third column of table 2. The fourth column shows the distribution of elements in the Sun (Cameron, 1970). The elements Al, Ca, K, Mn and Zn are seen from this table to be in considerable excess over solar values. It is the inclusion of such elements in the silicate structures that effectively provides absorption over the 8–9 $\mu$m waveband, the deficient region in the curves of figure 1. Restricting the mixture only to those minerals possessing cosmically plausible proportions of elements leads inevitably to a return to the difficulty seen in figure 1.

Another problem associated with models confined to silicates alone is that both the 10 $\mu$m and 20 $\mu$m emission features of the Trapezium nebula could not be simultaneously matched. The points in figure 3 represent the data of Forrest et al. (1976).

The curves show the situation for amorphous silicate particles (Krätschmer, 1986) with two different values of the temperature. Neither a mixture of minerals nor any reasonable distribution of temperatures can be found to remove the discrepancies shown in figure 3.

**Table 1:** Mineral spectra that were averaged.

| | |
|---|---|
| (1) Sillimanite | (46) Pink Microcline |
| (2) Staurolite | (47) Microcline VAR. Amazonstone |
| (3) Idocrase | (48) Clevelandite |
| (4) Hemimorphite | (49) Analcite |
| (5) Melilite | (50) Nepheline |
| (6) Beryl | (51) Heulandite |
| (7) Benitoite | (52) Diatomite |
| (8) Biotite | (53) Natrolite |
| (9) Muscovite | (54) Smithsonite |
| (10) Biotite flake | (55) Gypsum |
| (11) Pyrophyllite | (56) Pyrite |
| (12) Talc | (57) Galena |
| (13) Prochlorite | (58) Chert |
| (14) Selenite | (59) Opal |
| (15) Amblygonite | (60) Hematite |
| (16) Apatite | (61) Ilmenite |
| (17) Olivine | (62) Garnet |
| (18) Willemite | (63) Augite |
| (19) Kyanite | (64) Actinolite |
| (20) Andalusite | (65) Hornblende |
| (21) Magnesite | (66) Wollastonite |
| (22) Siderite | (67) Albite |
| (23) Rhodachrosite | (68) Oligoclase |
| (24) Calcite | (69) Anorth |
| (25) Dolomite | (70) Nepheline |
| (26) Aragonite | (71) Kaolinite |
| (27) Strontianite | (72) Illite |
| (28) Cordierite | (73) Montmorillonite |
| (29) Tourmaline | (74) Louisa loam |
| (30) Diopside | (75) Cecil clay loam |
| (31) Hypersthene | (76) Loess |
| (32) Anthophyllite | (77) Residual soil from granite |
| (33) Tremolite | (78) Flint clay |
| (34) Muscovite | (79) Plastic clay |
| (35) Witherite | (80) Fullers earth |
| (36) Cerussite | (81) Diatomaceous earth |
| (37) Calcite aragonite | (82) Reef limestone |
| (38) Calcite magnesite | (83) Limy oil shale |
| (39) Barite | (84) Mahogany shale |
| (40) Celestite | (85) Calcium oxalate |
| (41) Anhydrite | (86) Magnesium oxide |
| (42) Rock crystal | (87) Calcium phosphate |
| (43) Rose quartz | (88) Silica gel |
| (44) Orthoclase | (89) Barium sulfate |
| (45) Orthoclase VAR. Adularia | |

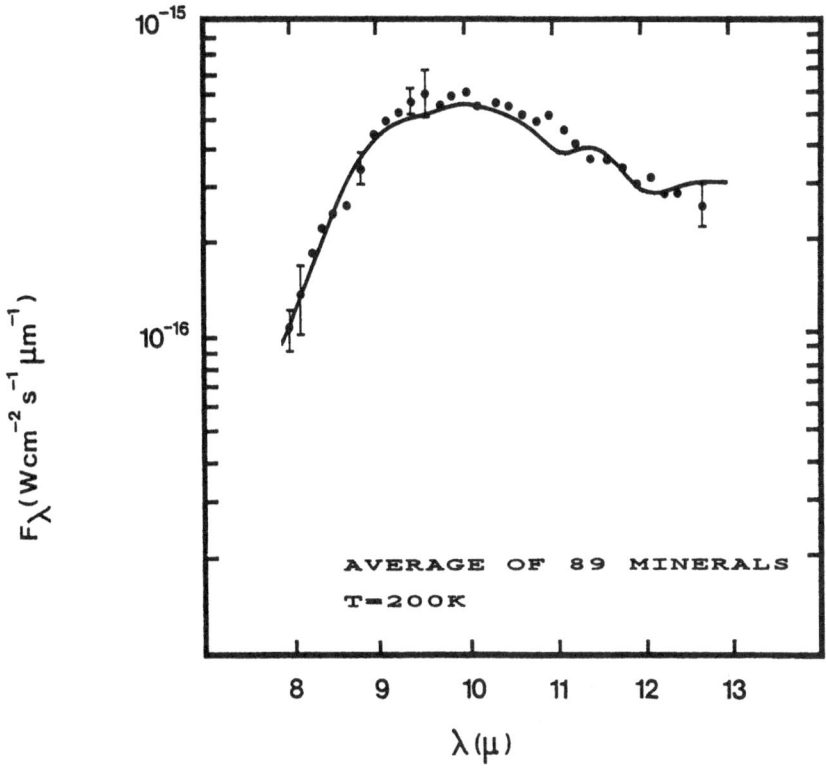

**Figure 2:** The calculated normalised flux curve for a mixture of minerals including high contents of Al and other cosmically underabundant elements. The points are the data for the Trapezium nebula (Forrest et al., 1975a,b). The particle temperature is taken to be 200 K.

**Table 2:** Elements in mineral mixture and in the cosmos.

| Element | AW | log A | log A |
|---------|----|-------|-------|
| Si | 28 | 7.52 | 7.52 |
| Al | 27 | 7.30 | 6.39 |
| Fe | 56 | 7.09 | 7.60 |
| Mg | 24 | 6.99 | 7.42 |
| Ca | 40 | 6.88 | 6.30 |
| Na | 23 | 6.61 | 6.25 |
| F | 19 | 6.57 | 4.60 |
| K | 39 | 6.47 | 4.95 |
| Mn | 55 | 6.17 | 5.40 |
| Zn | 65 | 6.12 | 4.20 |

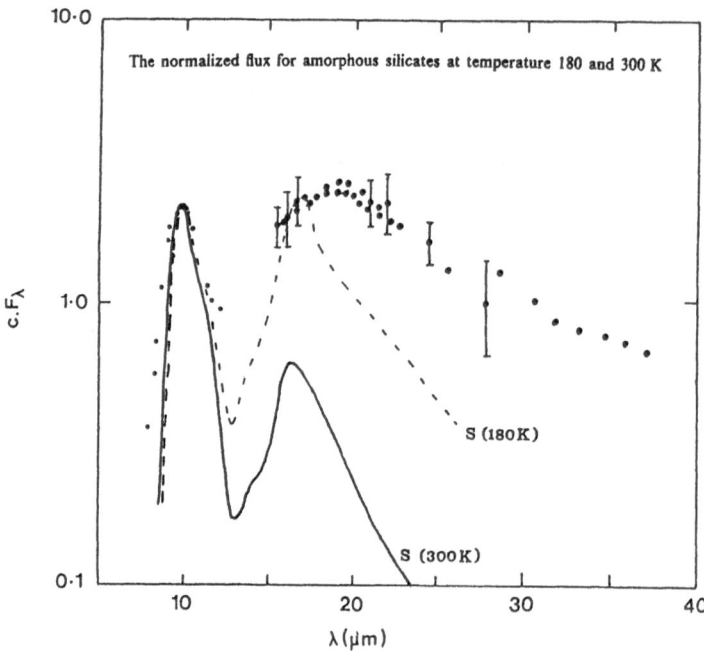

**Figure 3:** Normalised flux curves for amorphous silicate particles at temperatures of 180 and 300 K. The points are the observations for the Trapezium nebula (Forrest et al., 1976).

# 3   Mineral/organic combinations

Our search for a unified grain model which combines silicate material and organics in acceptable "cosmic" proportions, and which produces a close agreement with the Trapezium data in the 8–30 $\mu$m spectral region led eventually to a model that involved a mixture of diatoms (Hoyle and Wickramasinghe, 1984; Hoyle et al., 1982). The agreement with the data for such a biological model in the 8–12 $\mu$m waveband is shown in figure 4. The essential features of this agreement stem from the presence of a silica frustule comprised of $SiO_4$ tetrahedra in more or less random arrangement, along with C-O and C-C bonds in biomaterial. Subsequently, we showed that the same type of fit could be obtained by combining inorganic silica grains ejected from stars with organic particles whose infrared properties are represented by bacterial material (Wickramasinghe et al., 1989).

The model we favour at the present time is depicted schematically in figure 5.

The interior is comprised of porous organic material with an average visual refractive index n = 1.2 in the shape of a hollow oblate spheroid with axial ratio 2.5:1. The exterior is a thin dusting of amorphous silica dust with individual particles of size $\sim 10^{-6}$ cm. The organic cores arise from regions of star formation; the silica particles are supplied mainly from Mira stars. We consider a mass ratio organic/silica = 6.4/1.5 consistent with cosmic abundances of constituent elements (Wickramasinghe et al., 1989).

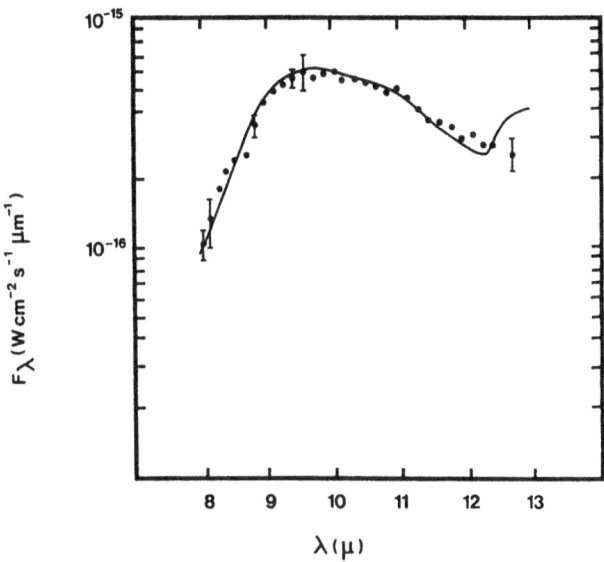

**Figure 4:** Observed flux of the Trapezium nebula compared with the behaviour of the grains of a mixed culture of diatoms heated to a temperature of 175 K.

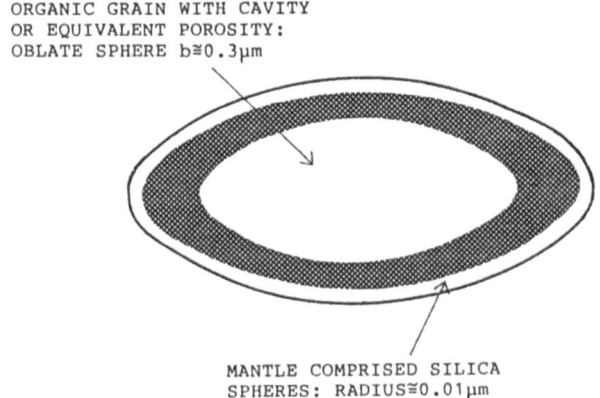

**Figure 5:** Schematic depiction of a composite organic-mineral grain model.

For the organic material we adopt the average optical properties of a bacterium as depicted by the heavy solid curve in the lower panel of figure 6.

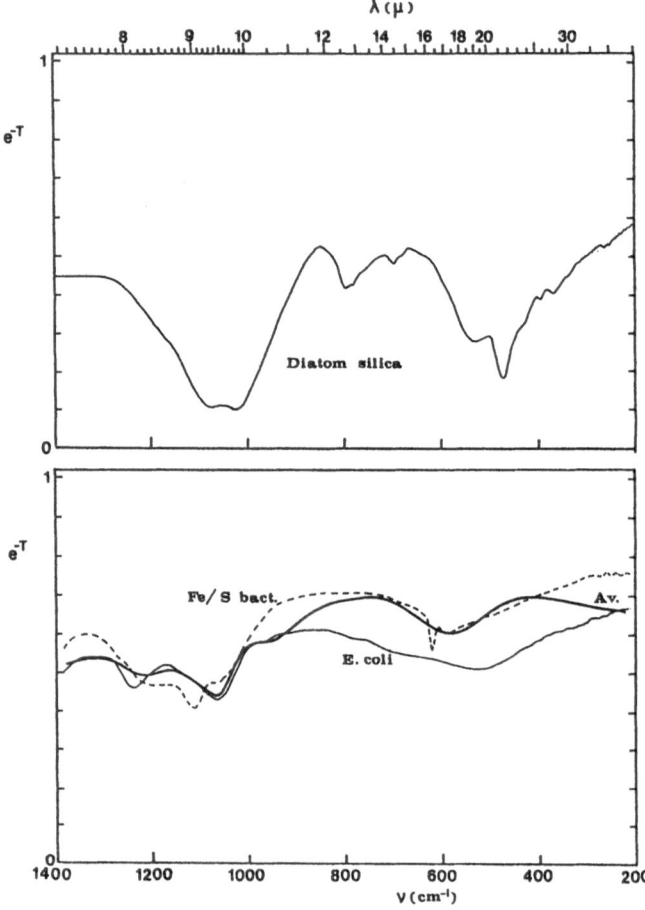

**Figure 6:** Upper panel: Transmittance curve for diatom silica. Lower panel: Transmittance curves for E. Coli (fine solid line), for a mixed culture of iron and sulphur bacteria (dashed curve), and for an "average" interstellar organic particle (heavy solid line).

For the silica component we use the optical properties of diatom silica shown in the upper panel of figure 6. We have argued earlier that this material contains arrangements of $SiO_4$ tetrahedra which may provide a good representation of the structure of silica particles condensing in stellar mass flows. Figure 7 shows the agreement which this combination yields with the entire 8–30 $\mu$m spectrum of the Trapezium nebula for a temperature T = 170 K. The close fit seen here would appear to be strong supportive evidence for the organic/mineral grain model considered here.

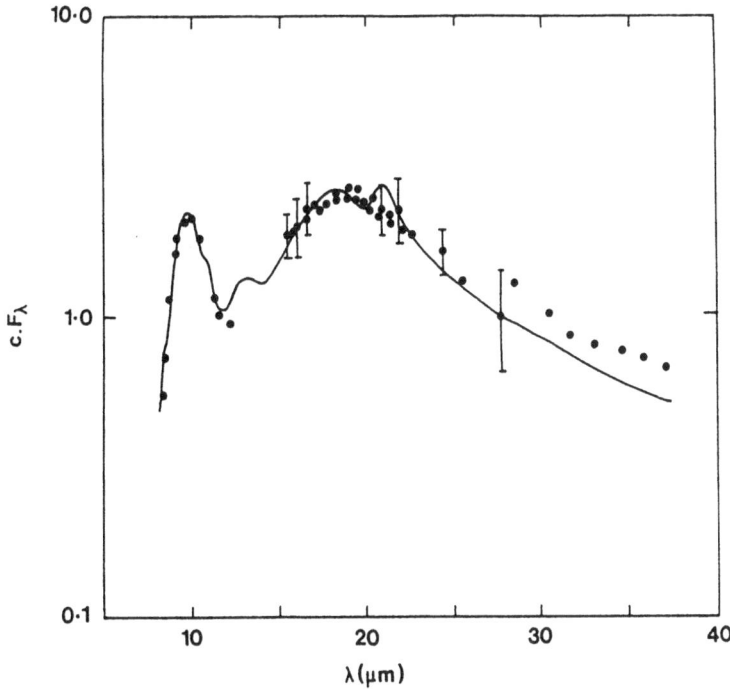

**Figure 7:** Flux curve for the "average" bacterium-diatom silica mixture normalised to accord with the observational point at 9.7 $\mu$m. The data is from Forrest et al. (1976).

## 4  Polarization in the 10 $\mu$m absorption band

Many galactic infrared sources are found to show a broad absorption band in the 8–12 $\mu$m region. Observations of the infrared source GC-IRS 3 and the BN object both show significant changes in polarization across this 10 $\mu$m absorption feature (Aitken et al., 1988). Such data can be used to discriminate between pure silicate and organic/mineral models because polarization curves are sensitive to the average $n(\lambda)$, $k(\lambda)$ values near the absorption peak (Martin, 1975). Whereas amorphous silica has a maximum $k$ value of 2.5 the organic/mineral combination proposed here has an average $k$ value at the 10 $\mu$m peak of 0.5, and a pure organic model has a maximum $k$ value $\sim 0.1$. In the calculations that follow we use $k(\lambda)$ curves determined directly in the laboratory to generate corresponding $n(\lambda)$ curves using a routine that computes the Kramers-Krönig dispersion relations, assuming the generally accepted values of n for these materials in the optical spectral region. For the case of the mineral-organic combination as depicted in figure 5 we use an average $\bar{k}(\lambda)$ and the appropriate average $\bar{n}$ value in the optical wavelength region. The computed $\bar{n}(\lambda)$, $\bar{k}(\lambda)$ functions are then used in the Mie formulae for homogeneous cylinders to compute extinction efficiencies $Q_{eE}$, $Q_{eE}$ for light with electric vectors respectively parallel and perpendicular to the cylinder axes. Computations for infinite cylinders might be assumed to hold approximately for oblate spheroidal particles

with an axial ratio ~2.5/1. The ratio of polarization to optical depth arising from a
column of partially aligned grains in the line of sight of a source is given by

$$\frac{P}{\tau} \propto \frac{Q_{eE} - Q_{eH}}{Q_{eE} + Q_{eH}} \tag{2}$$

the constant of proportionality depending on the degree of alignment in a particular case.

Figure 8 shows the observational data for the galactic centre source GC-IRS 3, and
the BN object compared with theoretical calculations for three types of mineral grains
of radii $r = 0.3$ $\mu$m.

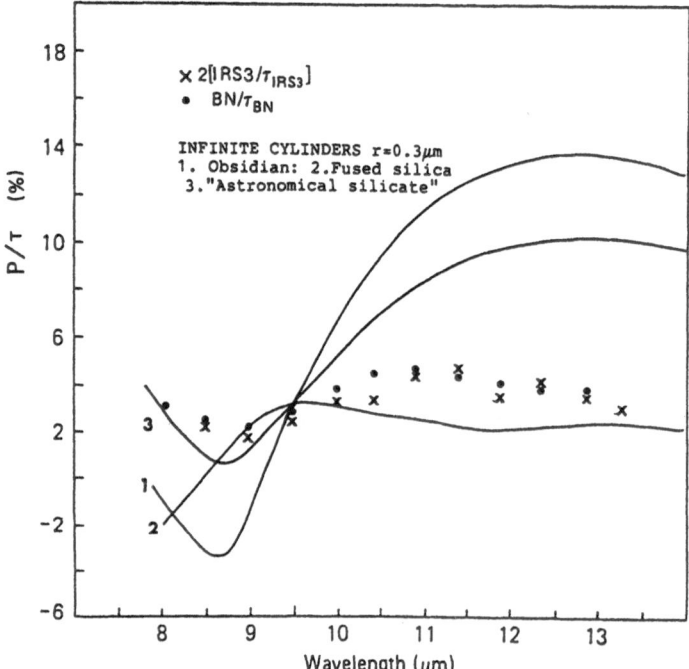

**Figure 8:** The $P/\tau$ ratio expressed as a percentage for the sources GC-IRS 3 and the
BN object compared with the corresponding quantity calculated for partially aligned cylinders comprised of three different types of silicate material. The degree of alignment is an
arbitrary parameter in the model.

The theoretical curves are normalised to agree roughly with the data at a wavelength
close to $\lambda = 9.5$ $\mu$m. In all three cases the theoretical curves differ markedly from
the trends in the observations. These results are not sensitive to our choice of the
radius $r$. The implication must be that mineral models are not viable. Figure 9 shows
the corresponding calculations for hollow organic grains (characterised by the data for

bacteria) and for organic/silica mixtures with a mass ratio 4.3/1.

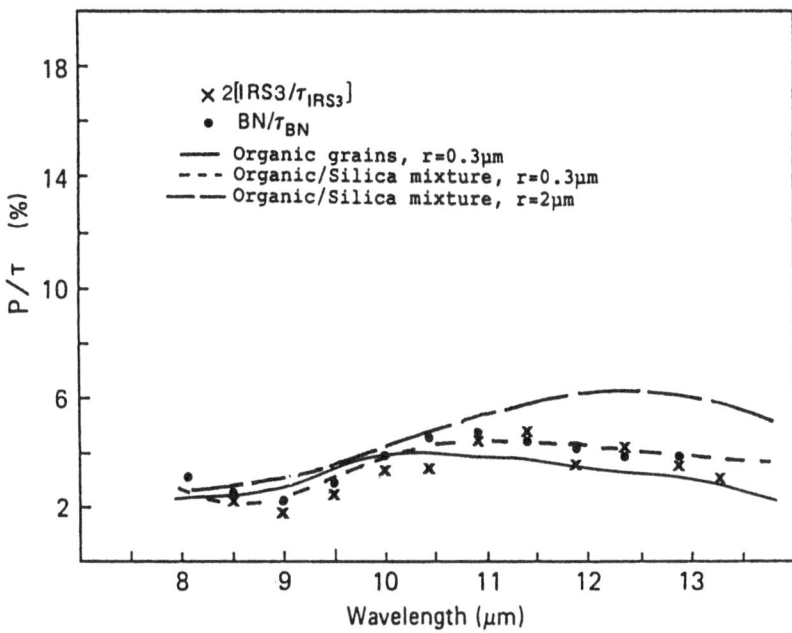

**Figure 9:** Same as figure 8 for organic grains and organic/silica composite grains.

Such particles with typical radii $\sim 0.3$ $\mu$m are seen to produce excellent agreements with the observational data.

# References

Aitken, D.K., Roche, P.F., Bailey, J.A., Briggs, G.P., Hough, J.H., Thomas, J.A.: 1986, M.N.R.A.S., **218**, 363.

Cameron, A.G.W.: 1970, Space Sci. Rev., **15**, 121.

Cohen, M., Tielens, A.G.G.M., Bregman, J.D.: 1989, Astrophys. J., **344**, L13-L16.

Cooke, A.R.: 1977, Ph.D. Thesis, University College Cardiff, University of Wales.

Cooke, A., Wickramasinghe, N.C.: 1977, Astrophys. Space Sci., **50**, 43–53.

Draine, B.T.: 1985, Astrophys. J. Suppl., **57**, 587.

Draine, B.T., Lee, H.M.: 1984, Astrophys. J., **285**, 89.

Forrest, W.J., Gillett, F.C., Stein, W.A.: 1975a, Astrophys. J., **192**, 351.

Forrest, W.J., Gillett, F.C., Stein, W.A.: 1975b, Astrophys. J., **195**, 423.

Forrest, W.J., Houck, J.R., Reed, R.A.: 1976, Astrophys. J., **208**, L133.

Hoyle, F., Wickramasinghe, C.: 1984, *"From Grains to Bacteria"*, Univ. Coll. Cardiff Press.

Hoyle, F., Wickramasinghe, N.C.: 1986, Q. Jl. R. Astr. Soc., **27**, 21.

Hoyle, F., Wickramasinghe, N.C.: 1989, Proc. 22nd Eslab Symposium on Infrared Spectroscopy in Astronomy, Salamanca, Spain, 7–9 December 1988, ESA SP-290, 67.

Hoyle, F., Wickramasinghe, N.C., Al-Mufti, S.: 1982, Astrophys. Space Sci., **86**, 63.

Hunt, J.M., Wisherd, M.P., Bonham, L.C.: 1950, Analytical Chemistry, **22**, 1478.

Keller, W.D., Spotts, J.H., Biggs, D.L.: 1952, Am. J. Sci., **250**, 453.

Krätschmer, W. 1986: Interrelationships Among Circumstellar, Interstellar and Interplanetary Dust, NASA Conf. Publ. 2403, A-13.

Martin, P.G.: 1975, Astrophys. J., **202**, 393.

Wickramasinghe, N.C.: 1975, M.N.R.A.S., **170**, 11P.

Wickramasinghe, N.C.: 1974, Nature, **252**, 462.

Wickramasinghe, N.C., Hoyle, F., Majeed, Q.: 1989, Astrophys. Space Sci., **158**, 335.

Woolf, N.J., Ney, E.P.: 1969, Astrophys. J., **155**, L197.

# NOVEL ION CHEMISTRY OF INTERSTELLAR POLYCYCLIC AROMATIC HYDROCARBONS

D.K. BOHME
*Department of Chemistry and*
*Centre for Research in Earth and Space Science*
*York University*
*North York, Ontario*
*Canada M3J 1P3*

ABSTRACT. Results of laboratory measurements are presented for gas-phase reactions of ground-state atomic silicon ions in the absence and presence of naphthalene, the simplest polycyclic aromatic hydrocarbon molecule. Naphthalene is observed to attach to $Si^+$ and to alter its chemistry. The observed gas-phase reactions of the $Si^+$.naphthalene adduct are viewed as "surface reactions". They provide a novel route for the synthesis of Si-bearing molecules. The results have implications for the role of polycyclic aromatic hydrocarbons in promoting surface reactions of $Si^+$ and other atomic ions in interstellar environments, both in the gas phase and on graphitic grains.

## 1. Introduction

Free neutral and positively charged polycyclic aromatic hydrocarbon (PAH) molecules, either completely or partially hydrogenated, have been invoked to account both for the observed infrared emission features in nebulae[1] and the observed diffuse interstellar absorption bands.[2] The possible presence of these molecules in these environments has led to a strong interest in their physics and chemistry.[3] Other authors have discussed how PAH molecules may attach electrons, ions or atoms and molecules and how ensuing chemistry may provide new routes for neutralization and the synthesis of simple and complex molecules in the interstellar gas.[3,4,5,6] However, this discussion has been largely speculative as there is very little supporting evidence available from laboratory measurements. The low vapour pressure of PAH molecules makes such measurements difficult. Here we present the results of recent measurements of reactions of atomic silicon ions with selected interstellar molecules occurring both in the absence and presence of naphthalene, the simplest PAH molecule and the one with the highest vapour pressure.[7,8,9,10,11] The results of these measurements provide experimental evidence for the attachment of atomic silicon ions to naphthalene and for a novel chemical role for naphthalene in the synthesis of silicon-bearing molecules. Also, the results have implications for the role of PAH molecules in promoting "surface reactions" of $Si^+$ and other atomic ions in interstellar environments, both in the gas phase and on graphitic grains.

### 1.1 ATTACHMENT TO PAH MOLECULES

Electrons, ions and atoms or molecules may attach to PAH molecules in the manner indicated in reactions [1] to [3]. In the gas phase at low pressures this attachment will be bimolecular

*E. Bussoletti and A. A. Vittone (eds.), Dusty Objects in the Universe, 69–76.*
© *1990 Kluwer Academic Publishers.*

[1] $e + PAH$ -----> $PAH^-$

[2] $X^+ + PAH$ -----> $(XPAH)^+$

ct

-----> $PAH^+ + X$

diss ct

-----> $P + AH^+ + X$

[3] $X + PAH$ -----> $(XPAH)$

diss

-----> products

and stabilization must occur by the emission of radiation. Collisional stabilization will become increasingly effective with increasing pressure. It should be noted that bimolecular channels may compete with the addition reactions with ions and atoms or molecules. For example, atomic ions with high electron-recombination energies such as $He^+$ and $H^+$ may charge transfer (ct) or dissociatively charge transfer (diss ct) with PAH molecules. We can also expect O atoms to release CO and $CO_2$ from PAH molecules and for H atoms to produce hydrocarbons.[4] Such reactions are important because they represent possible destruction processes for free PAH molecules.

### 1.2 NOVEL ROUTES FOR NEUTRALIZATION

The formation of positively and negatively charged PAH adduct ions can have important consequences for the neutralization of atomic ions with electrons which normally is not very efficient. The following two reaction sequences illustrate how PAH molecules may catalyze such recombinations. Route [4] has been discussed by Lepp and Dalgarno who have shown

[4] $e + PAH$ -----> $PAH^-$

$X^+ + PAH^-$ -----> $X + PAH$

---

$e + X^+$ -----> $X$

[5] $X^+ + PAH$ -----> $(XPAH)^+$

$e + (XPAH)^+$ -----> $X + PAH$

---

$X^+ + e$ -----> $X$

with calculations that the density of free electrons in the interstellar gas is reduced substantially by the presence of PAH molecules which take up most of the negative charge.[6]

### 1.3 NOVEL ROUTES FOR CHEMISTRY

We propose that novel routes leading to chemical bonding may be envisaged in analogy with the neutralization routes catalyzed by PAH molecules. For example, the following reaction sequence [6] could act to catalyze the association of the species $X^+$ and Y. In this sequence the positive charge is associated with the atom X in the adduct $(XPAH)^+$ as it reacts with Y.

[6]    $X^+$ + PAH         -----> $(X^+..PAH)$

   $(X^+..PAH)$ + Y    -----> $XY^+$ + PAH

   _____

   $X^+$ + Y           -----> $XY^+$

The reaction may be viewed to occur between $X^+$ and Y in the vicinity of the neutral PAH molecule, perhaps above the "surface" of the latter.

A derivative of scheme [6] is scheme [7] in which the charge on $X^+$ is transferred to the PAH molecule in the first step and a neutral reaction occurs in the second step in the vicinity, perhaps above the "surface", of a charged PAH molecule as follows:

[7]    $X^+$ + PAH         -----> $(X..PAH^+)$

   $(X..PAH^+)$ + Y    -----> $XY$ + $PAH^+$

   _____

   $X^+$ + Y + PAH     -----> $XY$ + $PAH^+$

This scheme is not catalytic since the PAH molecule is not regenerated as a neutral molecule. But, from the point of view of molecular synthesis, it has the attractive feature that it leads directly to the formation of a bonded neutral molecule from the reacting ion and molecule and it does so without requiring a separate neutralization step. The positive charge ends up on the departing PAH molecule.

The relative ionization energies of Si and naphthalene are such that scheme [7] is likely to be more appropriate for the description of the influence of naphthalene molecules on the gas-phase chemistry initiated by atomic silicon ions.

## 2. Experiment

The measurements were performed with the Selected-Ion Flow Tube (SIFT) apparatus in the Ion Chemistry Laboratory at York University which has been described in detail elsewhere.[12,13] Atomic silicon ions were produced from a 2-3% mixture of tetramethylsilane in deuterium by electron impact at 50-100 eV. The deuterium was added to scavenge the metastable $Si^+(^4P)$ ions in the ion source with the following reaction:

[8]    $Si^+(^4P)$ + $D_2$    -----> $DSi^+$ + D

$Si^+$ ions were selected from the source with a quadrupole mass spectrometer and injected into flowing helium gas at 0.35 Torr. Neutral reagents were added downstream. Reactant and product ions were monitored still further downstream with a second quadrupole mass spectrometer as a function of the added neutral reagent. Established methods of analysis were used to derive rate constants and product distributions from these observations. The experiments were performed with and without addition of naphthalene vapour upstream of the point of addition of neutral reagents. The operating temperature was 296 ± 3K.

## 3. Results and Discussion

The experiments with added naphthalene vapour provided evidence for the capture of $Si^+$ by the naphthalene molecules and for the transfer of an electron which leaves the naphthalene ionized:

[9a]   $Si^+$ + $C_{10}H_8$        $----->$   $(SiC_{10}H_8)^+$

[9b]                        $----->$   $C_{10}H_8^+$ + Si

The addition reaction [1a] was favoured by seven to one at 0.35 Torr of helium buffer gas. Presumably the addition occurs in a termolecular fashion but a contribution from bimolecular radiative association cannot be excluded. The electron-transfer channel is nearly energy-resonant since the ionization energies of Si ($8.15172 \pm 0.00003$ eV) and naphthalene ($8.14 \pm 0.01$ eV) are nearly identical, but the charge transfer is exothermic by $0.01 \pm 0.01$ eV. The observation of the electron transfer channel suggests that the $(SiC_{10}H_8)^+$ adduct ion might best be described as a charge-transfer adduct of the type $(Si..C_{10}H_8^+)$ in which a silicon atom is bonded to a charged naphthalene "surface".

Both the bare atomic silicon ion and its adduct with naphthalene were found to be unreactive with deuterium and carbon monoxide. For interstellar environments dominated by $H_2$ and CO, this result implies that $Si^+$ and its naphthalene adduct will be available to react with less abundant constituents.

Ground-state atomic silicon ions were seen to react with $H_2O$, $NH_3$, $C_2H_2$ and $C_4H_2$ in a bimolecular fashion while $O_2$ forms the adduct ion with $Si^+$, albeit at a rate immeasurably small. Rate constants, products and product distributions which were observed and measured are given in Table 1. An H atom is eliminated in the reactions with $H_2O$, $NH_3$ and $C_2H_2$ while hydride transfer predominates with $C_4H_2$. We have shown that with ammonia the H atom which is eliminated is that from the N-H bond undergoing insertion so that $SiNH_2^+$ is produced rather than $HSiNH^+$.[8] There are no major surprises in these results, although the difference in the nature of the reaction channels for acetylene and diacetylene is curious. The results have recently have been included in an extended pseudo-time-dependent model of the chemistry of silicon in dense interstellar clouds.[14]

The results obtained for the reactions with the $Si^+$.naphthalene adduct ion were more unexpected. The predominant feature of the chemistry was the production of ionized naphthalene according to the bimolecular reaction [10]. The reaction with ammonia was

[10]   $(SiC_{10}H_8)^+$ + M   $----->$   $C_{10}H_8^+$ + (SiM)

an exception, as only adduct formation was observed in this case. The results are summarized in Table 2. It is interesting to note that, with the exception of the reaction with $O_2$ which has a slightly enhanced rate, the reactions of $Si^+$ are uniformly slower in the presence of naphthalene. The neutral product of reaction [10] was not detected. It cannot be Si + M since the collisional dissociation is endothermic by an amount equal to the binding energy of Si to ionized naphthalene.

Table 1. Summary of rate constants (in units of $10^{-9}$ cm$^3$ molecule$^{-1}$ s$^{-1}$) and product distributions for reactions of Si$^+$($^2$P) with selected interstellar molecules at 296 ± 3 K.

| Reactant | Products | Distribution | Rate constant |
|----------|----------|--------------|---------------|
| $H_2$ | none | | < 0.0002 |
| $D_2$ | none | | < 0.0001 |
| CO | none | | < 0.00002 |
| $O_2$ | $SiO_2^+$ | | < 0.0001 |
| $H_2O$ | $SiOH^+$ + H | | 0.23 |
| $NH_3$ | $SiNH_2^+$ + H | | 0.64 |
| $C_2H_2$ | $SiC_2H^+$ + H | 0.7 | 0.35 |
| | $SiC_2H_2^+$ | 0.3 | |
| $C_4H_2$ | $C_4H^+$ + SiH | | 1.6 |

The observed production of ionized naphthalene and the likelihood of charge transfer within the adduct to form (Si..$C_{10}H_8^+$) suggest that the mechanism of reaction [10] may be viewed as a "surface reaction" of atomic silicon with the molecule M where the "surface" is provided by the charged naphthalene molecule. In this mechanism SiM may depart as a bound molecule if the reaction exothermicity is sufficiently large. Alternatively, bimolecular products may be formed at the surface and then be detached. Which of these neutral products is preferred is not known. Some insight can be gained from related experiments. Ground-state silicon atoms in the gas phase are known to react with $O_2$ and $C_2H_2$ and not to react with $H_2$.[15,16] The products of the reactions with $O_2$ and $C_2H_2$ have not been identified. There is disagreement in the rate constant for the reaction of Si with $O_2$ but in one study a large rate for this reaction has been rationalized in terms of the availability of symmetry-allowed exothermic routes to SiO + O($^3$P, $^1$D). The reaction of Si with acetylene has been attributed to efficient addition with ring insertion to form 3-silacyclopropenylidene.[15] A matrix isolation study of the reaction of silicon atoms with water has shown that a silicon-water adduct (SiOH$_2$) is formed which rearranges spontaneously to HSiOH, viz. the silicon atom inserts into the O-H bond.[17]

It is reasonable to expect the formation of 3-silacyclopropenylidene in the reaction of the Si$^+$.naphthalene adduct ion with acetylene if it proceeds by the "surface" mechanism. Formation of the isomers vinylidenesilene, CH$_2$CSi:, and silylenylacetylene, HSiC$_2$H, requires an additional 17 and 22 kcal mol$^{-1}$ of energy. Also, the possible bimolecular channels are probably endothermic: H-atom abstraction is endothermic by 63 kcal mol$^{-1}$ but the energetics for H-atom elimination with formation of SiC$_2$H are not known. The SiC$_{12}$H$_{10}^+$ adduct ion is then most reasonably viewed as 3-silacyclopropenylidene attached to the "surface" of an ionized naphthalene molecule.

Table 2. Summary of rate constants (in units of $10^{-9}$ cm$^3$ molecule$^{-1}$ s$^{-1}$) and product distributions for reactions of the adduct ion $(SiC_{10}H_8)^+$ with selected interstellar molecules at $296 \pm 3$K.

| Reactant | Products | Distribution | Rate Constant |
|---|---|---|---|
| $D_2$ | none | | < 0.00035 |
| CO | none | | < 0.0003 |
| $O_2$ | $C_{10}H_8^+$ + $(SiO_2)$ | | 0.00037 |
| $H_2O$ | $C_{10}H_8^+$ + $(SiOH_2)$ | | 0.0055 |
| $NH_3$ | $(SiC_{10}H_8NH_3)^+$ | | 0.41 |
| $C_2H_2$ | $C_{10}H_8^+$ + $(SiC_2H_2)$ | 0.9 | 0.063 |
| | $SiC_{12}H_{10}^+$ | 0.1 | |
| $C_4H_2$ | $C_{10}H_8^+$ + $(SiC_4H_2)$ | | 1.0 |

In analogy to the reaction with acetylene we can expect the reaction of Si$^+$.naphthalene with diacetylene to lead to the cyclic attachment of Si to a triple bond of diacetylene.

## 4. Conclusions and Suggestions

The experiments have shown that ground-state atomic silicon ions attach to naphthalene in the gas phase at 298 K in ambient helium at 0.35 Torr. Presumably the attachment occurs in a termolecular fashion under these conditions with helium acting as the stabilizing molecule. Attachment to naphthalene substantially alters the chemistry of Si$^+$.

The Si$^+$.naphthalene adduct is unreactive toward the abundant interstellar molecules hydrogen and carbon monoxide but reacts in a novel fashion with various other interstellar molecules at 298 K in bimolecular reactions of the general type [10]. The mechanism of these reactions may involve a reaction between neutral Si and the molecule M proceeding on a positively-charged naphthalene "surface". The identity of the neutral product (SiM) is not known but formation of silicon-bearing molecules can be expected.

The occurrence of reactions of type [10] suggests analogous reactions of type [11] involving larger PAH molecules or ions and reactions of type [12] involving other atomic ions, X$^+$, or

[11]   $(SiPAH)^+$ + M -----> PAH$^+$ + (SiM)

[12]   $(XPAH)^+$ + M -----> PAH$^+$ + (XM)

neutrals, X. In principle, the $(XPAH)^+$ ion may be formed either by the capture of the atomic ion X$^+$ by the PAH molecule or the capture of the neutral atom X by a positively-charged PAH ion. In the interstellar gas these adducts must be formed by radiative stabilization. With 7 eV as a representative ionization energy for the larger PAH molecules, it appears that within the adduct ion intramolecular charge transfer to the larger PAH molecules may occur with the atomic ions of S (10.360), Zn (9.394), Si (8.151), Fe (7.870) and Mg (7.646) but not with the atomic ions of Ca (6.113) and Al (5.986) (the numbers in parentheses are ionization energies in eV). We suggest that reactions of type [11] and [12] are not constrained to occur on the "surface" of PAH molecules in the interstellar gas but that they should also occur on PAH molecules or graphite, the end member of the PAH series, appended to interstellar grains.

Nothing is known about the spectroscopy of adduct ions of the type $(XPAH)^+$ so that their identification in the interstellar gas is not yet possible.

**References**

1.  Leger, A., Puget, J.L.: 1984, Astron. Astrophys. **137**, L5. Allamandola, L.J., Tielens, A.G.G.M., Barker, J.R.: 1985, Astrophys. J. **290**, L25.

2.  Van der Zwet, G.P., Allamandola, L.J.: 1985, Astron. Astrophys. **146**, 76. Leger, A., d'Hendecourt, L.B.: 1985, Astron. Astrophys. **146**, 81. Crawford, M.K., Tielens, A.G.G.M., Allamandola, L.J.: 1985, Astrophys. J. **293**, L45.

3.  Omont, A.: 1986, Astron. Astrophys. **164**, 159.

4.  Duley, W.W., Williams, D.A.: 1986, Mon. Not. R. astr. Soc. **219**, 859.

5.  Lepp, S., Dalgarno, A.: 1988, Astrophys. J. **324**, 553.

6.  Lepp, S., Dalgarno, A., van Dishoeck, E.F., Black, J.H.: 1988, Astrophys. J. **329**, 418.

7.  Wlodek, S., Fox, A., Bohme, D.K.: 1987, J. Am. Chem. Soc. **109**, 6663.

8.  Wlodek, S., Rodriquez, C.F., Lien, M.H., Hopkinson, A.C., Bohme, D.K.: 1988, Chem. Phys. Lett. **143**, 385.

9.  Bohme, D.K., Wlodek, S., Fox, A.: 1988, in *Rate Coefficients in Astrochemistry*, eds. T.J. Millar, D.A. Williams, Dordrecht, Kluwer, p. 193.

10. Wlodek, S., Bohme, D.K.: 1989, J. Chem. Soc., Faraday Trans. 2, **85**, 1643.

11. Bohme, D.K., Wlodek, S., Wincel, H.: 1989, Astrophys. J. **342**, L91.

12. Mackay, G.I., Vlachos, G.D., Bohme, D.K., Schiff, H.I.: 1980, Int. J. Mass Spectrom. Ion Phys. **36**, 259.

13. Raksit, A.B., Bohme, D.K.: 1983, Int. J. Mass Spectrom. Ion Phys. **55**, 69.

14. Herbst, E., Millar, T.J., Wlodek, S., Bohme, D.K.: 1989, Astron. Astrophys. **222**, 205.

15. Husain, D., Norris, P.E.: 1978, J. Chem. Soc. Faraday Trans. II **74**, 106.

16. Swearengen, P.M., Davis, S.J., Niemczyk, T.M.: 1978, Chem. Phys. Lett. **55**, 274.

17. Ismail, Z.K., Hauge, R.H., Fredin, L., Kauffman, J.W., Margrave, J.L.: 1982, J. Chem. Phys. **77**, 1617.

# ASTROPHYSICAL IMPLICATIONS OF THE INFRARED SPECTRUM OF CORONENE AT ELEVATED TEMPERATURES

G. C. Flickinger, T. J. Wdowiak, D. A. Boyd
Physics Department, University of Alabama at Birmingham, U.S.A.

ABSTRACT. Coronene encapsulated in a KBr pellet has been heated over a range of temperatures up to 788 K and its mid–infrared absorption spectrum obtained at selected temperatures. Changes in the aborbance spectum at elevated temperatures include a 3–fold strengthening of coronene's 3.29 $\mu$m band which assumes a profile very similar to the 3.29 $\mu$m UIR band. Coronene's 8.8 $\mu$m feature which is also correlated with a UIR also shows substantial increases in intensity at elevated temperatures. The 10.5 $\mu$m feature of coronene which does not correlate with the UIR decreases sharply as the temperature is increased. These results as well as some possible implications are discussed.

## 1. Introduction

The principle unidentified infrared emission bands (UIR) are generally taken to be 3.3 $\mu$m, 6.2 $\mu$m, 7.7 $\mu$m, 8.6 $\mu$m, and 11.2 $\mu$m. The origin of these emissions has been widely speculated upon and a general hypothesis now exists involving emission by vibrationally excited interstellar polycyclic aromatic hydrocarbons (PAH). The PAH hypothesis originated in 1981 with the observation by Duley and Williams that some of the UIR bands were coincident in frequency with absorption bands characteristic of PAH suggesting that the UIR bands might be due to the relaxation of thermally excited aromatic molecules as constituents of dust grains. Leger and Puget (1984), and Allamandola, Tielens, and Barker (1985) expanded on the initial PAH hypothesis by proposing that the aromatics responsible for emission might be in free molecular form instead of composite dust grains.

Although many aromatic molecules such as coronene and hexabenzocoronene have been shown to have absorption spectra which match some of the UIR bands, some of the PAH features do not have observed celestial counterparts. Also, there is difficulty in correlating the intensities of some of the laboratory bands with interstellar bands. A continuing endeavour to strengthen the PAH hypothesis for the UIR bands has been to look for particular a PAH or a group of PAH species which would have an absorption spectrum that would match more closely the observed UIR. A good example of the inability to match a specific PAH spectrum with a UIR interstellar emission spectrum is the coronene absorption spectrum. Because of its size and symmetric form, traits lending to survivability in the interstellar medium, coronene is one of the most widely studied PAH species in terms of the PAH–UIR hypothesis. It was the key molecule that led Leger and Puget to suggest initially that coronene and molecules like coronene might be responsible for the UIR. Coronene shows absorption at all the UIR bands except 11.2 $\mu$m. The 845 cm$^{-1}$ (11.8 $\mu$m) absorption feature of coronene due to the out–of–plane bending of pairs of hydrogens on an aromatic ring is the analogue to the 11.2 $\mu$m UIR thought to be due to the out–of–plane bending of isolated

77

E. Bussoletti and A. A. Vittone (eds.), Dusty Objects in the Universe, 77–84.

hydrogens of the periphery of interstellar PAH. A serious problem with obtaining a coronene—UIR match is the profile of the 3.29 $\mu$m band. Coronene has a split band in this region with two peaks of comparable intensity while the 3.29 $\mu$m UIR is a single well defined peak, and has been dismissed as a major interstellar component.

Laboratory spectra are usually obtained by mixing the sample with an infrared transparent salt such as KBr or CsI and then pressing this mixture into a pellet. With this procedure the sample is in the solid state and thus extensive inter—molecular interaction is inevitable. The hypothesis that interstellar PAH are in free molecular form implies that the interstellar emission from such molecules would not involve molecular coupling, and thus differences in laboratory and celestial spectra is to be expected. Another difference in conditions in the terrestrial laboratory and the conditions in the interstellar medium is the presence of UV radiation in the interstellar medium. Absorption of a UV photon by an aromatic molecule results in an vibrationally excited molecule. Since the time required for redistribution of much of the photon's energy is much shorter than the time required for infrared photon emission, the concept of an internal molecular temperature can be applied. If conditions in the laboratory were such that the internal temperatures of laboratory PAH were identical to the internal temperatures of interstellar PAH, the energy inconsistency would be eliminated. The most obvious and simple way of doing this is to heat the sample pellet to temperatures such that the average internal energy of the laboratory molecules is similar to that expected of UV excited interstellar molecular species. Blanco et al. (1988) have performed experiments by heating both compact and noncompact PAH to temperatures of 240 C, and have observed temperature dependence of some of the absorption bands of these PAH. Bernard et al. (1989) have also obtained absorbance spectra of coronene at temperatures up to 500 K, and have concluded that coronene's mid—infrared absorbance spectrum does not strongly depend on temperature in the range between 300 K and 500 K. Our research involving heating experiments but with a greater range of temperature sought to eliminate this apparent ambiguity.

## 2. Experimental Technique

The mid—infrared absorbance spectrum of coronene at temperatures ranging from 25 C (298 K) to 515 C (788 K) were obtained by mechanically mixing 0.2 mg of coronene and 100 mg of KBr in a vibrating agate ball mill. A hydraulic press is then used to compress the mixture in an aluminum collar forming a pellet with cross—sectional area 0.5 cm$^2$. A threaded hole on the circumference of the aluminum collar allows for the coupling to a resistance heater. In order to minimize heat loss a ceramic casing encloses the heater and aluminum foil was wrapped around the collar. A support stand for the heater—sample unit was constructed and the assembly inserted into the spectrometer specimen chamber. A shroud encloses the sample chamber to allow for nitrogen purging prior to and while making measurements. The temperature of the heater—disk unit is controlled and varied through use of a variable transformer in series with the resistance heater with the temperature of the aluminum disk holding the KBr encapsulated sample monitored with a chromel—alumel thermocouple inserted into a small hole on the circumference of the aluminum disk.

The coronene sample was heated to temperatures as high as 515 C (788 K) with mid—infrared absorbance spectra of coronene obtained at temperatures of 25 C (298 K), 245 C (518 K), 360 C (633 K), 440 C (713 K), 500 C (773 K), and 515 C (788 K). A room temperature spectrum of coronene was taken initially. The pellet was then heated to 240 C and a spectrum obtained. The sample was then allowed to cool to room temperature and another spectrum was taken. This procedure was repeated several times and also repeated for the higher temperatures. This method permitted comparison of room temperature spectra before and after heating to determine if the effects of elevated temperature were elastic or if the thermal treatment of the sample had altered its molecular structure. All spectra were obtained using a Mattson Polaris FTIR spectrometer.

## 3. Spectra

The mid–infrared absorbance spectrum of coronene in KBr has definite temperature dependent aspects. Most notably the 3040 cm$^{-1}$ (3.29 $\mu$m) feature shows dramatic changes in both peak intensity and structure. The strength of the peak increases more than 3–fold as the temperature is increased to 788 K (Figures. 1, 2). The intensity increase appears to be nonlinear, indicated by relatively gross changes in intensity over a narrow temperature range (15 C) at the highest temperatures. The distinct dual peak feature at room temperature gradually loses its split character as the temperature is increased. The 3040 cm$^{-1}$ (3.29 $\mu$m) peak strengthens considerably. This effect becomes most marked at the higher temperatures with the 3040 cm$^{-1}$ (3.29 $\mu$m) peak dominating the 3020 cm$^{-1}$ (3.31 $\mu$m) peak resulting in a single intense peak at 3040 cm$^{-1}$. The "knee" this feature displays is coincident with its room temperature sister peak at 3020 cm$^{-1}$ (3.31 $\mu$m). The 3040 cm$^{-1}$ (3.29 $\mu$m) absorbance feature of coronene at 788 C, due to the in–plane C–H stretch, is very similar to the 3.29 $\mu$m (3040 cm$^{-1}$) UIR feature of many celestial objects. Figure 3. is a comparison of the coronene 3040 cm$^{-1}$ (3.29 $\mu$m) feature and the 3.29 $\mu$m (3040 cm$^{-1}$) emission band of the Orion nebula (Bregman et al. 1989). The two peaks are exactly coincident in wavelength within limits of resolution and both reveal a "knee" feature. A difference however is that the "knee" feature of coronene is more prominent than the coresponding "knee" feature of the Orion emission. This apparent ambiguity may be a result of telluric methane which interferes with ground–based observations in the wavelength region of the "knee" (Chris Sellgren: personal communication).

Other coronene absorption bands which exhibit significant variation with temperatures are the 1135 cm$^{-1}$ (8.8 $\mu$m) band and the 955 cm$^{-1}$ (10.5) $\mu$m band. The intensity of the 1135 cm$^{-1}$ peak increases more than 2–fold while the intensity of the 955 cm$^{-1}$ peak decreases 4–fold as the temperature is raised from 298 K to 788 K (Figure 4.). The 1135 cm$^{-1}$ (8.8 $\mu$m) coronene feature is correlated with one of the UIR while the 955 cm$^{-1}$ (10.5 $\mu$m) coronene feature is not. The fact that a UIR correlated band increases with increasing temperature while a non–correlated band decreases may explain the failure to obtain a detailed match between room temperature laboratory spectra of PAH and celestial emission spectra. Absorption bands of PAH at room temperature which are not prominent in celestial spectra may decrease or disappear as the temperature is increases giving a better match between laboratory and celestial spectra.

The 1620 cm$^{-1}$ (6.2 $\mu$m) and 1310 cm$^{-1}$ (7.6 $\mu$m) coronene features, both of which correlate well with the UIR, show no variation with temperature in the explored range. The 845 cm$^{-1}$ (11.8 $\mu$m) coronene feature also reveals negligible change with temperature increase. Although this band is not correlated with a UIR, it is associated with the 11.2 $\mu$m (890 cm$^{-1}$) UIR. The 11.2 $\mu$m band and the 11.8 $\mu$m band of aromatic species are both due to the out–of–plane bending of hydrogens on the periphery of the aromatic rings, the 11.2 $\mu$m band due to isolated hydrogens and the 11.8 $\mu$m band due to two isolated hydrogens. It is reasonable therefore to expect that both features will have similar temperature dependence and so if the

80

Figure 1. Composite of tracings of infrared absorption spectra in the 4000 cm⁻¹ − 400 cm⁻¹ range of coronene (3.1 mg/cm²) at temperatures of : 298 K, 518 K, 633 K, 713 K, 773 K, and 788 K. Peaks correlated with UIR bands are indicated in the room temperature spectrum. The spectra also show presence of atmospheric water and carbon dioxide in varying amounts as artifacts.

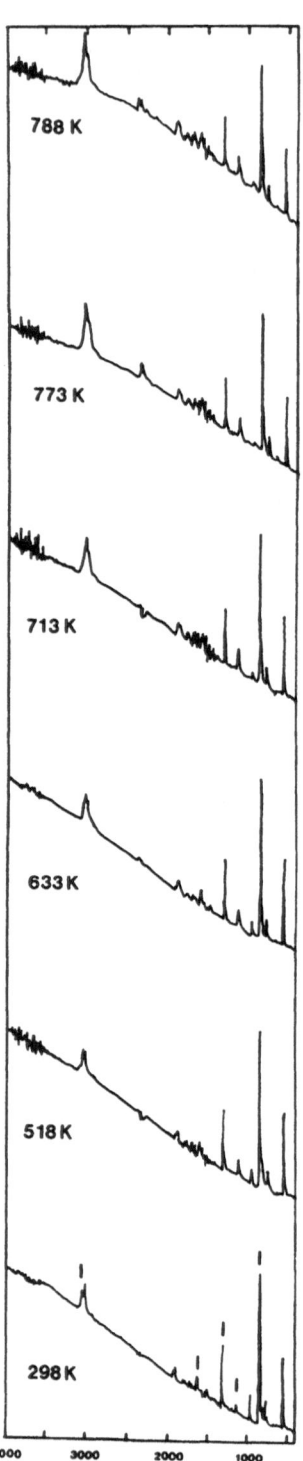

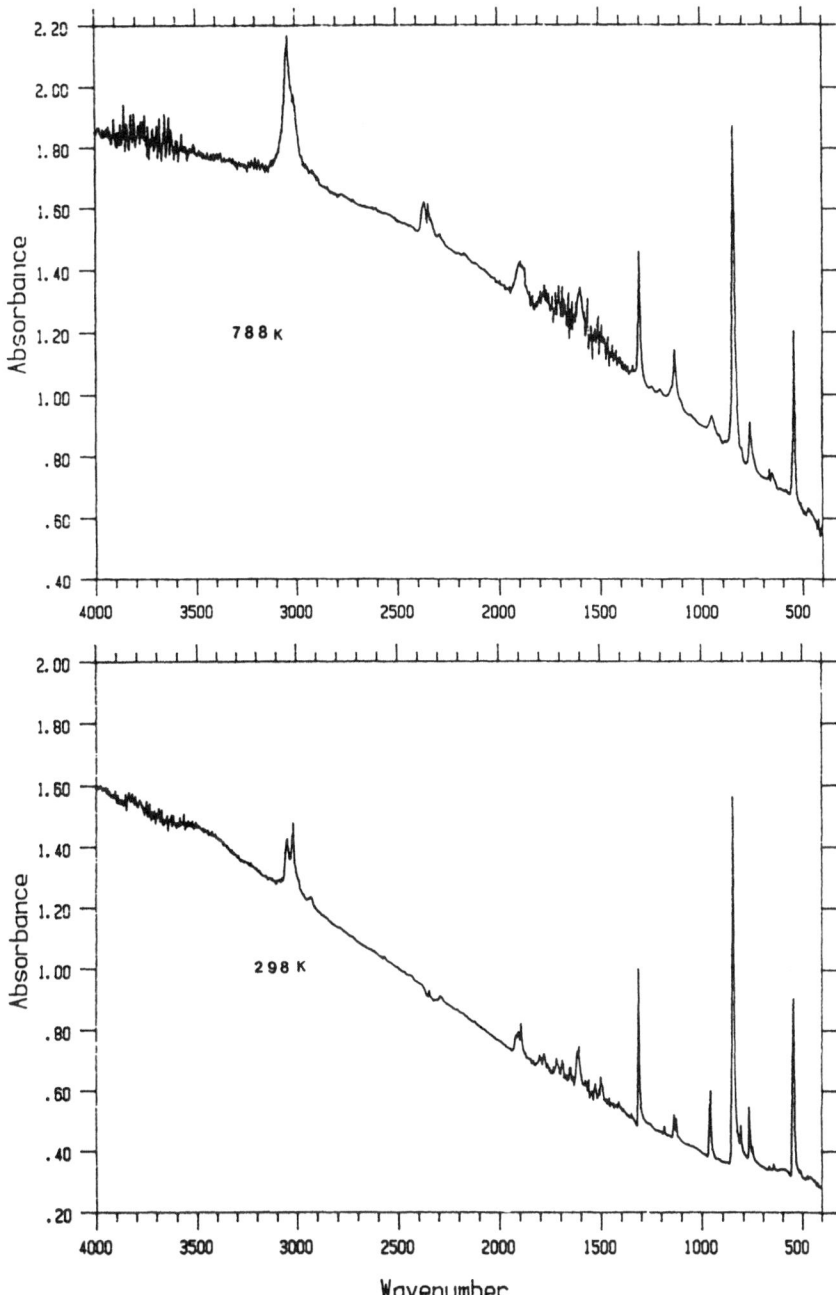

Figure 2. Spectra obtained at 298 K (bottom) and 788 K (top) (as in Figure 1.) showing details of the feature differences over the extremes of the temperature range studied. Note the 3X strengthening of the 3040 cm$^{-1}$ (3.29 $\mu$m) and 1135 cm$^{-1}$ (8.8 $\mu$m) features correlated with the 3.29 $\mu$m and 8.6 $\mu$m UIR bands, and the reduction in strength of the 955 cm$^{-1}$ (10.5 $\mu$m) band not correlated with a UIR.

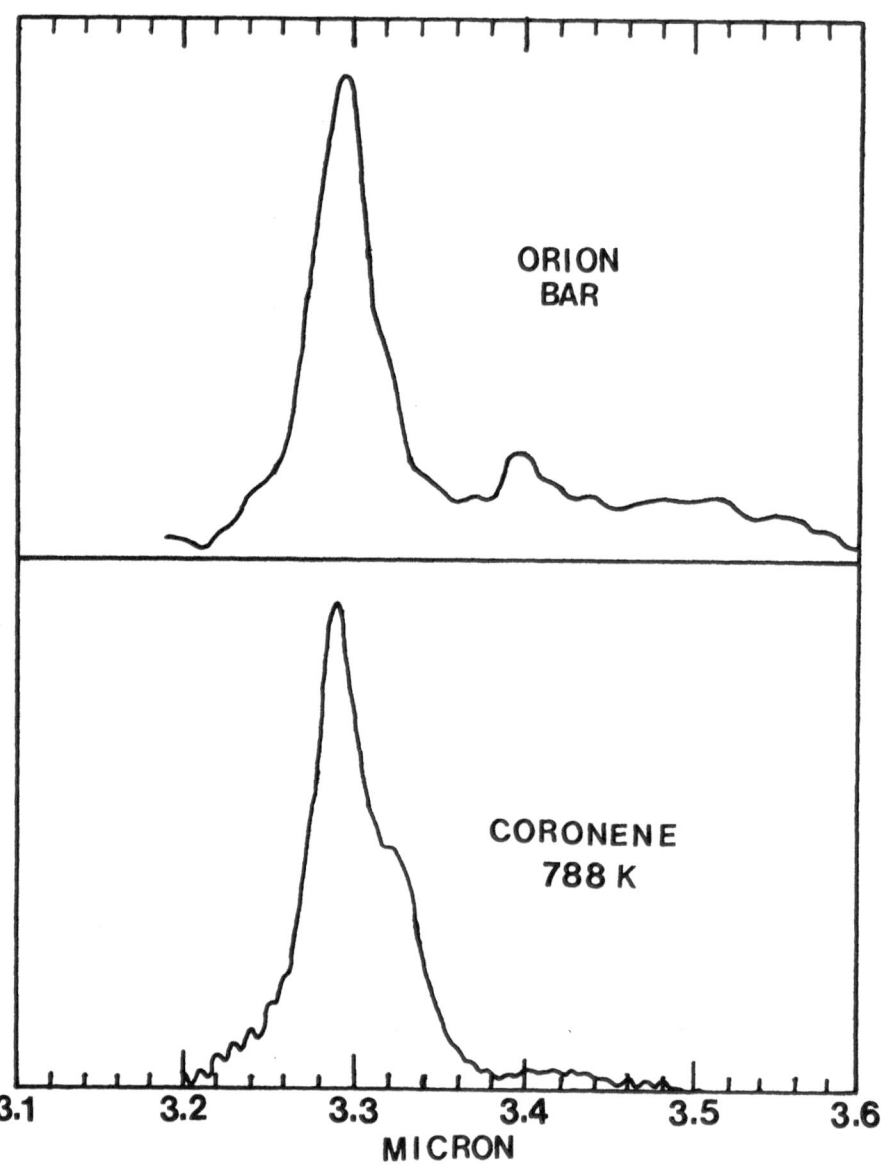

Figure 3. Comparison of the 3000 cm⁻¹ region feature of coronene (0.4 mg/cm²) at 788 K in a N₂ purge with the Orion Bar 3.29 μm UIR (Bregman et al. 1989).

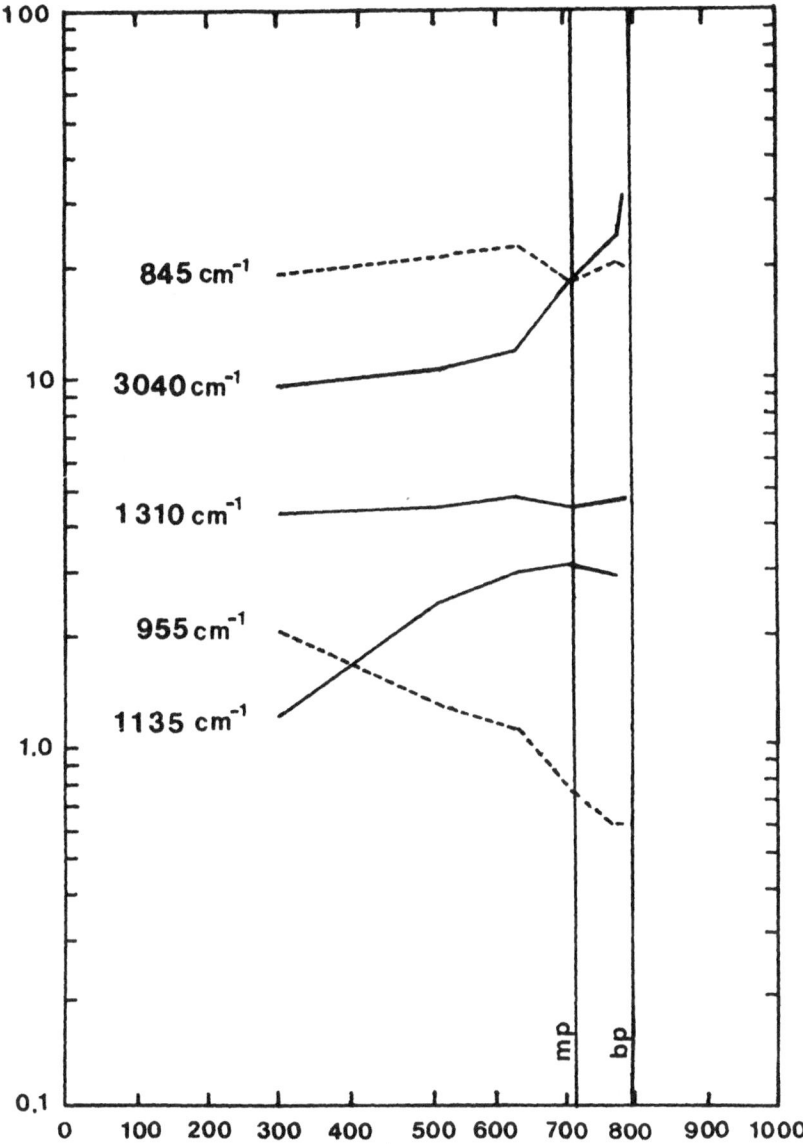

Figure 4. Variation in feature strength (Absorbance x Wavenumber) for five spectral features of coronene (3.1 mg/cm²) at temperatures of 298 K, 518 K, 633 K, 713 K, 773 K, and 788 K (except for absence of data for 1135 cm⁻¹ at 788 K). The two vertical lines indicate the melting and boiling points of coronene at STP.

interstellar emission is due to excited PAH, then the intensity of the 11.2 $\mu$m UIR band would be independent of the internal molecular temperature of the emitting species. If this is the case then a ratio of the temperature dependent 3.29 $\mu$m (3040 cm⁻¹) aromatic band to the temperature independent 11.2 $\mu$m (890 cm⁻¹) aromatic band could be an indicator of the temperature of the molecule.

## 4. Conclusion

The temperature dependence of the mid–infrared absorbance spectrum of coronene may have interesting implications. Some of the difficulties with the assignment of coronene as a major interstellar emitter seems to have been resolved. Most notably, the 3040 cm⁻¹ (3.29 $\mu$m) coronene feature which at room temperature shows little similarity to the 3.29 UIR feature, when heated to 788 K assumes an almost identical profile to the interstellar band. The weakening of coronene's 955 cm⁻¹ (10.5 $\mu$m) band not correlated with a UIR coupled with the strengthening of its 1135 cm⁻¹ (8.8 $\mu$m) band that is correlated with a UIR also gives more credibility to the hypothesis that coronene or coronene–like species are responsible for the UIR. The apparent independence of coronene's 845 cm⁻¹ (11.8 $\mu$m) band of temperature hints that the 890 cm⁻¹ (11.2 $\mu$m) absorption band of other PAH may also be independent of temperature. If this is the case then the ratio of the intensities of the 3.29 $\mu$m (3040 cm⁻¹) UIR and the 11.2 $\mu$m (890 cm⁻¹) UIR could be used as an astrophysical thermometer for UIR emitting excited molecules.

The spectrum of coronene at higher temperatures up to 1000 K, at which point coronene will likely undergo thermal alteration (Sakata, 1989, personal communication) needs to be obtained to determine if the temperature dependent coronene bands continue to alter their profiles. High temperature gas phase measurements are also desirable. Further, higher resolution spectra of the interstellar 3.29 $\mu$m emission including measurements from space to eliminate the interference of telluric methane needs to be obtained for detailed comparison with the high resolution laboratory spectra. A greater understanding of the molecular physics involved in the emission and absorption of polyatomic molecules including the temperature dependent aspects is required so that a general relationship between peak position and intensity, and temperature can be found. This work supported by NASA grant NAGW–749.

## References

Allamandola, L. J., Tielens, A., and Barker, J. R.: 1985, Ap. J., **290**, L25.

Bernard, J. P., d'Hendecourt, L. B., and Leger, A.: 1989, Astron. Astrophys., **220**, 245.

Blanco, A., Borghesi, A., Fonti, S., Orofino, V., Bussoletti, E., and Colangeli, L.: 1988, in "Dust in the Universe", M. E. Bailey and D. A. Williams ed. (Cambridge University Press, Cambridge).

Bregman, J. D., Allamandola, L. J., Tielens, A. G. G. M., Geballe, T. R., and Witteborn, F. C.: 1989, Ap. J., **344**, 791.

Duley, W. W., and Williams, D. A.: 1981, M.N.R.A.S., **196**, 269.

Leger, A., and Puget, J. L.: 1984, Astr. Ap., **137**, L5.

# THE 3.3 MICROMETER FEATURE: THE LARGE SCALE GALACTIC EMISSION

M. GIARD[1], N. SALES[2], F. PAJOT[1], J.M. LAMARRE[1] and G. SERRA[2]
[1] I. A. S. , Campus d'Orsay, Bat. 120, 91405 Orsay-cedex
[2] C. E. S. R. , 9 Av. du Colonel Roche, BP 4346, 31029 Toulouse-cedex

ABSTRACT. The AROME balloon borne experiment was built in order to perform measurements of IR emission features in extended sources (Field of view = 0.5°). It has shown that the 3.3 micrometer feature is present in the spectrum of the diffuse galactic light. This means that the feature emitting material (PAHs, HACs, QCCs, etc.) is to be considered as a standard component of interstellar grains models. Our aim here is to emphasize the ubiquity of the 3.3 micrometer feature. Historicaly, the infrared emission features were first observed in nebulae of intense UV excitation just because these are the brightest infrared astronomical objects. Actually, the intensity of the 3.3 μm feature (relative to the total far infrared radiation ) increases in the most diffuse and less UV-excited regions of the interstellar medium.

## 1. Introduction

The IRAS infrared survey has widely shown the ubiquity of the diffuse galactic emission both in the spectral ($\lambda$ = 12, 25, 60 and 100 micrometer) and the spatial (cirrus clouds) domains. It is now well established that this emission originates in interstellar grains whose sizes range from a micrometer to a few angströms (Draine and Lee 1984, Desert et al. 1989). The smaller grains (i.e. large molecules) are responsible for the shorter wavelength emission because they are subject to thermal fluctuations. Their general existence was established by the IRAS detection of the 12 and 25 micrometer diffuse galactic emission. It was necessary to perform the identification of those grains by the detection of selected spectral features. In the hypothesis of Polycyclic Aromatic Hydrocarbon large molecules (Puget et al. 1985) the characteristic emission bands at 3.3, 6.2, 7.7, 8.6 and 11.3 micrometer should be present in the spectrum of the diffuse galactic light. In terms of surface brightness, the diffuse galactic emission is 10 to 100 times fainter than the emission of nebulae (HII, planetary and reflection) in which the aromatic infrared bands (AIBs) are currently observed with ground-based or airborne (KAO) instruments. We started the AROME program in order to achieve this peculiar goal of detecting the AIBs in the diffuse galactic light.

85

E. Bussoletti and A. A. Vittone (eds.), Dusty Objects in the Universe, 85–88.

## 2. Instrument

The instrument consists in two 15 cm Cassegrain telescopes and cooled (60 K) detectors and optics. The measured flux has been done as large as possible by maximising both the transmittance ($\approx 0.5$) and the troughput ($= 0.4\ 10^{-7}\ m^2$ sr, 0.5 degree field of view) of the optics. The feature is detected by comparison of the integrated fluxes measured in a narrow and a wide photometric band, both centered on the feature's wavelength. A spatial modulation is provided by oscillating secondary mirrors (20 Hz) and a slow azimutal scanning at a constant elevation angle. The 3.3 μm feature was choosen as our first aim because of the reduced thermal background at this wavelength. The sensitivity of the instrument is $0.8\ 10^{-7}\ W/m^2/sr/\sqrt{Hz}$ in terms of the integrated energy in the 3.3 μm feature.

## 3. Results

Up to now, two flights of the instrument have been performed. They led to the first detection of the 3.3 μm feature in the diffuse galactic light (Giard et al. 1988). The first flight (Sicily , August 1987) allowed to map a large area of the galactic plane between 10 and 35 degrees of galactic longitude (Giard et al. 1989). The coverage of the second flight was larger, $310° < l < 60°$, and it fully confirmed and extended the previous results. The integrated feature's intensity (continuum corrected) is as high as $10^{-7}$ W/m2/sr in the direction of the galactic disk. It is clearly correlated with the far infrared emission and has a spatial distribution very different from the underlying continuum. As a consequence, we concluded that the 3.3 μm feature finds its source in the interstellar medium whereas the continuum at the same wavelength is of stellar origin.

## 4. Ubiquity of the 3.3 μm feature

The 3.3 μm emission feature is observed at large scales in the interstellar medium and is not specific of UV irradiated regions.

All the known HII-molecular complexes which are gathered in the galactic plane are strong emitters of 3.3 μm feature radiation. The brightest sources are of course those associated with the ionized nebulae M17, NGC 6334, NGC 6357, M42 (Orion A) and NGC 2024 (Orion B), but a few tens of fainter objects can be isolated on our maps. Despite our low angular resolution (0.5°) , the comparison of our flux with previous smaller beam measurements shows that the 3.3 μm feature flux does not principally arise from the HII regions but from the whole associated complexes. Among the sources detected with AROME, observations of the 3.3 μm feature in a 1 arc minute beam were available only for the very bright HII nebulae M42 and M17.

In the case of Orion A (the infrared source associated with M42) the feature's flux in the 30' AROME beam is 7 time as high as the flux integrated over the central 5'x5' map of Sellgren 1981 ( see the poster by N. Sales et al. in these proceedings).

In the case of M17 the feature's AROME flux is three orders of magnitude higher than the 1 arc minute beam value of Grasdalen and Joyce 1976.

Actually, small beam measurements of such extended sources are to be taken with care since they are altered down by the finite chopping amplitude (a few times the field of view), which implies a non-zero flux in the reference beam. However, the relatively high feature's intensities measured in the AROME beam demonstrates that this emission extends over the whole HII-molecular complex.

Outside the complexes, the AROME instrument was also able to measure a noticeable 3.3 μm feature galactic background which was detected up to galactic latitudes of ±5°. The emission observed at such a galactic latitude originates in interstellar diffuse clouds similar to the local 'cirrus', but which are much further away from the sun: d = 5 kpc.

A remarkable fact is that the 3.3 μm emission is less concentrated in HII regions than is the far infrared (FIR) radiation. This is shown in Fig. 1 where we have plotted the ratio of the feature's intensity to the total FIR radiation (3.3 to FIR) versus the total FIR emission, for different lines of sight .

--------------------------------------------

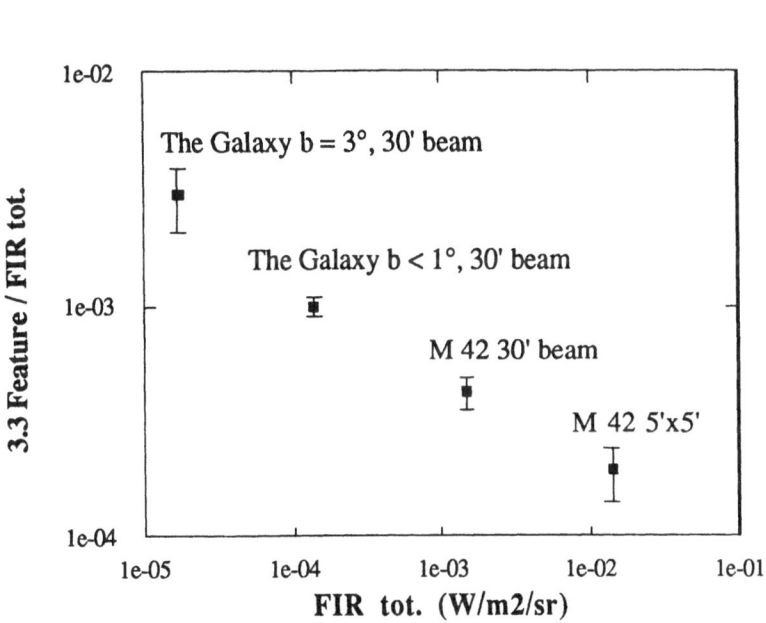

**Fig. 1.** Ratio of the 3.3 μm feature's intensity to the total FIR radiation versus the total FIR. Averaged 30' FIR galactic fluxes are from IRAS; the FIR for M42 is from Harper 1974; and the 3.3 μm 5'x5' value is from Sellgren 1981. Averaged 30' 3.3μm energies are from AROME data.

The 3.3 μm measure has not been corrected for interstellar extinction. However, we have shown (Giard et al. 1989) that absorption by dust is only able to account for half of the decrease of the 3.3 to FIR ratio in the case of the averaged galactic emission between b = 3° and b= 0°. The 3.3 to FIR ratio is very sensitive to the very small grains composition (abundance, hydrogenation rate and minimum size) and not much to the strength and spectrum of the exciting radiation field. (See Desert et al. for a complete description of grain models including very small grains). Thus, It is likely that both extinction and variations in the nature and composition of the very small grains contribute to the observed variations in the 3.3 to FIR ratio.

More observations of the other emission features and whith a better angular resolution are necessary and will certainly give very valuable information and new constraints on the nature and abundance of the interstellar very small grains.


## 5. Conclusion

Up to now, the observations of the infrared emission features were limited to astronomical objects in which a strong UV flux irradiates interstellar dust. This was essentially due to a limitation in sensitivity, since these objects are the brightest infrared emitters in the sky. Actually, AROME observations have shown that the 3.3 μm feature emitting material (i.e. very small grains) is present in the interstellar medium at the scale of the galactic plane, like CO, HI and dust. We have first indications of variations in the abundance and composition of this material under different interstellar environments. This point has to be investigated by more experimental work, since the exact nature of the interstellar very small grains will have important consequences on the understanding of the chemistry and the energetics of the interstellar medium.


## References

Desert X., Boulanger F. and Puget J.L.: 1989 in preparation.
Giard M., F. Pajot, J.M. Lamarre, G. Serra, E. Caux, R. Gispert, A. Léger, D. Rouan: 1988, A&A **201**, L1.
Giard M., F. Pajot, J.M. Lamarre, G. Serra and E. Caux: 1989, A&A **215**,92.
Grasdalen G.L. and R.R. Joyce: 1976, Ap.J. **205**, L11.
Harper D.A.: 1976, Ap. J. **192**, 557.
Perault M.: 1987, Thèse d'Etat, Université de Paris VII.
Puget J.L., Léger A., Boulanger F.: 1985, A&A **142**, L19.
Sellgren K.: 1981, Ap.J. **245**, 138.
Thronson H.A.: Ap.J. **204**, 420.

# SEARCH FOR THE UV AND IR SPECTRA OF $C_{60}$ IN LABORATORY-PRODUCED CARBON DUST

**W. Krätschmer[1], K. Fostiropoulos[1], D.R. Huffman[2]**
[1]*Max-Planck-Institut für Kernphysik, Heidelberg, F.R.G.*
[2]*University of Arizona, Tucson, Arizona, U.S.A.*

ABSTRACT. Carbon dust samples were prepared by evaporating graphite in an atmosphere of an inert quenching gas (Ar or He). Changes of the spectral features of the carbon dust were observed when the pressure of the quenching gas was increased. At low pressures (order 10 torr), the spectra show the familiar broad continua. At high pressures (order 100 torr), narrow lines in the IR and two (or possibly three) broad features in the UV emerge. The four strongest IR features are located at 1429, 1183, 577, and 527 $cm^{-1}$. Their positions are close to that predicted for the $C_{60}$ molecule. The UV features are observed at 340, 270, and, less distinct, at 220 nm. One of these features may be related to the known 368 nm transition of $C_{60}$. It thus appears that at high quenching gas pressures $C_{60}$ is produced along with the carbon dust.

# 1 Introduction

Interest in the mechanism for forming carbon molecules —e.g. in circumstellar shells— led to a group of interesting experiments which employed mass spectroscopy. In such a study, the extraordinary stability of the $C_{60}$ molecule was discovered and assigned to the truncated icosahedron or soccer ball structure of this molecule (Kroto et al., 1985). It has been speculated that $C_{60}$ may be the carrier of interstellar features, like e.g. the 220 nm absorption, the diffuse lines in the visible and others. In order to provide convincing evidence for $C_{60}$ in interstellar space, the spectral features of this molecule must be known. Until now, however, the molecule has not been produced in sufficient quantity to permit conventional spectroscopy to be done. In the UV-VIS domain longwards of 335 nm only one weak (f= 0.004) absorption feature located at 386.0 nm, with a width of about 1 nm, has been detected (Heath et al., 1987). The technique used in that work, namely laser depletion spectroscopy of van der Waals complexes between $C_{60}$ and other molecules, has been applied to $C_{60}$ molecules in the gas phase. Calculations of the electronic eigenstates of $C_{60}$ indicate the existence of strong lines at 250 nm and shorter wavelength (Larsson et al., 1987). So far, no experimental data on the spectral features in the IR domain exist. There are, however, theoretical predictions of the IR spectra which are based on certain assumptions concerning the $C_{60}$ molecular shape and the interatomic forces. Provided the molecular geometry of $C_{60}$ in fact is that of a soccer-ball, i.e. that of a truncated icosahedron (point group $I_h$), four IR active vibrations should

*E. Bussoletti and A. A. Vittone (eds.), Dusty Objects in the Universe, 89–93.*
© 1990 Kluwer Academic Publishers.

exist. The force fields and constants assumed in the calculation of the actual vibration frequencies are either adopted from smaller carbon-containing molecules or obtained by quantum-chemical calculations. According to these calculations, the four IR-active frequencies should be located at about $1600\pm200$ cm$^{-1}$, $1300\pm200$ cm$^{-1}$, $630\pm100$ cm$^{-1}$, and $500\pm50$ cm$^{-1}$. The ranges indicate roughly the spread found in the literature (see e.g. Weeks and Harter, 1989; Stranton and Newton, 1988; Cyvin et al., 1987; Wu et al., 1987). Stereoscopic views of the atomic displacements at the fundamental vibrations of a 60-atomic soccer-ball molecule are displayed in an article by Weeks and Harter 1989.

During the course of our work on carbon smokes we produced by chance samples showing peculiar UV absorption features in addition to the familiar 220 nm absorption. We were surprised when we recently found that these smokes exhibited unusual IR spectra as well. We believe that in this cases we produced $C_{60}$ molecules along with the smoke particles.

## 2  Experimental procedures

Carbon dust was produced within a conventional bell-jar carbon evaporator which was first evacuated to $10^{-4}$ torr by either an oil diffusion pump or a turbo pump (both equipped with LN$_2$ cold-traps) and then filled by an inert quenching gas. We used He or Ar for this purpose. The pressures ranged between 1 and 200 torr (He) and 1 to 100 torr (Ar). As the next step, the graphite rods were evaporated by resistive heating. The smoke which formed in the vicinity of the evaporating rods was collected on two kinds of substrates: Si or Ge for the IR, and quartz-glass for the UV-VIS spectral range. The substrates were placed within a distance between 5 and 10 cm from the rods. We usually tried to produce similar optical densities in the IR and the UV-VIS. This required a much larger dust coverage for the IR —as compared to the UV-VIS samples. The evaporator was opened after a cool-down period of 10–30 min and the smoke sample taken out. Finally, the spectrum of the collected smoke was measured by ratioing the transmission of the dust-coated substrate to that of the empty substrate. UV-VIS spectra were taken by a grating spectrometer (type PE 330; 2 nm resolution) and IR spectra by an evacuable Fourier-transform spectrometer (type Bruker 113 V; resolutions between 2.0 and 0.5 cm$^{-1}$, range between 5000–400 cm$^{-1}$, occasionally 5000–20 cm$^{-1}$).

## 3  Results and discussion

Typical UV-VIS and IR spectra of carbon dust are shown in figures 1 and 2. For low quenching gas pressures (i.e. in the order 10 torr), the spectra show the expected featureless continua on which broad humps are superposed, e.g. at 220 nm in the UV, and 1600–1200 cm$^{-1}$ as well as 900–500 cm$^{-1}$ in the IR. The 220 nm absorption hump originates from an electronic transition, the 1600–1200 cm$^{-1}$ hump from streching modes, and the 900–500 cm$^{-1}$ hump from bending modes of carbon atoms within distorted graphitic structures. The featureless continuum is due to scattering of the light by the dust particles as well as to absorption by the free conduction-electrons in the graphitic material.

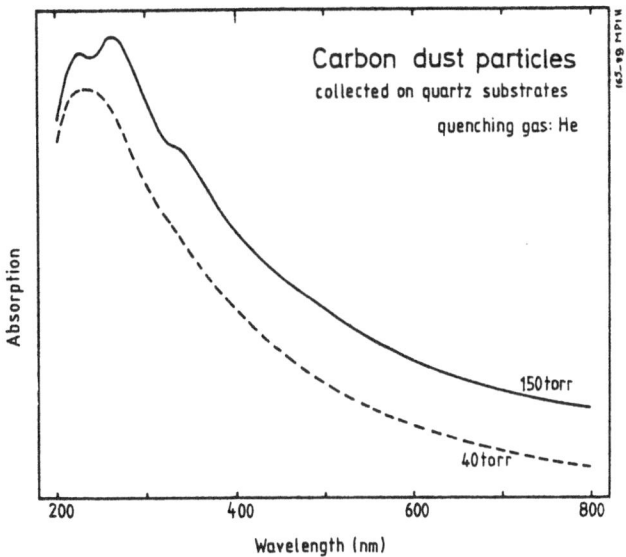

**Figure 1:** The UV-VIS spectrum of carbon smokes produced at a quenching gas pressure of 40 torr He (dotted line) and at 150 torr He (solid line). The absorption ($-\log_{10} T$, were T is the transmission) of both spectra ranges from about 0.2 (minimum) to 2.0 (maximum). The carbon dust was collected on UV transparent quarts-glass. The broad feature at about 220 nm originates from graphitic carbon dust. Note the additional humps at 270, 340 nm, and —less distinct— at 220 nm in the upper spectrum. The absorption at 340 nm may be related to the 368 nm band of $C_{60}$.

For high quenching gas pressures (i.e. in the order 100 torr), additional spectral features appear in the carbon smokes. Under these conditions, apparently new components are formed and become admixed to the "regular" dust. Especially the appearance of the narrow IR-lines is striking; it suggests that these new components are molecules. The lines are located at 1539, 1429, 1183, 1135, 675, 577, and 527 $cm^{-1}$. Four lines are much stronger than the others; these are the lines at 1429, 1183, 577, and 527 $cm^{-1}$. Their intensity is a few percent above the continuum. The additional UV features are centered at about 270 and 340 nm and have a width of about 25 nm. In the vicinity of 220 nm an additional UV feature may exist as well since the shapes of the two spectra of figure 1 are slightly different in this wavelength range.

The IR lines and UV bands always appear together. They therefore seem to belong to the same molecule(s). However, since UV and IR measurements on the same smoke samples are difficult, a strict intensity correlation of both kinds of features could not yet be established.

As far as the appearence of the features is concerned, we observed that a kind of transition pressure of the quenching gas exists above which the features appear regularly and below which they usually do not. For He, this pressure is about 50 torr and for Ar

92

it seems to be smaller, i.e. about 30 torr.

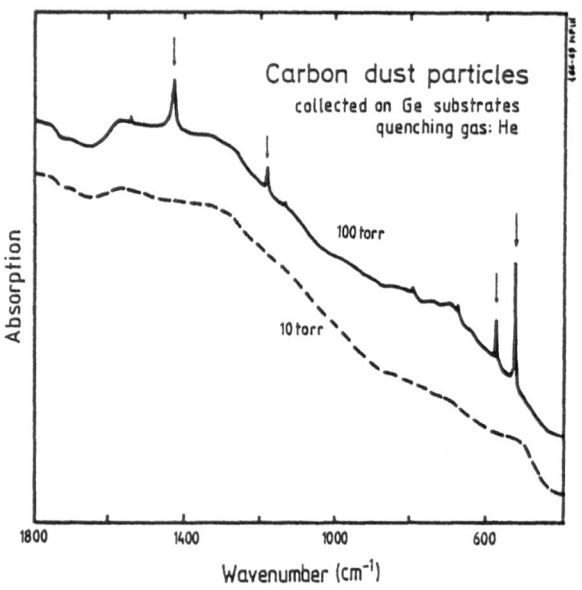

**Figure 2:** The IR spectrum of carbon smokes produced at a quenching gas pressure of 10 torr He (dotted line) and 100 torr He (solid line). The absorption of both spectra ranges from about 0.2 (minimum) to 0.3 (maximum). The broad humps in the range 1600–1200 $cm^{-1}$ and 900–500 $cm^{-1}$ in both spectra originate from graphitic dust. Notice the lines which are absent in the lower spectrum. The four most intense lines, marked by arrows, may originate from $C_{60}$.

We first thought the IR line features were produced by a molecular contamination of the carbon smoke which, for example, might have taken place within the vacuum system, e.g. by pump-oil. However, when we replaced the ($LN_2$ baffled) oil-diffusion pump by a turbo-pump, equipped with a rough-pump oil-absorber, and an $LN_2$ cooled baffle, we observed no change in the spectral patterns. Thus, pump-oil is unlikely to be responsible for the features. Hydrocarbons as carriers can be ruled out as well since the characteric C-H bond absorptions in the vicinity of e.g. 3000 $cm^{-1}$ are usually weak or absent. Also a change of the quenching gas supply to a higher degree of cleanliness ($10^{-6}$ ppV) had no effect. We thus believe the carrier of the IR lines (and UV features) is a carbon molecule. However, definite proof whether this is in fact the case requires the same experiments to be performed with $^{13}C$ graphite. We plan to do this in the future.

As it appears to us, the only likely carrier for the IR and UV features is the $C_{60}$ molecule. This assumption mainly rests on our IR data. The number of IR features agrees with what is predicted for this species: Four IR bands (in the case of icosahedric symmetry) are expected and this is what we observe. In a search down to 20 $cm^{-1}$ we have found no other prominent IR lines. The positions of the four lines are all within the ranges predicted by theory. For a rough estimate of the $C_{60}$ concentration in the dust, the ratio of the equivalent widths of the lines to that of the broad features in the continuum may be used. The ratio and thus the $C_{60}$ concentration seems to range in the order of percent. In the UV, our data seem to be in conflict with that of Heath et al., 1987: The reported 1 nm wide UV band at 386 nm does not show up in our spectra. Either this feature is too weak to be detected at our level of $C_{60}$ concentration, or this band is related to the broad feature we find at 340 nm. In the latter case there is an obvious mismatch in the line width and position. However, the $C_{60}$ molecules in our experiment are embedded in a kind of matrix of graphitic particles on which they probably stick. The interaction in this environment may shift and broaden the UV features as well as the IR lines. The additional weak lines present in the IR spectra may in fact indicate distortions (e.g. in the molecular symmetry) due to such interactions. However, their degree of influence on the spectra must be rather limited. Otherwise, the main IR bands which, due to the symmetry of the molecule, are highly degenerate, should show a substantial broadening or splitting into subbands. We do not see such an effect. The additional UV absorptions we see in the vicinity of 220 and 270 nm are not far away from the wavelengths positions of strong electronic transitions predicted by theory (Larsson et al., 1987).

If our conjecture is correct, then the $C_{60}$ molecule in fact has the expected icosahedric symmetry and is formed in macroscopic quantities along with graphitic smoke particles at high quenching gas pressures. It may be possible in the future to extract micro-gram quantities of pure $C_{60}$ from the smoke.

In case the kind of carbonaceous material we have produced in this work exists in interstellar space, it should be observable in astronomical sources. It would likely show up either as structure associated with the 220 nm interstellar absorption feature or as absorption or emission band at the positions of the stronger IR absorptions (e.g. at 1429 and 527 $cm^{-1}$). In detailed examinations of the 220 nm interstellar band shape to date, we are not aware of any evidence for the kind of structure observed here.

# References

Cyvin, S.J., Brendsdal, E., Cyvin, B.N., Brunvoll, J.: 1988, Chem. Phys. Letters, vol. 143, no. 4, 377.

Heath, J.R., Curl, R.F., Smalley, R.E.,: 1987, J. Chem. Phys., vol. 87, no. 7, 4236.

Kroto, H.W., Heath, J.R., O'Brien, S.C., Curl, R.F., Smalley, R.E.: 1985, Nature, vol. 318, 162.

Larsson, S., Volosov, A., Rosen, A.: 1987, Chem. Phys. Letters, vol. 137, no. 6, 501.

Stanton, R.E., Newton, M.D.: 1988, J. Phys. Chem., vol. 92, 2141.

Weeks, D.E., Harter, W.G.: 1989, J. Chem. Phys., vol. 90. no. 9, 4744.

Wu, Z.C., Jelsky, D.A., George, T.F.: 1987, Chem. Phys. Letters, vol. 137, 291.

# CIRCUMSTELLAR DUST ENVELOPES OF M GIANT STARS

O. Hashimoto[1*], Y. Nakada[1], T. Onaka[1], T. Tanabé[2], F. Kamijo[1]
[1]Department of Astronomy, University of Tokyo, Japan
[2]Institute of Astronomy, University of Tokyo, Japan

ABSTRACT. IRAS data of M giant stars are studied using a spherical dust envelope model. The infrared colors and the strengths of the 10 $\mu$m silicate emission band of the IRAS M stars which show the emission band in LRS spectra are well explained by the model with dirty silicate grains. However, some additional grains which have no feature at 10 $\mu$m are required in order to explain the spectra of the M stars which do show no features or very weak band emission at 10 $\mu$m.

## 1 Introduction

Most M giant stars are losing their mass and are surrounded by the circumstellar envelopes of the ejected matter. The absorption, the scattering and the thermal emission by the dust grains in the circumstellar envelopes affect the spectra of the mass losing stars. Dusty oxygen-rich objects such as M giant stars often show the 10 $\mu$m silicate emission band in the IRAS Low Resolution Spectrometer (LRS) spectra. The study of the IRAS data of the late type stars should lead to an understanding of the dust envelopes and mass loss phenomena of these objects. IRAS observations of M stars whose LRS spectra are available are studied using a dust envelope model.

## 2 IRAS data

LRS spectra with no features are classified as *1n*, those which show the 10 $\mu$m emission band as *2n* (IRAS Science Team, 1986). The IRAS M stars mostly belong to the LRS class *1n* or *2n*. Only M stars whose fluxes are of good quality in all three bands at 12, 25, and 60 $\mu$m are selected from IRAS point source data base; 473 *1n* M stars and 188 *2n* M stars are selected. Two colors of $\log(S_{25}/S_{12})$ and $\log(S_{60}/S_{25})$ are used for our analysis. The strengths of the 10 $\mu$m emission band in the LRS spectra are examined in the case of the M stars with the emission band (LRS *2n*). Photometric data of IRC catalog are also used in the discussion.

---

*Present address: Department of Applied Physics, Seikei University, Musashino, Tokyo 180 Japan

*E. Bussoletti and A. A. Vittone (eds.), Dusty Objects in the Universe, 95–102.*
© 1990 *Kluwer Academic Publishers.*

## 3 The dust envelope model

Equations of radiative transfer in a spherical dust envelope are solved using a generalized two-stream Eddington approximation method (Unno and Kondo, 1976, 1977; Unno, 1989).

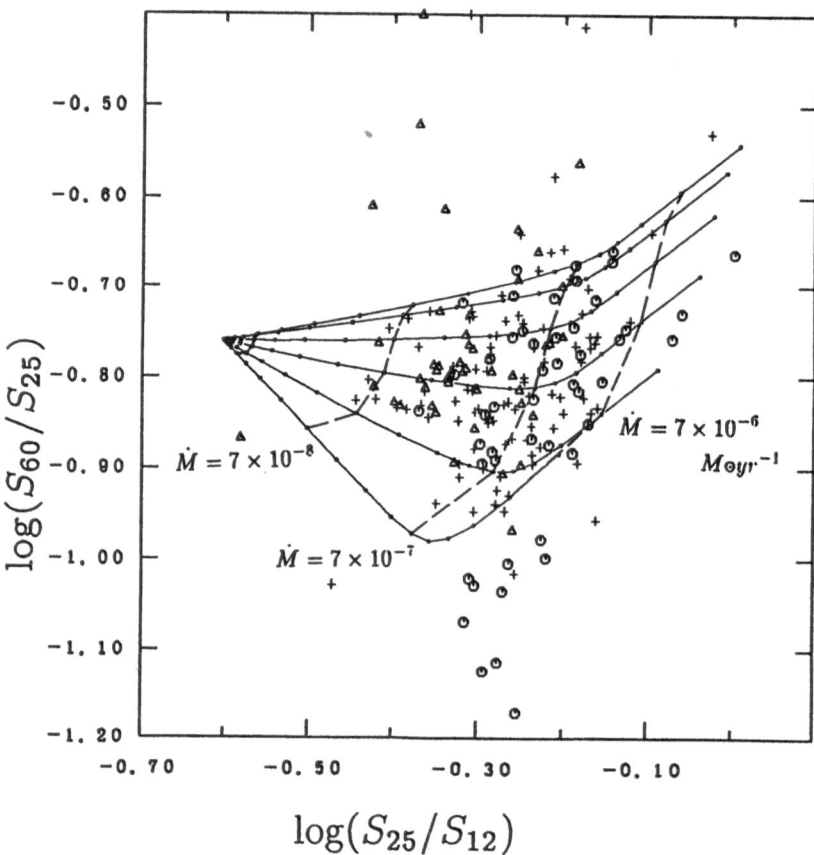

**Figure 1:** Two-color diagram of $\log(S_{25}/S_{12})$ vs. $\log(S_{60}/S_{25})$ for M stars with the 10 $\mu$m silicate emission band (*2n* M stars). The IRAS data are plotted as triangles (weak band emission; *21-22*), circles (strong band emission; *27-29*) and crosses (others). The best fit model grid is also presented.

Several models are calculated with various combinations of the mass loss rate, the condensation temperature of the dust grains and the time which has passed since the mass loss started.

Steady mass flow is assumed for our model. The expansion velocity of the envelope is fixed as $V(\text{expansion}) = 15$ km s$^{-1}$ from radio observations (Knapp and Morris, 1985). The stellar radius is assumed as $R_* = 400R_\odot$. The spectrum of the central star is approximated by a 3000 K black body. The gas to dust mass ratio is set to be 200.

## 4  M stars with the 10 μm emission band

IRAS point source data of the M stars with the 10 μm silicate emission band (LRS ℓn) are examined by the models. Two IRAS colors of $\log(S_{25}/S_{12})$ and $\log(S_{60}/S_{25})$ and the strengths of the 10 μm emission band in the LRS spectra of the selected 188 M stars with the emission band are well explained by the models with the dirty silicate grains (Jones and Merrill, 1976) whose opacity is modified to be proportional to $\lambda^{-1.5}$ in the wavelength range longer than 28 μm (Hashimoto et al., 1989). The condensation temperatures of the dirty silicate grains at the inner boundary of the best fitted models are about 900 K.

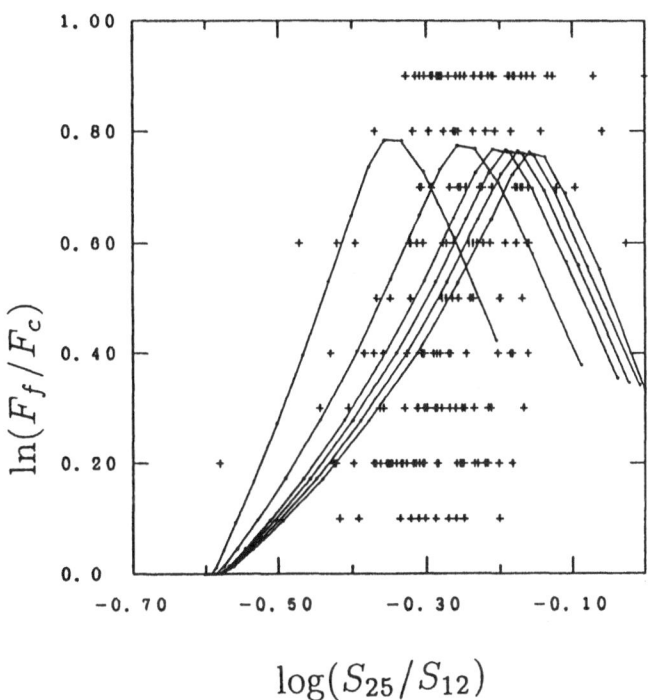

**Figure 2:** Diagram of $\log(S_{25}/S_{12})$ vs. the strength of the 10 μm emission band. Models and IRAS data presented in figure 1 are plotted.

The model fit in the IRAS two-color diagram is presented in figure 1. The IRAS data are plotted as triangles(weak band emission), circles(strong band emission) and crosses (others). Dashed lines in the model grid represent the models with mass loss rate $7 \cdot 10^{-8}$, $7 \cdot 10^{-7}$ and $7 \cdot 10^{-6} M_\odot$ yr$^{-1}$ from the left to the right, and solid lines show those with the duration of the mass loss 50, 170, 580, $1.8 \cdot 10^3$, $5.9 \cdot 10^3$ and $1.8 \cdot 10^4$ years from the bottom. The strength of the 10 μm emission band shows a maximum when $\log(S_{25}/S_{12})$ $\sim -0.3$ to $-0.2$ (see figure 2) for both our models and IRAS objects. However, it is difficult to explain the objects with red $\log(S_{25}/S_{12})$ and weak emission feature.

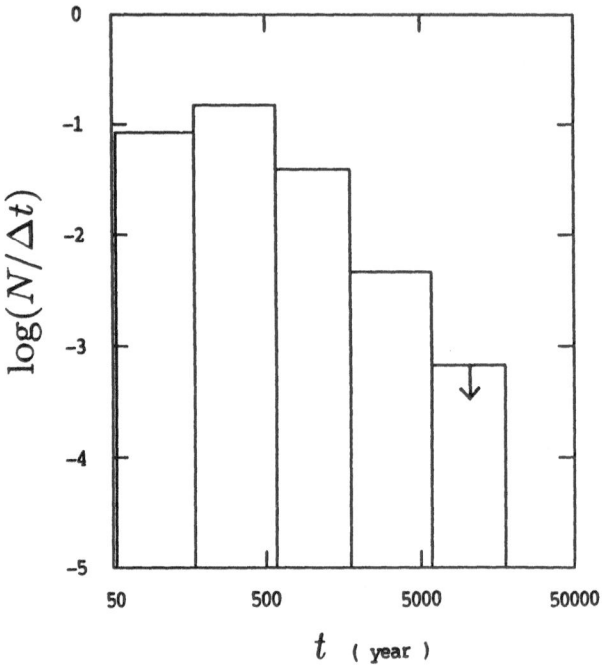

**Figure 3:** A histogram of the distribution of the duration time of the steady mass loss.

In the two-color diagram very few objects are located in the region where the duration of the mass loss of the model is longer than a few of $10^4$ years. A histogram of the distribution of the duration time of the steady mass loss is presented in figure 3. The value of $N/\Delta t$ in the ordinate is the mean number of IRAS objects per year which are located between two solid lines of constant duration time in the model grid in figure 1. It is suggested that the characteristic time of the steady mass loss is about 2000 years.

There are no IRAS $\ell n$ objects which correspond to the models with $\dot{M} < 7 \cdot 10^{-8} M_\odot$ $\mathrm{yr}^{-1}$ in figure 1. Because our model indicates that the 10 $\mu$m emission band becomes strong enough to be observed as $\ell n$ when $\dot{M} > 1.4 \cdot 10^{-8} M_\odot \ \mathrm{yr}^{-1}$ (see figure 4), the critical value of $7 \cdot 10^{-8} M_\odot \ \mathrm{yr}^{-1}$ can be interpreted as the minimum mass loss rate necessary for the formation of the silicate grains.

The strength of the 10 $\mu$m emission band is well correlated with the mass loss rate in our model (figure 4), irrespective of the value of $t$ (mass loss). The suggested correlation is supported by the mass loss rates obtained from radio observations (Knapp and Morris, 1985; Knapp, 1985), which are also plotted in figure 4. Therefore, the strength of the 10 $\mu$m emission band is thought to be a direct indicator of the mass loss rate (cf. Skinner and Whitmore, 1988).

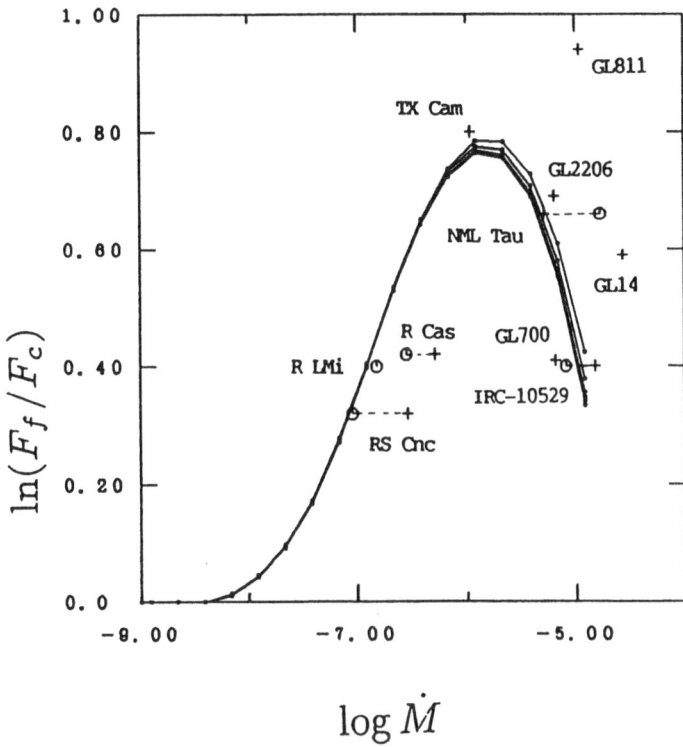

$$\log \dot{M}$$

**Figure 4:** Relation between the mass loss rate and the strength of the 10 $\mu$m emission band. Mass loss rates obtained by radio observations (Knapp and Morris, 1985, Knapp, 1985; crosses and circles respectively) are compared.

## 5  M stars without the 10 $\mu$m feature

The two-color diagram of the selected 473 *1n* M stars is presented in figure 5. Two separated groups of the *1n* objects are seen in the diagram. One is a group of blue M stars with $\log(S_{25}/S_{12})$ less than $-0.5$, and the other is that of red objects with $\log(S_{25}/S_{12})$ > $-0.5$. Most of the blue objects are M stars earlier than about M4 (triangles), while most of the red objects are later than M5 (circles). The IRAS color of the blue objects can be explained by the photosphere of the star. However, it is impossible for our model with the modified dirty silicate grains to reproduce the spectra of the red objects without showing the 10 $\mu$m emission feature. Some reasons other than the dirty silicate grains are necessary to explain the colors of the red objects.

Another two-color diagram of $\log(S_{25}/S_{12})$ vs. $\log(S_{12}/S_K)$ of the M stars is presented in figure 6. Photometric data of K-band are taken from the IRC catalog. The M stars with the weak and strong 10 $\mu$m band emission are denoted by crosses and circles

respectively,

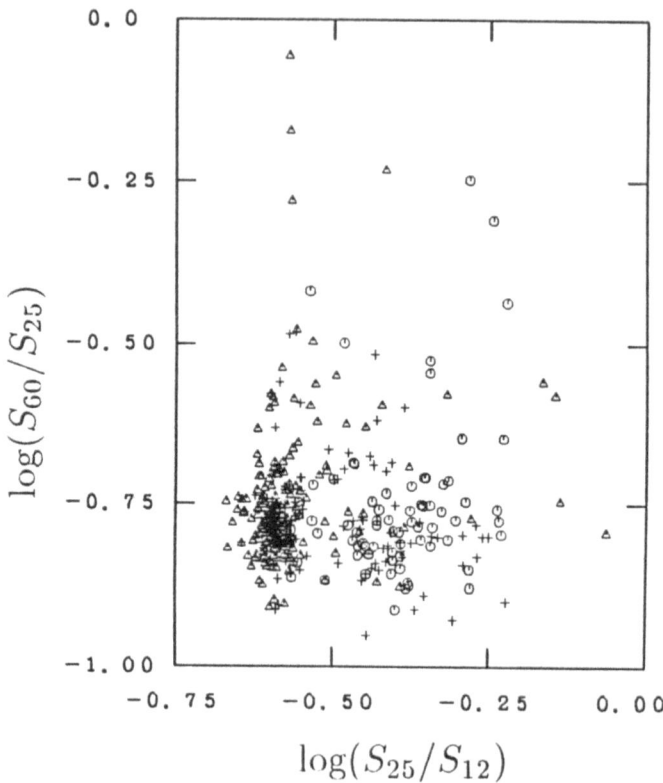

**Figure 5:** Two-color diagram of $\log(S_{25}/S_{12})$ vs. $\log(S_{60}/S_{25})$ for IRAS $1n$ M stars. M stars from M1 to M3 are plotted as triangles, M4–M5 as crosses and M6–M9 as circles.

while the $1n$ M stars are denoted by triangles. The red $1n$ objects with $\log(S_{25}/S_{12})$ $> -0.5$ have the different colors of $\log(S_{12}/S_K)$ from those of the blue objects with $\log(S_{25}/S_{12}) < -0.5$. It is noted that the distributions of the red $1n$ objects and of the $2n$ objects with the weak 10 $\mu$m emission band overlap each other. The $1n$ M stars with $\log(S_{25}/S_{12}) > -0.5$ may be the similar objects to the $2n$ M stars with the weak band emission. The red colors of $2n$ M stars are due to the dust grains in the circumstellar envelopes. The red $1n$ M stars may be reddened by the circumstellar dust grains having no features in the wavelength region of LRS. If there are such grains in the envelope, they could also account for the red color of some $2n$ M stars with the weak band feature. Radio emission of CO, SiO and $H_2O$ are observed for some $1n$ M stars (Engels and Heske, 1989, etc.). It means that these objects are undergoing mass loss. It is very likely that they are surrounded by the dust envelopes with some featureless grains.

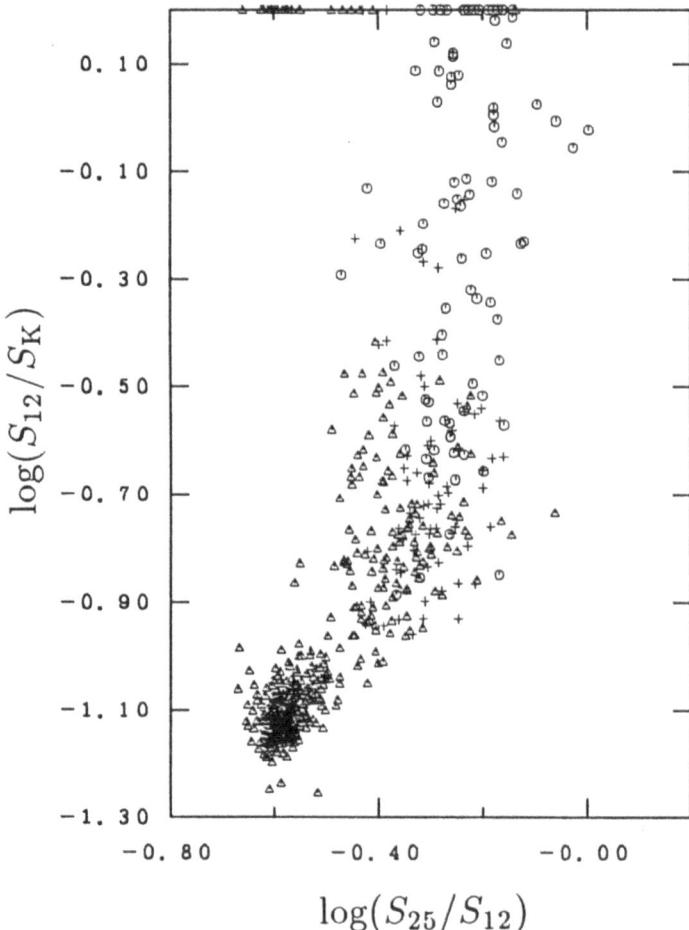

**Figure 6:** Two-color diagram of $\log(S_{25}/S_{12})$ vs. $\log(S_{12}/S_K)$. IRAS $1n$ M stars are plotted as triangles, the M stars with the weak 10 $\mu$m band emission ($21$–$24$) as crosses and those with the strong band emission ($25$–$29$) as circles.

Of cource, reddening of the stellar atmosphere itself, or effect by the molecules in the extended atmosphere cannot be ruled out to explain the color of the redder $1n$ M stars. Further study is necessary to solve the problem.

# 6 Conclusions

1. The IRAS data of the M stars with the 10 $\mu$m emission band are well explained by the models with the dirty silicate grains whose opacity is proportional to $\lambda^{-1.5}$ at $\lambda > 28$ $\mu$m. The condensation temperature of the dirty silicate grains is about 900 K.

2. The strength of the 10 $\mu$m emission band is a direct indicator of the mass loss rate of the $\mathcal{L}n$ M stars (see figure 4).

3. The lower limit of the mass loss rate $7 \cdot 10^{-8} M_\odot$ yr$^{-1}$ indicates a critical lower limit for the grain formation around M stars.

4. The characteristic time of the duration of the steady mass loss of the $\mathcal{L}n$ M giant stars is about 2000 years.

5. Late M stars without the 10 $\mu$m feature show reddened colors of $\log(S_{25}/S_{12})$ and $\log(S_{12}/S_K)$.

6. There might be some featureless grains in the envelopes of the red $1n$ M stars and the red M stars with the weak 10 $\mu$m emission band.

# References

Engels, D., Heske, A.: 1989, Astron. Astrophys. Suppl., in press.

Hashimoto, O., Nakada, Y., Onaka, T., Tanabé, T., Kamijo, F.: 1989, Astron. Astrophys., in press.

IRAS Science Team: 1986, Astron. Astrophys. Suppl., **65**, 607.

Jones, T.W., Merrill, K.M.: 1976, Astrophys. J., **209**, 509.

Knapp, G.R.: 1985, Astrophys. J., **293**, 273.

Knapp, G.R., Morris, M.: 1985, Astrophys. J., **292**, 640.

Skinner, C.J., Whitmore, B.: 1988, M.N.R.A.S., **231**, 169.

Unno, W.: 1989, P.A.S.J., **41**, 211.

Unno, W., Kondo, M.: 1976, P.A.S.J., **28**, 374.

Unno, W., Kondo, M.: 1977, P.A.S.J., **29**, 693.

# INTERSTELLAR PLANETESIMALS

W.M. Napier
*Royal Observatory, Blackford Hill, Edinburgh, Scotland*

ABSTRACT. The dust to gas ratio in a nebula is unstable in the presence of a radiation field. A radiative analogue of the Jeans' critical mass for gas exists, from which it appears that in a molecular cloud environment the dust tends to accumulate into discrete concentrations of mass $\sim 10^{15} - 10^{23}$ g. Solid bodies of cometary or asteroidal dimensions may therefore comprise a significant component of the interstellar medium. The implications of this interstellar population for theories of star formation are briefly discussed. Evidence is presented that some proportion of the comet population, including Halley's Comet, may have been recently captured during passage of the Solar System through the Scorpio-Centaurus association.

## 1 Solid body growth in molecular clouds

It is believed that the interstellar medium is prone to a number of instabilities. Jeans' collapse, giving rise to star formation, is the best known of these, but in addition molecular clouds may attain states of chemistry or temperature in which small-scale structures, of dimensions $\sim 10 - 1000 AU$, are formed. The detection of clumpy structure within molecular clouds, limited apparently only by the resolving power of the telescope employed, is well known, molecular cores of characteristic dimension $\sim 0.1$ pc and molecular number density $n(H_2) \sim 10^4 - 10^5$ being found. Such structures are found even in the IRAS high-latitutde cirrus clouds (e.g. Turner et al. 1989).

Inhomogeneities in the dust distribution, over and above those created by the gas instabilities, appear to be created by the action of radiation on dust grains which, shielding each other in an ambient radiation field, would tend to be forced together by the differential radiation pressure created by the mutual shadowing. This phenomenon has been confirmed in a number of theoretical studies (e.g. Flannery and Krook 1978). Gerola and Schwartz (1976) pointed out that the photodesorption of molecules from the surfaces of grains might enhance this instability by an order of magnitude and predicted that all diffuse interstellar clouds should have an inhomogeneous dust distribution. They considered that the growth rate of the perturbations was almost scale-free.

This radiative driving, with or without photodesorption, mimics gravity in some respects, and Napier and Humphries (1986) pointed out that, arising out of it, a radiative analogue of the Jeans' critical mass exists for the dust component of the interstellar medium. Starting with a homogeneous gas in an ambient radiation field, one applies a spectrum of perturbations in the dust-to-gas ratio. Left to itself, a perturbation would

*E. Bussoletti and A. A. Vittone (eds.), Dusty Objects in the Universe, 103–110.*

disperse through the Brownian motion of the dust grains, unless this motion were offset by the inward driving of the radiation (with or without photodesorption). Thus a generalised virial theorem can be constructed and the condition for marginal instability found:

$$2K_E + \Omega + \Psi = 0 \tag{1}$$

where

$$\Omega = \int_V \rho_g \mathbf{r}.\mathbf{g} dV \tag{2}$$

represents the gravitational potential, and

$$\Psi = \int_V \rho_g \mathbf{r}.\mathbf{f} dV \tag{3}$$

represents the radiative one. Here f, g are the accelerations on a dust grain due to radiation pressure and gravity respectively, while $\rho_g$ represents the density of an individual grain. The integration is taken over the volume V of a prescribed perturbation. The acceleration f may be enhanced by photodesorption, while g refers to the self-gravity of the perturbation. This self-gravity will be due to an enhanced dust level over some region: an overdensity of gas would disperse at the sound speed, whereas the thermal motion of the dust is only $\sim 1$ cm s$^{-1}$ at 10 K.

To apply the criterion (1), the acceleration acting on a grain within a dust concentration must be found. Radiation pressure in the interstellar medium acts most effectively on 'classical' grains of diameter $\sim 10^{-5}$ cm, less so on small grains or large aromatic molecules; and the incident radiation is assumed to be absorbed rather than scattered. The acceleration, assuming photodesorption of molecules from an icy mantle, turns out to take a simple form:

$$f = \frac{\pi a^2}{m_g} KUG \tag{4}$$

where $\pi a^2$ represents the cross-section of a grain and $m_g$ is its mass. The factor K represents the momentum transfer from an ejected molecule or impinging photon, U is the energy density of the radiation field outside the perturbation, and G is a geometrical factor representing the anisotropy of the radiation field within the perturbation. This latter takes the form

$$G = \delta \tau \tag{5}$$

for a homogeneous slab perturbation, or

$$G = \frac{1}{3} \delta \tau \tag{6}$$

for a spherical one. Here $\delta \tau$ represents the optical depth enhancement due to the perturbation measured from its mid-plane or centre; it can be related to the enhancement $\delta n_g$ in grain density through

$$\delta \tau = \pi a^2 \delta n_g r \tag{7}$$

For a dust-to gas ratio $10^{-2}$ by mass and a local gas density $n_H$, one has

$$n_g = 1.6 \times 10^{-12} \epsilon n_H \qquad (8)$$

where $\epsilon = \delta n_g / n_g$ is the strength of the imbedded perturbation.

The efficiency of photodesorption in interstellar conditions is uncertain, although there is a theoretical basis for expecting a photodesorption yield $Y \sim 0.02$ for loosely bound molecules on the mantles of UV-irradiated grains (Napier and Humphries 1986). Adopting this in the first instance, and assuming that molecules of molecular weight 15 are dislodged from grain mantles with a kinetic energy 0.2–0.5 eV by near-UV photons, one finds

$$K \sim 225 - 350$$

Putting these factors together, and noting that the kinetic energy $K_E$ of the dust perturbation is given by

$$K_E = \frac{3}{2} M_{cr} \frac{\mathcal{R}}{\mu} T \qquad (9)$$

where $\mathcal{R}$ represents the gas constant, and $(\mu, T)$ refer respectively to the molecular weight and kinetic temperature of the grains, one obtains the critical mass $N_{cr}$ above which a dust perturbation would collapse in the presence of a radiation field. For a weak ($\epsilon \ll 1$) spherical perturbation

$$M_{cr} = \frac{1.4 \times 10^{15}}{(K/250)^{3/2}(U/10^{-10})^{3/2}(\epsilon/10^{-1})^{3/2}(n(H_2)/10^3)^{1/2}} \text{ g} \qquad (10)$$

the collapse time $t_c$ is determined by the terminal speed of the grains through the gas, which is reached rapidly. One finds

$$t_c \sim \frac{630}{(K/250)^{3/2}(U/10^{-10})(\epsilon/10^{-1})} \text{ yr} \qquad (11)$$

for a nebula of temperature 10 K, independently of the dimensions of the perturbation or the density of the nebula. In an OB association, with $U \sim 10^{-10}$ erg cm$^{-3}$, collapse will occur strongly in $10^3 - 10^5$ yr with or without photodesorption, yielding bodies with $M_{cr} \sim 10^{15} - 10^{19}$ g. For $U \sim 10^{-13}$ erg cm$^{-3}$, more appropriate to the mean stellar background, collapse will occur marginally in the presence of photodesorption and not at all under the action of radiation pressure only. A turbulent velocity gradient $\sim 0.04$ m s$^{-1}AU^{-1}$ has been assumed.

The case of a uniform slab or sheet perturbation is likely to be common, arising from radiation pressure from nearby individual stars, the interaction of clouds and so on. In this case the distribution of forces acting on the dust is the same as the distribution of self-gravitational forces acting within an isothermal gas layer. The evolution of the radiative perturbations may therefore be analogous to that of a collapsing sheet of gas (Miyama et al. 1987). Linear and second order perturbation analyses for this case have revealed that a collapsing sheet breaks into slender cylinders or filaments, whose subsequent evolution and break-up are controlled by opacity to radiation. The random break-up of a sheet would yield fragments with a power law mass distribution, of index −1.75. Thus by number small bodies would be most common, whereas by mass the rare, large bodies would predominate.

In the case of a spherical perturbation, sub-fragmentation may not occur (the most unstable mode having an infinite wavelength in the gravitational analogue). An initial blob may thus, in principle, form a single planetesimal. The finest structures so far detected in the interstellar medium are clumps of interstellar gas with characteristic dimensions $\sim 10 - 20 AU$. Their existence has been inferred from rapid radio variability in extragalactic radio sources, (Quirrenbach et al. 1989), from extreme scattering events occurring in quasars (Fiedler et al. 1987), and from VLBI observations of HI variation in a number of compact extragalactic radio sources (Diamond et al. 1989). Each of these types of phenomenon has been attributed to the passage of small foreground nebulae in the galactic disc. Diamond et al. require $\sim 10^6$ such mini-nebulae per cubic parsec to exist in the disc. If the dust in such a clump were to aggregate into a single solid body, it would have a characteristic mass $\sim 4 \times 10^{22}$ g and diameter 40 km, about that of a medium-sized comet. For the clumps predicted by Gilden (1984) to occur in the interstellar medium, arising out of a thermal instability in cool molecular clouds, objects of diameter $\sim 3000$ km would form, about that of the largest satellites or smallest planets in the Solar System.

The limiting factor in the process may be turbulence. In a star-forming environment, say with U $\sim 10^{-10}$ erg cm$^{-3}$ and $n(H_2) \sim 10^{-3}$ cm$^{-3}$, the critical radius for collapse is $\sim 0.2 AU$ and the infall speed is $\sim 8$ m s$^{-1}$. Any velocity shear due to turbulence must then exceed $\sim 40$m s$^{-1}AU^{-1}$ to destroy the process. The cold clumps contained within the dark clouds of the star-forming regions in Taurus have dimensions 0.1 pc and ammonia profiles which are virtually thermal (Myers 1985), implying a velocity shear two orders of magnitude less than that required to destroy a modest perturbation. Sufficiently quiescent conditions for the growth of such perturbations do seem to occur.

It appears therefore that there may be a tendency for the dust grains in dense molecular cores to aggregate into bodies of $\geq$ km dimensions. There thus seem to be two 'Jeans collapse' regimes in the interstellar medium, one associated with the gas and yielding stellar critical masses, and an analogous one associated with the dust and yielding critical masses which are more asteroidal or cometary, and sometimes even planetary. Planetesimals, if so formed by this process, would be concentrated in dense star forming regions where the UV radiation density is relatively high and collapse is rapid ($10^3$ yr). Of course they would then disperse throughout the galactic disc.

# 2 The origin of comets

The existence of a significant population of icebergs in the interstellar medium is likely to have implications for, inter alia, theories of star and planet formation. The problem, of course, is that of detecting them. One approach would be to trawl a dense nebula with a large net, and observe the catch. Such a net may exist in the form of three–body encounters when the Solar System passes through a comet field: some proportion of the planetesimals would lose sufficient energy to convert their orbits from hyperbolic, in the Sun's frame of reference, to elliptic. The third body might be one of the giant planets, or the molecular cloud itself. Capture might have taken place in the star cluster within which, possibly, the Sun formed, or it might be a continuing process, some proportion of the present comet population being captured from the ambient interstellar environment. The expected mass range of the interstellar planetesimals, at least when derived from a collapsing sheet of dust, is about that of the observed comets of the Solar System, and their probable constitution, that of a cold aggregation of interstellar grains, likewise seems to correspond to that observed (e.g. Greenberg, these proceedings). It is therefore tempting to speculate that the Solar System comets may have originated in a molecular cloud environment rather than, as popularly assumed, a planetary one.

Significant dynamical constraints on theories of cometary origin have emerged in recent years, arising from the observed constancy of the cratering record over the past 3.9 Gyr (Baldwin 1985), the discovery that the classical Oort cloud is dynamically unstable in the galactic environment, a general upward revision of cometary masses, and so on. The upshot of these new constraints is that the question of comet origins has, after a long period of relative stagnation, once again been thrown open (a comprehensive review of theories of comet cosmogony is given by Bailey et al. 1990).

Also, in recent years, evidence has emerged of various imbalances in the Oort cloud, indicative of a recent disturbance or origin. Lindblad's Ring, the material associated with the young blue stars of Gould's Belt and the dust clouds of the Great Rift, appears to be an expanding and rotating ring of material through which the Solar System passed $10 \pm 2$ Myr ago (Olano 1982; see figure 1).

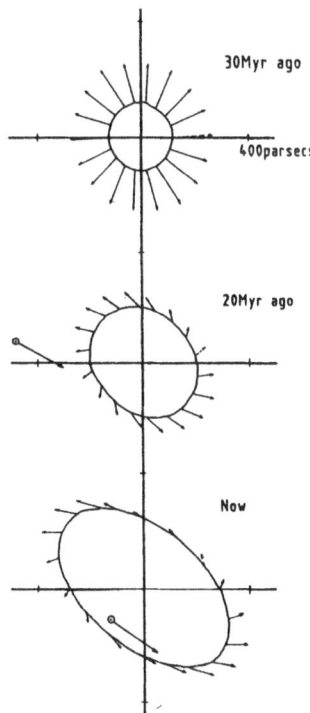

**Figure 1:** Lindblad's Ring and Gould's Belt, after Olano (1982).

In figure 2 is shown the optical equator of Gould's Belt against a recent map of dark nebulae. The solar antapex is seen to coincide with the belt, confirming that the Solar System has passed through it. Specifically, the solar antapex also coincides with the convergent point of the Scorpio-Centaurus association. The isochrones of this star-forming region reveal three main subgroups, with ages 3–6 Myr, 10–13 Myr and 13–16 Myr (de Geus and de Zeeuw 1987). It therefore appears that in the relatively recent past the Solar System passed through a molecular cloud in which star formation was taking place and in which, according to the theoretical arguments above, a dense field of comets might have been located.

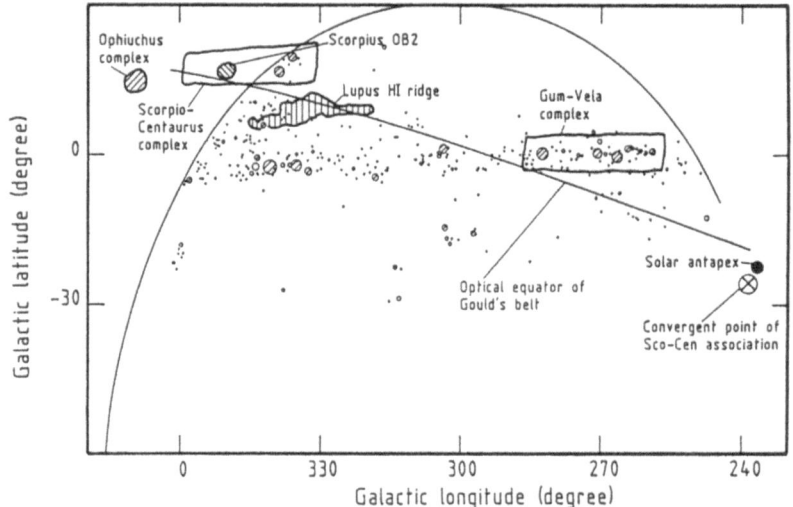

**Figure 2:** The distribution of nebulae catalogued by Hartley et al. (1986). Also shown is Lindblad's Ring, which is seen edge-on as Gould's Belt. Note that the track of the Sun implies a recent passage through the Sco-Cen association.

The question of whether a population of comets adequate to replenish the entire Oort cloud might be captured during passage through a molecular cloud is at present unsolved, although rough analytic arguments suggest that it might be possible (Clube and Napier 1984). The capture of interstellar comets using Jupiter as a third body has been investigated by Valtonen and Innanen (1982) and Valtonen (1983), who find the process to be exceedingly inefficient.

However from an analytic discussion by Radzievskii and Tomanov (1977), it appears that a few $10^4$ comets might be captured into the short-period comet system during passage through a comet field of moderate density (Bailey et al. 1990). Given that such a recent encounter has occurred, an interstellar component of the present system of short-period comets ($\sim 120$ known) is a reasonable expectation.

Consistently with this viewpoint, there is evidence that the Oort cloud has been recently disturbed, if not captured. The rate of injection of comets into the planetary system is expected to vary as $sin2bcosb$, under the influence of the galactic tidal field, where $b$ is galactic latitude. This is roughly as observed for the aphelion directions of the long-period comets (there being zones of avoidance around poles and equator), but Yabushita (1989) has pointed out that the detailed orbital configurations of the long-period comets are not consistent with a steady state distribution, there being a significant excess of new comets currently arriving into the region of the inner planets. Since the relaxation time of the Oort cloud is of the same order as the orbital period of a long-period comet, a disturbance within the last few Myr is implied. Further, there appears to be a second plane of symmetry, defined by a discrete group of 38 comets moving in a largely retrograde direction (Delsemme 1989). This second plane includes both the solar apex (or antapex, depending on whether perihelia or aphelia are plotted) and the galactic centre. Now the solar apex moves over the sky at 1.5 degrees per million years, and this coincidence again implies that, at a minimum, strong perturbation of the Oort cloud has taken place within the last few Myr. These and other considerations are discussed more fully by Bailey et al. (1990).

Significant chemical constraints have also emerged in recent years and seem to imply a formation environment for comets closer to that of molecular clouds than, say, planets (references may be found in loc. cit.): (1) the elemental composition of Halley's Comet is known to be closer to that of the Sun than even the most primitive meteoritic material; (2) the extremely low abundances of ammonia and methane in Comet Halley are unlike those of the outer planets, relatively and absolutely; (3) several indicators ($S_2$ in Comet IRAS-Iraki-Alcock, the ortho/para ratio in Halley etc.) seem to require a temperature, at formation, appropriate to a molecular cloud environment in the range 20–50 K; and (4) likewise the deuterium abundance.

These considerations are equally consistent with an origin in a protosolar nebula lying beyond the present positions of the outer planets, provided that sufficient shielding from solar radiation occurred (although the presence of $S_2$ apparently requires there to have been UV irradiation of the grains which accreted into comet Iras-Araki-Alcock, as required also by the mechanism discussed above). Further, for the most part the isotope ratios determined for Comet Halley cannot discriminate between a Solar System or an interstellar cosmogony. The carbon isotope ratio, however, is an exception: it has been determined for Comet Halley by ground-based observations, and its value ($65 \pm 8$) is that of the current galactic environment ($62 \pm 7$, although uncertain), not that of 4.5 Gyr ago ($89 \pm 2$) –see Bailey em et al. for references. To preserve a primordial Solar System origin for Halley's Comet, therefore, one must postulate some mechanism acting on the ratio in the comet, which mechanism, acting 4.5 Gyr ago, coincidentally produced a ratio equal to that of the interstellar medium at the present time.

# 3  Conclusion

The strong depletion of metals in dense nebulae may arise from the progressive locking up of dust into larger bodies, and the slow homologous collapse of a molecular cloud seeded with planetesimals, say of mean mass $10^{18}$ g, would yield a concentration of such bodies in dense nebulae. For example a nebula with $n(H_2) = 10^4$ cm$^{-3}$, collapsing from an original dimension 0.1 pc to 1 AU, would yield a protostar of solar mass surrounded by $\sim 10^{14}$ comets orbiting within 5 AU of the protostar. The potential for constructing a planetary system out of such an assembly is clear. The idea that nebulae may contain many bodies of subplanetary dimensions, collisions between which created stars and star clusters, is originally due to Krat (1952). If such bodies do exist in abundance in nebulae, it seems there may be considerable implications for theories of star and planet formation as well as for comet cosmogony.

# References

Bailey, M.E.B., Clube, S.V.M., Napier, W.M.: 1990, *The Origin of Comets*, Pergamon, Oxford.

Baldwin, R.B.: 1985, Icarus, **61**, 63–91.

Clube, S.V.M., Napier, W.M.: 1984, M.N.R.A.S., **208**, 575–588.

de Geus, E., de Zeeuw, T.: 1987, in *Star Forming Regions* (eds. M. Peimbert and Jugaku), IAU Symp., **115**, 205–206. D. Reidel, Dordrecht, Holland.

Delsemme, A.H.: 1989, Sky Tel., **77**, 26–264.

Diamond, P.J. et al.: 1989, NRAO preprint 89/85.

Fiedler, R.L. et al.: 1987, Nature, **326**, 675–678.

Flannery, B.P., Krook, M.: 1978, Astrophys. J., **223**, 447–457.

Gerola, H., Schwartz, R.A.: 1976, Astrophys. J., **206**, 452–457.

Gilden, D.H.: 1984, Astrophys. J., **283**, 679–686.

Hartley, M. et al.: 1986, Astron. Astrophys. Suppl., **63**, 27–48.

Krat, V.A.: 1952, Voprosy Kosmogonii, **1**, 34–91.

Miyama, S.K., Narita, S., Hayashi, C.: 1987, Prog. Theor. Phys., **78**, 1051–1064.

Myers, P.C.: 1985, in *Protostars and Planets II* (eds. D.C. Black and M.S. Matthews), Univ. of Arizona, Tucson.

Napier, W.M., Humphries, C.M.: 1986. M.N.R.A.S., **221**, 105–117.

Olano, C.A.: 1982, Astron. Astrophys., **112**, 195–208.

Quirrenbach, A. et al.: 1989, Nature, **337**, 442–444.

Radzievskii, V.V., Tomanov, V.P.: 1977, Sov. Astron., **12**, 218–223.

Turner, B.E., Rickard, L.J., Xu, L.: 1989, Astrophys. J., **344**, 292–305.

Valtonen, M.J.: 1983, The Observatory, **103**, 1–4.

Valtonen, M.J., Innanen, J.A.: 1982, Astrophys. J., **225**, 307–315.

Yabushita, S.: 1989, M.N.R.A.S., **240**, 69–72.

# THE ANATOMY OF DENSE MOLECULAR CLOUDS

P.G. Mezger, R. Zylka
Max-Planck-Institut für Radioastronomie,
Auf dem Hügel 69, 5300 Bonn 1, F.R.G.

ABSTRACT. At mm/submm wavelengths, where the Rayleigh-Jeans approximation applies and dust optical depths are small, the observed surface brightness of dust emission is $S_\nu/\Omega_A \propto N_H \sigma_\nu^d T_d$. Here $N_H$ is the hydrogen column density, $\sigma_\nu^d$ is the dust absorption cross section per H-atom at the frequency $\nu$ and $T_d$ is the dust temperature. For volume densities $n_H \lesssim 1E5-1E6 cm^{-3}$ molecular spectroscopy and dust continuum observations in general yield comparable results. But at higher volume densities and low gas temperatures it is found that molecular spectroscopy becomes an unreliable tracer of hydrogen column densities. Continuum observations of dust emission, on the other hand, provide the most reliable method of investigating especially very compact cores of molecular clouds, since $\sigma_\nu^d$ and $T_d$ are usually known within factors of 2 or better. As examples recent observations of star forming cloud cores in the Orion A and B clouds and of two giant molecular clouds in the galactic center region will be summarizeded in this review.

## 1. A (Historical) Introduction

After the discovery of the $\lambda 21 cm$ line of atomic hydrogen in the early 50ies most astronomers expected that the basic problems related to the interstellar matter (ISM), such as star formation and the spiral structure of the Galaxy, would be solved within the next ten years. It took about just that time, however, to realize that at densities $n_H \gtrsim 10^2 cm^{-3}$ atomic hydrogen very abruptly transforms into molecular hydrogen which has no transition in the radio range.

Today we know that about one half of the total mass of the ISM (or quantitatively $\sim 1-2 \cdot 10^9 m_\odot$) exists in the form of molecular clouds. The cores of these dense clouds are the places where stars form, and high-mass star forming Giant Molecular Clouds (GMCs) with typical masses of $\sim 10^5 m_\odot$ outline the spiral structure. Most of this knowledge has been obtained through molecular radiospectroscopy, which started to gain momentum in the late 60ies through the discovery of spectral lines of $NH_3$ and $H_2O$. Since the number of molecular transitions increases with decreasing wavelength, mm astronomy has evolved during the last two

111

*E. Bussoletti and A. A. Vittone (eds.), Dusty Objects in the Universe*, 111–121.
© 1990 *Kluwer Academic Publishers.*

decades as one of the most important branches of the still young science of radioastronomy.

A new road for the investigation of dense molecular clouds was opened, when in 1978 R. Hildebrand and his Chicago bolometer group published their first map of 400$\mu$m dust emission from two GMCs close to the galactic center. The underlying physics is simple and straightforward. Dust grains absorb stellar photons, get heated and reemit the absorbed energy as quasi black-body (bb) radiation in the FIR. The absorption cross section of dust at UV/optical wavelengths is much larger than at FIR wavelengths. Dust grains, therefore, attain considerably higher temperatures than the equivalent bb temperatures which correspond to the mean intensity of the Interstellar Radiation Field (ISRF), and which are ~3K in the solar vicinity and ~8K in the galactic center. Typical temperatures of dust in the intercloud gas are ~20K in the solar vicinity and ~27K in the galactic center region. This fact was first noted by van der Hulst (1949).

Since 1982 the MPIfR bolometer group observes cold dust emission with its $^3$He cooled Ge-bolometer (Kreysa, 1985) using first optical/NIR telescopes and – since 1986 – both the IRAM 30m MRT on Pico Veleta, Spain, and the 15m SEST on La Silla, Chile. Especially these latter high angular resolution observations ($\theta_A$~7"-30") showed the existence of dense ($n_H$≥$10^6$cm$^{-3}$) and cold ($T_d$~10-20K) condensations of gas and dust which are barely detectable through their molecular line emission. Line opacity could be ruled out as an explanation of this discrepancy by using as probes very rare isotopic transitions such as $C^{18}O$. Depletion of molecules, which freeze out and form ice mantles around refractory grain cores, is a plausible explanation of the observed molecular underabundances (Mezger et al. 1987). Our observations suggest that most molecular transitions become unreliable tracers of hydrogen column densities at low gas (and dust) temperatures and at volume densities $n_H$≥$10^6$cm$^{-3}$. Due to this rather fundamental limitation high angular resolution observations of cold dust emission at mm/submm wavelengths point out the most promising – if not the only – road to investigate the early stages of star formation. In sect.4 we report on observations of two cloud cores.

By combining observations at NIR and submm wavelengths one can observe dust both in absorption and emission. This allows to place stars and dust clouds along the line of sight, as will be demonstrated in sect.5 for the case of two Galactic center GMCs.

## 2. Determination of hydrogen column densites

The observable surface brightness of a cloud, which is more extended than the telescope HPBW $\theta_A$, is related to dust temperature $T_d$ and hydrogen column density $N_H$ by

$$S_\nu/\Omega_A = B_\nu(T_d) \ (1-e^{-\tau_\nu})$$

Here $\Omega_A = 1.133\theta_A^2$ is the beam solid angle, $B_\nu$ is the Planck function,

$$\tau_\nu = \sigma_\nu^H \, N_H$$

is the dust optical depth at frequency $\nu$ and $\sigma_\nu^H$ is the dust absorption cross section per H-atom, which can be approximated by

$$\sigma_\nu^H(\lambda) = (Z/Z_\odot)\,b \begin{cases} 7 \cdot 10^{-21}\ \lambda_{\mu m}^{-2} & \lambda_{\mu m} \gtrsim 100 \\ 7 \cdot 10^{-22}\ \lambda_{\mu m}^{-1.5} & 40 \leqslant \lambda_{\mu m} \leqslant 100 \end{cases}$$

$Z/Z_\odot$ is the metallicity relative to that of the solar vicinity and b is an empirical parameter accounting for dust in the diffuse ISM (b~1), in molecular clouds (b~1.9) and in high density cores (b~3.4). For $\tau_\nu \ll 1$ and in the Rayleigh–Jeans approximation ($1.44\ 10^4/\lambda\mu m T_d \ll 1$) is $S_\nu/\Omega_A \propto \sigma_\nu^H$ $N_H T_d$. This relation allows a straightforward estimate of hydrogen column densities as long as gas and dust are well mixed. For further details see the Appendix in Mezger, Wink and Zylka (1989b). From the above relation one sees that $\tau_{1300} < 1$ for $N_H < 10^{26} \text{cm}^{-2}$. The shape of the dust emission spectrum is $\nu^m B_\nu(T_d)$ for optically thin emission, with m=-2 for $\lambda_{\mu m} \gtrsim 100$ and -1.5 for $40 \leqslant \lambda_{\mu m} \leqslant 100$; furthermore m=0 whenever $\tau_\nu > 1$. The fit of one or more of such model spectra to an observed dust emission spectrum usually allows meaningful estimates of average dust temperatures $T_d$. Note that (within the same limits for m) Wien's displacement law can be generalized:

$$\lambda_{max} T_d \sim 5100[3/(3+m)] \qquad \mu m\ K$$

### 3. What is wrong with IRAS observations, where are the protostars?

IRAS provided us with a nearly complete survey of the sky in four IR wavelength bands. In the longest wavelength band (~100$\mu$m) dust opacity still is not yet a too serious limitation ($\tau_{100} < 1$ for $N_H < 10^{24} \text{cm}^{-2}$) but rather is the dust temperature, since $\lambda_{max} < 100$ $\mu$m for $T_d \gtrsim 35-51K$ (for m=2 and 0, respectively). Dust inside dense clouds without internal heating sources never attains such high temperatures and its contribution to the emission in the IRAS bands therefore is usually negligible. In this case dust in GMCs is heated solely by the ISRF, whose optical/UV photons get absorbed in the outer skin of the clouds (i.e. within $A_v \leqslant 3$mag counted from the cloud surface) and thus get transformed into FIR radiation. It is this radiation, together with the strong MIR emission contributed from small and very small grains, which prevent the dust temperature to fall below ~10K even in very dense cloud cores and protostellar condensations (Mathis et al., 1983; Mezger, 1989). IRAS is sensitive only to the emission from warmer dust in the cloud skins and completely ignores the cold dust in the interior of the clouds. This insensitivity to cold dust is also the basic reason why IRAS did not detect a single isothermal protostar.

There are two principal objections to estimate hydrogen column densities from dust surface brightnesses: Contamination of broadband continuum dust emission by a forest of molecular lines; and grain coagulation. In specific cases of high excitation temperatures of the molecular gas, such as in the Orion "hot core", line emission can in fact account for up to ~50% of the total broadband emission. In colder condensations, however, line contamination does not appear to be a serious problem (see Mezger et al., 1989b, and references therein). Grain coagulation, which almost certainly occurs in very dense and cool regions (Mezger et al., 1987) is no problem as long as the grain diameters are $a \ll \lambda$. In this case $\sigma_\lambda^H \propto a^3$, i.e. the observed surface brightness is proportional to the dust volume (and mass), independent on the grain size distribution. This pertains in general at FIR/mm wavelengths for dust particles associated with the ISM. Hence, in the submm/mm range, the quantity $\sigma_\lambda^H = \tau_\lambda / N_H$ is rather insensitive to the particular chemical constitution and size distribution of the dust, while the corresponding quantity at optical wavelengths, $A_V / N_H$, can vary over a large range.

## 4. Anatomy of massive star forming cloud cores

### 4.1 THE CLOUD CORE NGC2024

The Orion B cloud is part of the Orion GMC complex. A map in the CO(J=3-2) transition (Fig.1a; White et al. 1981) shows several warm cores embedded in an extended GMC of $M_H \sim 10^5 m_\odot$. One of these warm cores is associated with the compact HII region NGC2024, a well-known region of high-mass star formation.

We mapped this cloud core with the 3m IRTF on Mauna Kea, Hawaii, and with the 30m MRT on Pico Veleta, southern Spain (Mezger et al., 1988). At $\lambda 1300 \mu m$, with an angular resolution of 90" and after subtraction of the free-free emission from the HII region, we found an elongated compact cloud core of $\sim 800 m_\odot$ (Fig.1b). At $\lambda 350 \mu m$ and with $\theta_A = 30"$ this cloud core was resolved into two emission peaks (Fig.1c) which – observed at $\lambda 1300 \mu m$ with the 30m MRT($\theta_A = 11"$) – were finally resolved into six condensations of $T_d \lesssim 20K$ and masses $M_H \sim 10-60 m_\odot$ (Fig.1d). Comparison with model computations of pre MS evolutionary tracks indicates that most of these condensations are isothermal protostars on the brink of forming stellar cores. These are probably the first observed genuine protostars. There are no conspicuous counterparts of these objects observed in molecular line emission.

In contradiction to our interpretation Moore et al. (1989) derived for the dust associated with NGC2024 a single temperature of 47K by fitting a modified Planck curve of the form $\nu^{1.6} B_\nu (47K)$ to the observed dust emission spectrum. Based on this dust temperature the authors claim ~3 times lower dust (and gas) masses and central stellar cores for the condensations. This is a good example for the pitfalls of an uncritical interpretation of dust emission spectra. Firstly, all relevant observations favor an opacity with $m \sim 2$ rather than 1.6 for $\lambda \gtrsim 100 \mu m$. Secondly, the high resolution observations of NGC2024 clearly show that the dust emission at $60 \mu m$ (warm dust) and $1300 \mu m$ (cold dust) has quite different spatial distributions. This latter fact alone requires a two-temperature fit to the observed spectrum, which for m=2 yields T(wd)~45K and T(cd)~16K (see Mezger et al., 1988, and especially their Fig.7b).

# NGC2024

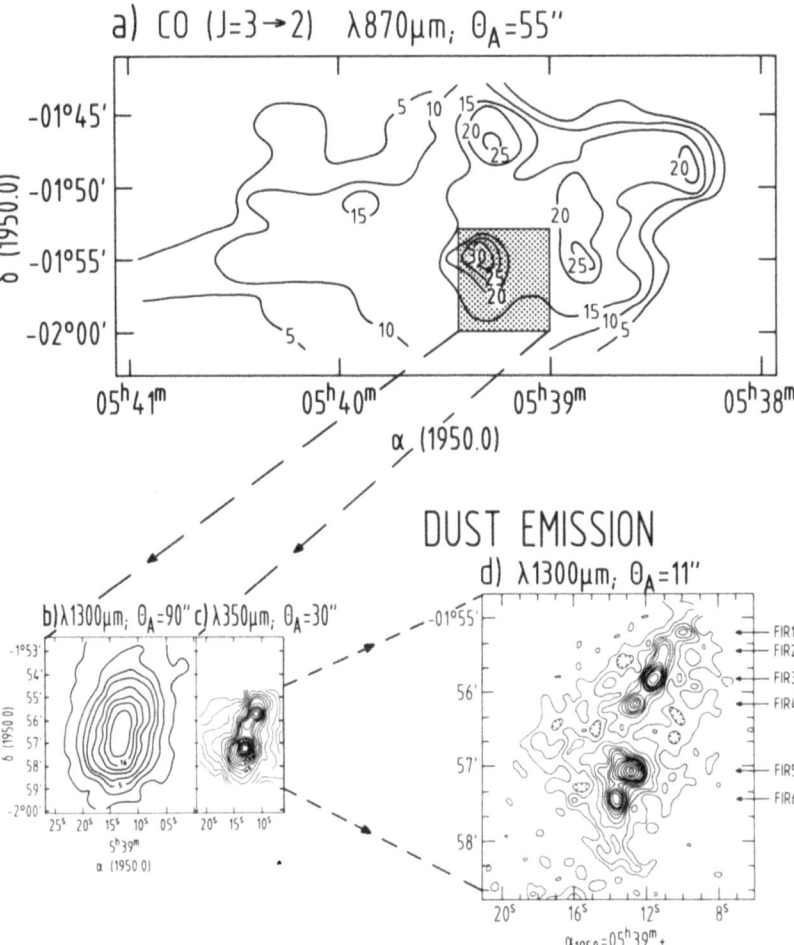

**Figure 1:** *Molecular line and dust emission observed in the direction of NGC2024. a) CO(3→2) map in a single velocity channel of width $\Delta V=1.8 kms^{-1}$ centered at $V_{LSR}=9.0 kms^{-1}$ (White et al., 1981). b) and c) Maps of the dust emission made with the 3m IRTF at $\lambda 1300 \mu m$ and $\lambda 350 \mu m$, respectively. Contours are in units of Jy/beam area. d) Map of the $\lambda 1300 \mu m$ dust emission made with the 30m MRT. Contour units, in Jy/beam area, increase from 0.2-2 in steps of 0.2 and above 2 in steps of 0.5 (Mezger et al., 1988a).*

## 4.2 THE CLOUD CORE OMC1

The Orion molecular cloud (OMC) has been most thoroughly investigated by means of radiospectroscopy and dust continuum emission. Recently Mezger et al. (1989b) mapped its dust emission at $\lambda 1300 \mu m$. With an angular resolution of 90" they observed OMC1 and OMC2 as part of a cigar-shaped cloud core of total mass $M_H \sim 3~10^3 m_\odot$. With an angular resolution of ~11", OMC1 and 2 were resolved into a number of components which are partly self-luminous, dust enshrouded pre MS stars, partly could still be massive protostars without stellar core. As in the case of NGC2024, the pre MS and protostars in OMC1 and 2 are located close to the projected central axis of the elongated cloud core, like beads on a string. This could indicate that magnetic fields play an important (if not dominant) role in the formation of massive stars.

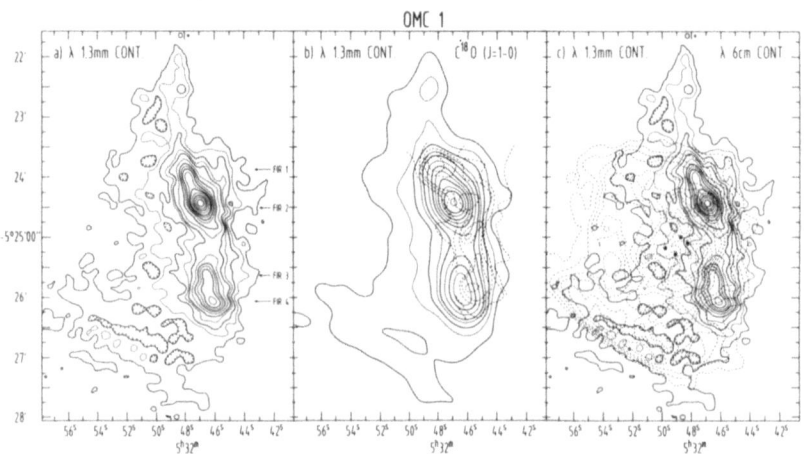

**Figure 2:** $\lambda 1.3mm$ maps of OMC1 a) As observed with the MRT ($\theta_A = 11"$). Solid contour levels are: 0.45, 0.89, 1.33, 1.87, 2.34, 3.33, 4.67. 6.00, 7.67, 10.0, 12.3, 16.7, 21.3, 26.0, 30.7 $Jy(11"-beam)^{-1}$. b) Map a) smoothed to $\theta_A = 22"$ and overlaid with the velocity-integrated $C^{18}O(J=1-0)$ MRT map of Wilson et al.(1986) observed with this angular resolution. Contour levels of dust emission are: 1.79, 3.55, 5.34, 7.47, 9.34, 13.3, 18.7, 24.0, 30.7, 40.0, 49.4, 66.7 $Jy(22"-beam)^{-1}$. c) Map a) overlaid with a $\lambda 6cm$ free-free emission VLA map obtained by Johnston et al.(1983) with a synthesized HPBWs of 10"x13".

Fig.2a shows OMC1 as observed with the MRT. The extended emission is mostly free-free emission from the HII region Orion A, as visible in Fig.2c which is an overlay of the 1300$\mu m$ map on a VLA 6cm map of free-free emission observed by (Johnston et al., 1983). None of the conspicuous compact HII regions in the Orion A complex coincides with the dust emission peaks OMC1, FIR1-4. Fig.2b is an overlay of the MRT 1300$\mu m$ map on the $C^{18}O(J=1-0)$ map of Wilson et al. (1986) smoothed to the same angular resolution of 22". Obviously, the peaks of $C^{18}O$ and dust emission do not coincide, indicating that this molecular line, although optically thin, is no longer a reliable tracer of hydrogen column density. The explanation of this discrepancy by Wilson et al. as a temperature effect assumes that gas and dust are thermally coupled, an assumption which is not borne out by observations. Moreover, conversion of velocity-integrated $C^{18}O$ emission yields considerably lower $H_2$ column densities if "standard" molecular abundances are adopted. Hence, molecular depletion in the densest condensations appears to be the most plausible explanation.

4.3 CONCLUSIONS REGARDING DUST EMISSION.

i) Dust absorption cross sections at mm/submm wavelengths: There are no observational objections against an opacity law $\sigma_\lambda^H \propto \lambda^{-2}$ for $\lambda \gtrsim 100\mu m$. Note that in the Rayleigh-Jeans regime, with this opacity law and for an optically thin, centrally heated dust emission source model computations have shown that – contrary to many statements – $S_\nu$ is not proportional to $\nu^4$ but rather to $\nu^3$, since the source solid angle widens with decreasing $\nu$. Failure to consider this effect has led to considerable confusion regarding the wavelength dependence of the dust opacity. ii) The critical density at which dust and gas get thermally coupled is higher than usually assumed and falls in the range $10^5 < n_H/cm^{-3} \lesssim 10^7$. iii) At low temperatures ($\lesssim 20K$) and high densities ($n_H \gtrsim 10^6 cm^{-3}$) most molecular transitions – even if they are optically thin – cease to be reliable tracers of hydrogen column densities, probably due to molecular depletion (for references to conclusions i) to iii) see Mezger et al., 1989b).
iv) A large fraction of dust in GMCs is cold ($<20K$) and does not contribute much to the emission in the IRAS bands $\lambda \lesssim 100\mu m$. Dust and gas mass determinations based on IRAS observations therefore tend to be severe underestimates (Cox and Mezger, 1989; Snell et al. 1989). v) There are indications of grain coagulation in the massive circumstellar disk of S106, which may decrease the "standard" visual extinction $A_v \sim 5.34 \ 10^{-22} N_H$ by as much as a factor of $\sim 20$ (Mezger et al., 1987).

Molecular depletion and deviations from the "standard" extinction curve due to grain coagulation explain in fact the low disk mass derived by Bieging (1984, who – by adopting the standard value $A_v/N_H$ – underestimated the actual disk mass by a factor of $\sim 20$) and the much more far reaching claim by Barsony et al. (1989) that there is no disk at all surrounding IRS4.

## 5. Anatomy of two Galactic Center clouds

5.1 THE INNER 50PC: AN OVERVIEW

The stellar density of the central cluster – which consists primarily of M and K giants with $T_{eff} \sim 4000K$ – increases rapidly towards the galactic center. A visual extinction of $A_v \sim 30mag$ along the line of sight caused by dust in the galactic disk prohibits its observation at optical wavelengths. At $\lambda 2.2\mu m$, however, extinction has decreased to $\sim 3mag$ and the central star cluster dominates the observed surface brightness. The inner part of a 3°x2°, $\lambda$2.2$\mu$m map obtained by Glass et al. (1987) shows nevertheless dark patches, one of which coincides with the intensity ridge of $\lambda$1.3mm dust emission (Fig.3a). In these regions dust clouds with $A_v \gtrsim 60$mag must be located in front of the bulk of the stars of the central cluster.

The two clouds labelled (according to their positions in galactic coordinates) M-0.02-0.07 and M-0.13-0.08 and shown in Fig.3a form part of a much more extended belt of molecular clouds. The dynamical center of the Galaxy probably coincides with the nonthermal radio point source Sgr A*, which is surrounded by the HII region Sgr A West. The more extended synchrotron source Sgr A East is shifted (with respect to Sgr A* and West) towards negative latitudes. Along the line of sight Sgr A* and West are located in front of Sgr A East.

Recent observations of dust and $^{13}$CO(J=2-1) emission made with an angular resolution of ~11"-12" (Mezger et al., 1989a; Zylka et al., in prep. Fig.3b,c) indicate that the two clouds mentioned above – one surrounding Sgr A East, the other comprising M-0.13-0.08 – form one entity. They are connected by a "streamer" of gas, which causes the strong 2.2$\mu$m absorption east of Sgr A West (see Fig.3a). This geometry is discussed in sect.5.2.

## 5.2 THE MORPHOLOGY OF THE COMPACT GMCS SGR A EAST AND M-0.13-0.08

We investigated the $\lambda$1.3mm dust emission of these two GMCs with the 30m MRT. The two maps obtained are shown in Figs.3b and c.

M-0.13-0.08 (Fig.3b), on the basis of a comparison of the 2.2$\mu$m and 1300$\mu$m maps (Fig.3a), is placed at least 30pc in front of the galactic center. Its morphology is that of a compact high-mass star forming GMC. It consists of an extended component of size ~15x7.5pc$^2$ with a clumpy structure, which accounts for ~90% of the total cloud mass of $M_H \sim 3 \cdot 10^5 m_\odot$. A central ridge containes two subcores of ~2-3 $10^3 m_\odot$, which exhibit all signposts of high-mass star formation.

The morphology of the cloud M-0.02-0.07 (Fig.3c) is more complex. It consists of two components. One is the Sgr A East cloud of size ~12.5x12.5pc$^2$ and total hydrogen mass of ~2 $10^5 m_\odot$, which surrounds the Sgr A complex of radio sources; the other is the streamer. A ring of dust emission surrounds the synchrotron source Sgr A East, which represents a massive (~6 $10^4 m_\odot$) shell of neutral gas and dust, which was probably piled up by the blastwave of the explosion that created Sgr A East (Mezger et al., 1989a, and references therein). The center of this ring and the peak dust emission (as seen e.g. in the $\lambda$1.3mm IRTF map Fig.3a) are separated by ~2'. This dust emission peak is due to the high column densities in directions where the streamer is seen tangentially.

The schematic Figs.4a and b outline this geometry. Fig.4a shows the central region as seen from the sun, Fig.4b shows the geometry as seen from the galactic pole. Radial velocities obtained from the $^{13}$CO(J=2-1) survey are indicated. The width of the features shown in Fig.4b is a measure of their hydrogen column density.

M.0.13-0.08 and part of the streamer are located in front of the galactic center. The streamer passes just ~1' below Sgr A*. This means

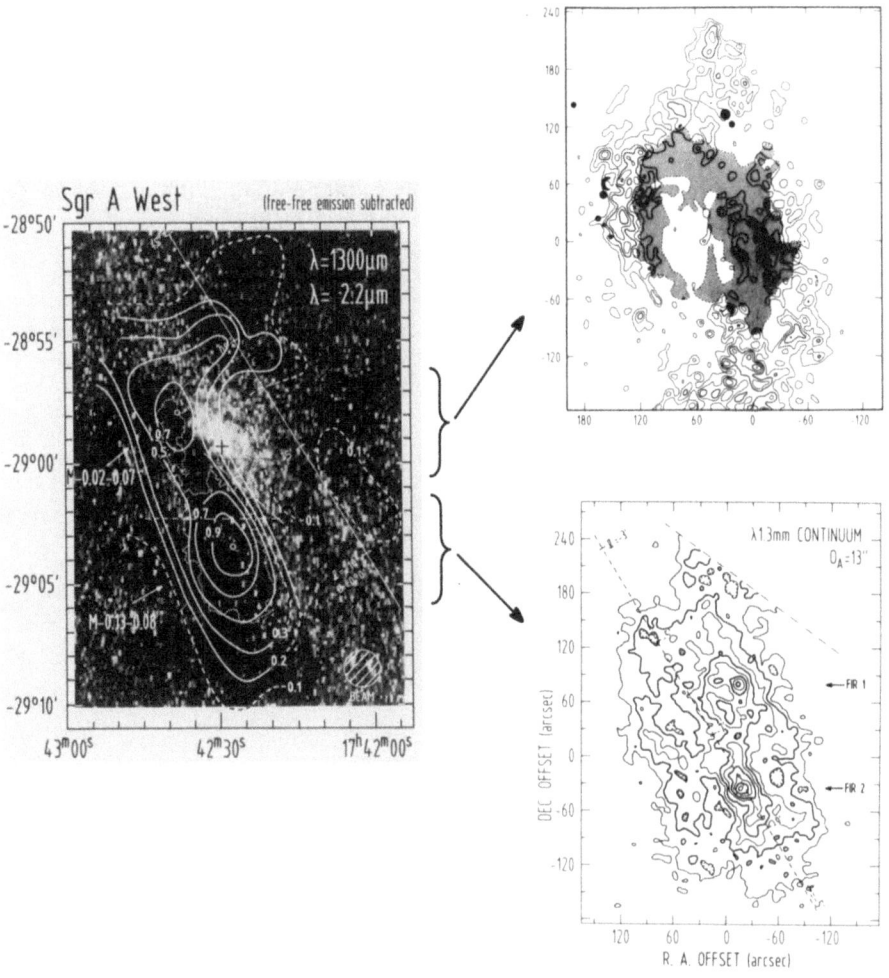

**Fig.3**: a)  λ1300μm,  θ_A=90" dust emission map of two galactic center GMCs (adapted from Mezger et al., 1986, corrected for free-free emission) overlaid on the λ2.2μm, 6"x12" map of Glass et al.(1987). b,c) MRT, λ1300μm  θ_A=11"-13" maps of the central regions of the two clouds (Mezger et al., 1989a; Zylka et al., in prep.). The lowest contours of the two maps joined together are shown in a) as dotted curves. In Fig. b) the contributions from Sgr A* and A West have been subtracted. The stippled area shows the extent of the synchrotron source Sgr A East.

120

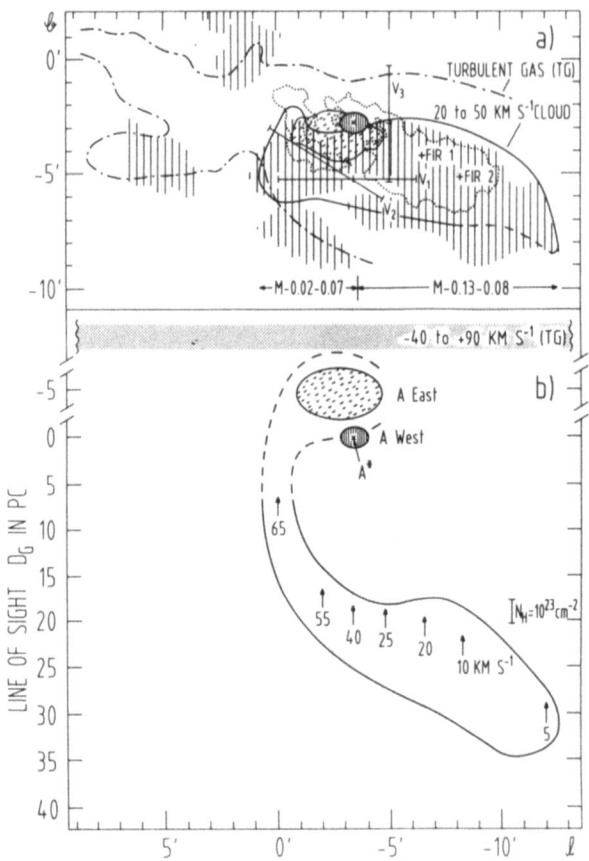

**Fig.4:** Schematic representation of geometry and dynamics of the Sgr A cloud and radio complex. a) Dash-dotted and solid contours outline the boundaries of the Turbulent Gas (TG) cloud and the 20 to 50 kms$^{-1}$ cloud. Dotted contours represent the lowest contours of the MRT dust emission maps of the cloud cores. M–0.02–0.07 and M–0.13–0.08 joined together. Hatch-marked areas outline regions of high 2.2μm absorption. The three components of the Sgr A radio complex are labelled in b. b) Geometry along the line of sight. Radial velocities of M–0.13–0.08 and the "streamer" as observed. The thickness of the clouds measures their hydrogen column densities. The distribution of gas surrounding the Sgr A radio complex is not yet well understood.

good luck for astronomy, since otherwise all NIR/MIR emission from the central region would be absorbed by this cloud. The bulk of the gas in M-0.13-0.08 has radial velocities of ~20kms⁻¹ and in the Sgr A East cloud of ~50-60kms⁻¹, while radial velocities of the streamer increase continuously from ~25 to 65kms⁻¹.

A possible explanation of the different morphology of the two cloud cores is given by Zylka et al. (in prep.). In this scenario the 20 to 50kms⁻¹ cloud originally was an elongated high-mass star forming cloud approaching the central region on an excentric orbit. The approaching end of the cloud was torn apart by tidal forces and the cloud was braked due to interaction with a relatively dense (some 100cm⁻³) intercloud gas which may be represented by the turbulent gas (TG) cloud indicated in Fig.4a and b. Due to this braking part of the cloud could reach – and interact with – the galactic center, resulting in the explosion which created Sgr A East. Most of the gas in the Sgr A East cloud is now moving away from the galactic center. The star forming GMC M-0.13-0.08, on the other hand, is still approaching the galactic center and possibly will undergo a similar phase of interaction within the next few million years.

**References**

Barsony, M., Scoville, N.Z., Bally, J., Claussen, M.J.: 1989,
    Astrophys.J. **343**, 212
Bieging, J.H.: 1984, Astrophys.J. **286**, 591
Cox, P., Mezger, P.G.,: 1989, The Astronomy and Astrophysics Review 1, 49
Glass I.S., Catchpole, R.M., Whitelock, P.A.: 1987, Monthly Notices
    Roy.Astron.Soc. **227**, 373
Hildebrand, R.H., Whitcomb, S.E., Winston, R., Stiening, R.F., Harper, D.A.
    Moseley, S.H.: 1978, Astrophys.J. **219**, L101
Johnston, K., Palmer, P., Wilson, T., Bieging, J.: 1983, Ap.J. **271**, L89-L93
Kreysa, E.: 1985, in : Int.Symp. on MM- and Submm Wave Radio Astronomy,
    URSI, Granada, p.153
Mathis, J.S., Mezger, P.G., Panagia, N.: 1983, Astron.Astrophys. **128**, 212
Mezger, P.G., Chini, R., Kreysa, E., Gemünd, H.P.: 1986,
    Astron-Astrophys. **160**, 324 (PaperI)
Mezger, P.G., Chini, R., Kreysa, E., Wink, J.E.: 1987,
    Astron.Astrophys. **182**,127
Mezger, P.G., Chini, R., Kreysa, E., Wink, J.E., Salter, C.J.: 1988,
    Astron.Astrophys. **191**, 44
Mezger, P.G., Zylka, R., Salter, C.J., Wink, J.E., Chini, R., Kreysa, E.,
    Tuffs, R.: 1989a, Astron.Astrophys. **209**, 337
Mezger, P.G.: 1989, Proc. IAU Symp. 139 (in press)
Mezger, P.G., Wink, J.E., Zylka, R.: 1989b, Astron.Astrophys. (in press)
Moore, T.J.T., Chandler, C.J., Gear, W.K., Monntain, C.M.: 1989,
    M.N.R.A.S. **237**, 1P
Snell, R., Heyer, M.R., Schloerb, F.P.: 1989, Astrophys.J. **337**, 739
van de Hulst, H.C.: 1949, "The solid particles in interstellar space"
    Recherches Astronomiques de l'Observatoire d'Utrecht II, No.2
White, J.G., Phillips, J.P., Watt, D.G.: 1981, M.N.R.A.S., **197**, 745
Wilson, T.L., Serabyn, E., Henkel, C., Walmsley, C.M.: 1986,
    Astron.Astrophys. **158**, L1

# S O F I A
## A Stratospheric Observatory For Infrared Astronomy

P.G. Mezger, H.P. Röser
Max-Planck-Institut für Radioastronomie
Auf dem Hügel 69, 5300 Bonn 1, FRG

## 1. Airborne facility

SOFIA will be an airborne observatory designed to address fundamental questions in galactic and extragalactic astronomy and in the origin and evolution of the Solar System. Much of the radiant energy in the Universe lies at infrared wavelengths which are not observable from ground-based telescopes. SOFIA will provide ready and frequent access to these wavelengths with spatial and spectral resolutions which will be unmatched in this century.

SOFIA will be a telescope with a physical diameter of 2.7 meters and an effective aperture diameter (entrance pupil) of 2.5 meters operating in a Boeing 747 aircraft at altitudes from 41,000 to 45,000 feet. It will provide astronomers routine access to infrared, far-infrared and submillimeter wavelengths unavailable from the ground, and with the means to observe transient astronomical events from anywhere in the world. SOFIA is currently being studied jointly by NASA and the German Science Ministry (BMFT). Phase B studies on the aircraft modification and the telescope system have been completed in 1989. The concept is based on 15 years of experience with NASA's Kuiper Airborne Observatory (KAO), which SOFIA will replace in the mid 1990's. SOFIA's wavelength range covers nearly four decades of the electromagnetic spectrum, from the visible, throughout the infrared and submillimeter, to the microwave region. Relative to the KAO, SOFIA will be roughly ten times more sensitive for compact sources, enabling observations of fainter objects and measurements at higher spectral resolution. Also, it will have three times the angular resolving power for wavelengths greater than 30 microns, permitting more detailed imaging at far infrared wavelengths.

## 2. Scientific objectives of SOFIA

The spectral regime from the infrared to submillimeter encompasses a multitude of rich and varied physical processes and is uniquely suited for study of the cosmic birth on all scales. SOFIA's high spectral and spatial resolution will exploit and extend the scientific legacy left by IRAS

123

E. Bussoletti and A. A. Vittone (eds.), Dusty Objects in the Universe, 123–130.
© 1990 Kluwer Academic Publishers.

(Infrared Astronomical Satellite) and will complement the capability of the coming space observatories like ISO (Infrared Space Observatory), SIRTF (Space Infrared Telescope Facility) and FIRST (Far Infrared Space Telescope). Questions that would be addressed by SOFIA include:

(1) Interstellar cloud dynamics and star formation in our galaxy:
  - Why and how do galactic clouds form stars?
  - How important are magnetic fields and rotation in this process?

(2) Proto-planetary disks and planet formation in nearby star systems:
  - How common are solar-systems? Under what conditions are they created?

(3) Origin and evolution of biogenic materials in the interstellar medium and in proto-planetary disks:
  - What environments are hospitable to pre-biotic molecules and compounds?
  - What is the chemical composition of the interstellar medium?

(4) Comets, planet atmospheres and rings in our solar system:
  - How did our solar-system evolve?
  - What was the composition of the solar nebula?

(5) Star formation, dynamics and chemical content of other galaxies:
  - How different are other galaxies?
  - Why do some (e.g. colliding galaxies) exhibit extraordinarily large infrared luminosities?
  - What is the origin of this luminosity?
  - What is the structure and composition of the interstellar medium of other galaxies and the relation to star formation?

(6) The dynamic activity in the center of our own galaxy:
  - What powers the highly luminous phenomena hidden at the center of the Milky Way — a compact star cluster or a black hole?
  - Is this region similar to the "Active Galactic Nuclei" seen in some other galaxies?

(7) The Evolution of the Universe:
  - What do very distant and very young galaxies look like?
  - How do galaxies evolve?
  - What is the form of the missing mass in the Universe?
  - What lies beyond the galaxies?

## 3. Observing Program

SOFIA will be operated by NASA with support from the BMFT as an international facility for astronomy throughout a 20 year lifetime. It will provide 120, 8 hour flight opportunities per year for approximately 40 principal investigator teams, selected by annual peer review. Evolution of state-of- the-art focal plane instrumentation is a key aspect of the SOFIA concept. Roughly half of the investigator teams will provide the observatory with a wide variety of specialized instruments, including array cameras, polarimeters, and several types of spectrometers. Thus, in addition to its wide-ranging scientific contributions, SOFIA will stimulate development of focal plane instrument technologies, and provide extensive science community involvement in support of future space astronomy missions.

## 4. International Observatory

SOFIA will be a joint project under U.S. leadership (NASA) with the Federal Republic of Germany. West Germany has tentatively agreed to provide a significant contribution to the development by supplying a major part of the telescope system as well as operating costs commensurate with the fraction of flight time the German science community would receive. It is expected that West Germany will take a share of about 20 % to build the observatory and to operate the airborne facility. Therefore by collaborative agreement, science teams from West Germany would be guaranteed a 20 % fraction of the observing time (~ 25 flights) on SOFIA. Teams from other countries could compete for time with U.S. proposers on the same basis that they have for time on the KAO. U.S. and West Germany also agreed that other European countries could share the German technical and financial contribution to SOFIA to receive an equivalent fraction of flight time.

## References

1) Chairman of the SOFIA Science Working Group
   E.F. Erickson
   NASA Ames Research Center
   Moffett Field, CA 94035, USA

2) SOFIA Technology Workshop
   May 14-15, 1986
   NASA Ames Research Center
   Edited by E.F. Erickson, P.M. Harvey

126

127

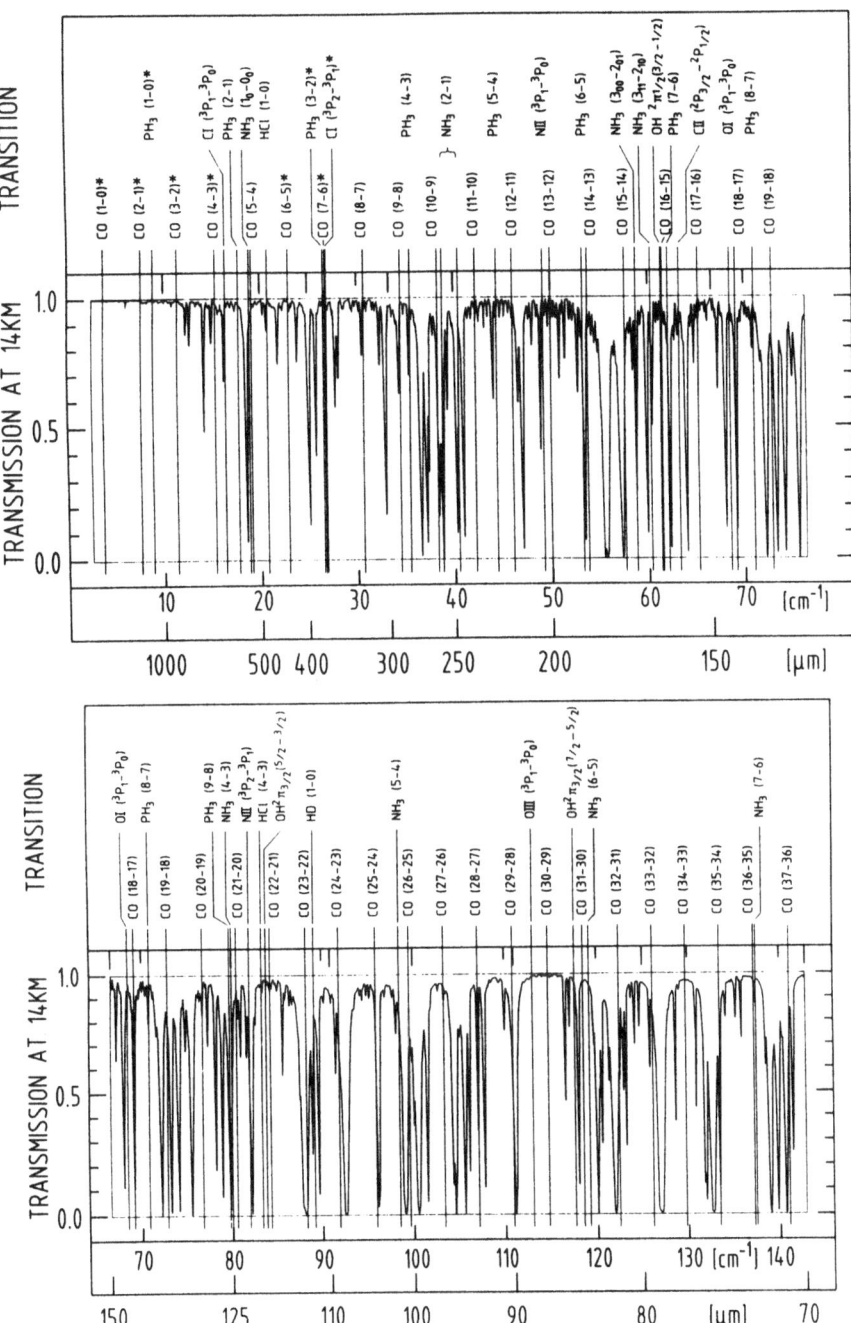

**Figure 2** Atmospheric transmission at 14 km in the wavelength range a) 1000μm to 150μm and b) 150μm to 70μm. Molecular and atomic transitions of astronomical interest are also indicated.

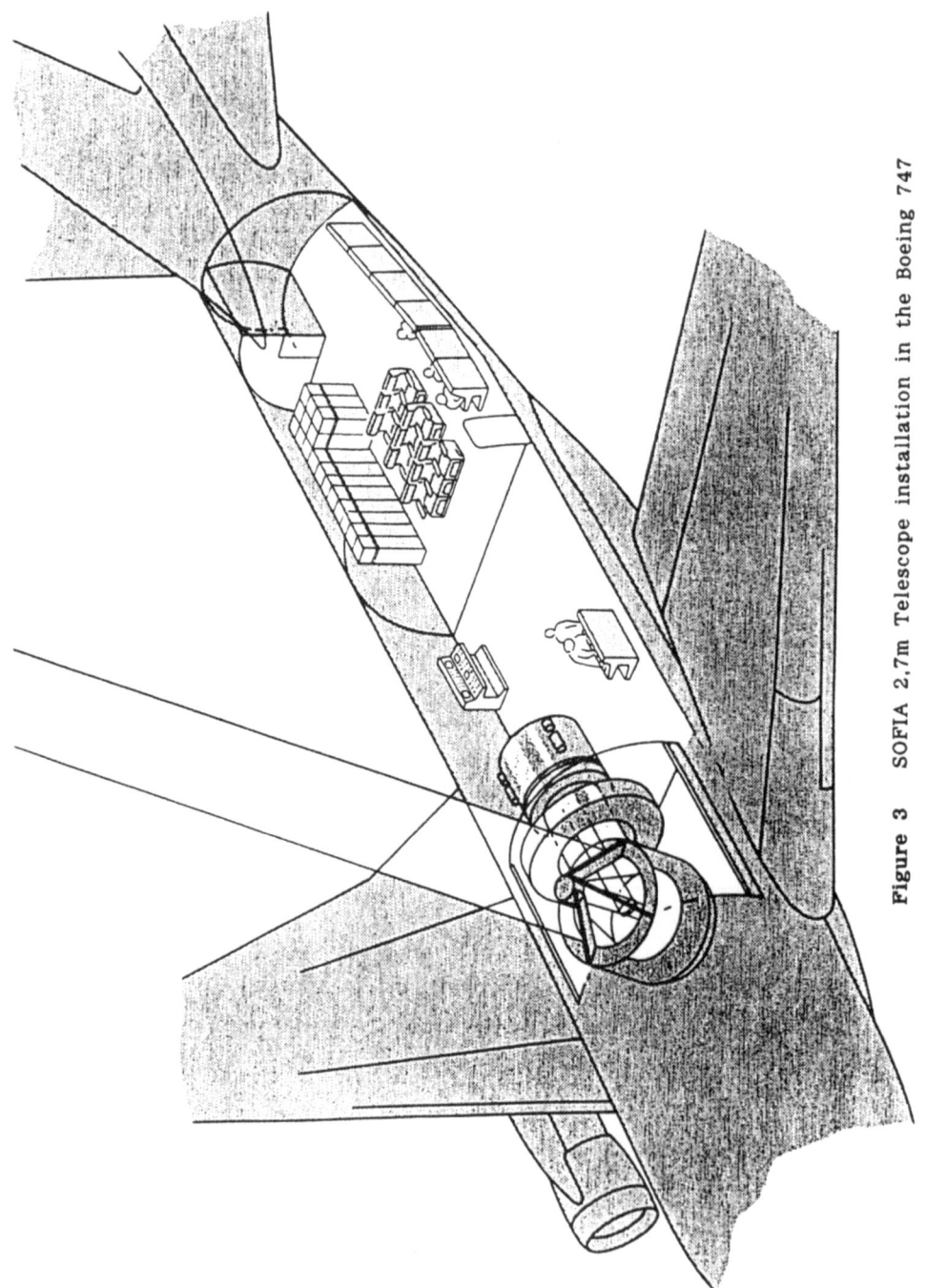

Figure 3    SOFIA 2.7m Telescope installation in the Boeing 747

# MISSION / SYSTEM CHARACTERISTICS

- **TELESCOPE PERFORMANCE**
  - EFFECTIVE APERTURE DIAMETER (ENTRANCE PUPIL) — 2.5 METERS
  - ELEVATION RANGE — 20-60 DEGREES
  - WAVELENGTH RANGE — 0.3-1600 MICRONS
  - CONFIGURATION — INSTRUMENT ACCESS IN CABIN
  - OPTICS IMAGE QUALITY (80% VISIBLE IMAGE DIA.) — 1 ARCSEC
  - POINTING ACCURACY / STABILITY — 0.2 ARCSEC
  - IR PERFORMANCE — CHOPPING, NODDING, SCANNING

- **AIRCRAFT PERFORMANCE**
  - AIRCRAFT TYPE — BOEING 747 SP
  - TIME AT ≥ 41,000 FEET — 5-6 HOURS
  - SHEAR LAYER "SEEING" — 2-3 ARCSEC

- **OBSERVATORY OPERATIONS**
  - OBSERVING FLIGHTS / YEAR — 120
  - PI TEAMS PER YEAR — 40

Figure 4   Mission/System Characteristics

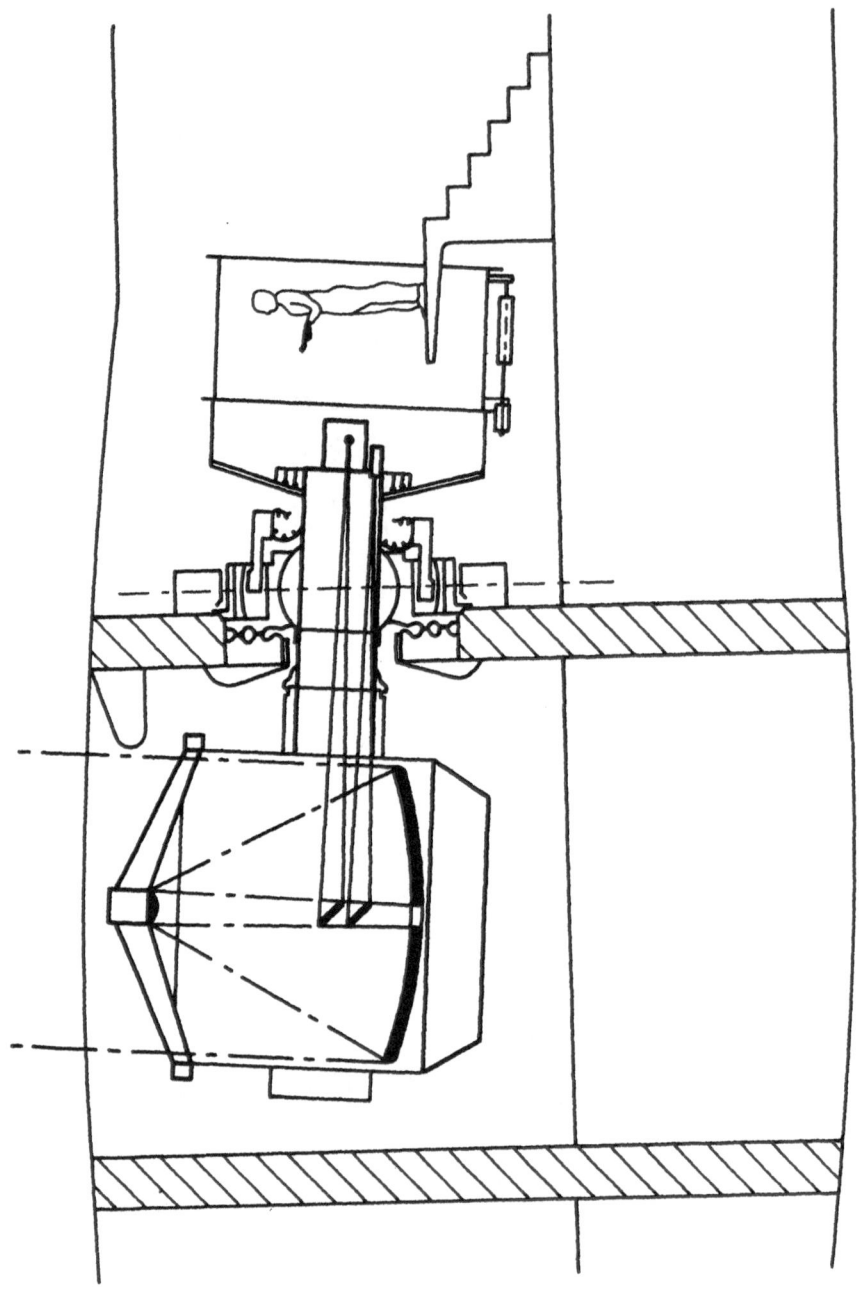

Figure 5   Telescope concept

# OBSERVATIONS OF THE COLD DUST AND GAS IN CENTAURUS A: PROBING AN EXTRAGALACTIC ISM

M. Cameron, A. Eckart, R. Genzel, W. Wild
*Max Planck Institut für Physik und Astrophysik,*
*Institut für Extraterrestrische Physik, F.R.G.*

ABSTRACT. We present SEST molecular line and IRAS CPC far-IR measurements of the cold interstellar gas and dust in NGC 5128 (Centaurus A). These measurements indicate that the bulk of the gas is confined to the optical dust lane but that the far-IR emission has a significantly greater scale height. We conclude from a comparison of these two data sets that there is a substantial amount of cool dust present in Centaurus A which is not intimately associated with star formation.

## 1 Introduction

Detailed comparisons of the distribution of dust and gas making up the cooler phase of the interstellar medium in external galaxies are severely hampered by the limited number of objects with sufficient angular extents that they can be spatially resolved by far-infrared detectors with beam sizes >1'. One of the best candidates for such a study is NGC 5128 (Centaurus A), one of the most frequently observed extragalactic sources which, from its peculiar optical morphology, appears to be an elliptical stellar distribution straddled by a prominent dust lane along its minor axis. Principally as a result of its proximity (3 Mpc, ESO Messenger no. 44, p. 1), Centaurus A exhibits highly extended optical, far-IR continuum, molecular line and radio emission on spatial scales ranging up to the 10° extent of its giant radio lobes. Extensive observations in recent years across most accessible wavebands have uncovered a multi-component system comprising a compact milliarcsecond core and IR hot spot at the nucleus, a dust lane which rotates about the major axis of the elliptical component and which is severely warped at the extremities, and a system of faint narrow shells around the elliptical galaxy. The many features that make up this, our closest radio galaxy, have been summarised by Ebneter and Balick (1983).

This investigation of the cold ISM in Centaurus A was motivated by a need to determine

1. the influence of star formation on the temperature of dust and gas in the dust lane
—from the relative amounts of material in these two phases it should be possible to determine the star formation efficiency (SFE) as a function of position in the galaxy.

131

*E. Bussoletti and A. A. Vittone (eds.), Dusty Objects in the Universe, 131–138.*

2. how various dust components with different temperatures correlate with the differing atomic and molecular gas distributions

3. what clues to the origin of this system can be extracted from a kinematic study of the molecular ISM

4. the similar and contrasting characteristics of the ISM in our own Galaxy and that in the disk of Centaurus A.

Various non-molecular phases of the ISM in Centaurus A have already been studied. These include the 21cm HI emission (van Gorkom 1987; van Gorkom et al. 1990), detection of the 158 $\mu$m [CII] fine structure line at the nucleus (Crawford et al. 1985) and the far-IR continuum emission, observed both with IRAS (DSD-mode data with lower spatial resolution in the far-IR than the CPC data presented here; Marston and Dickens 1988) and from the KAO (Joy et al. 1988). The most extensive far-IR data on Centaurus A, which probes the thermal emission from "standard" (i.e. non-PAH) dust particles, were obtained with the Chopped Photometric Channel (CPC) instrument onboard IRAS. Study of the neutral molecular interstellar medium in Centaurus A has begun only recently with the advent of large millimetre wave telescopes capable of accessing a declination of –43°. Phillips et al. (1987, 1988), using the Caltech Submillimeter Telescope on Mauna Kea, Hawaii, have partially mapped the $^{12}$CO(2–1) emission in the dust lane. Here we present a map of the $^{12}$CO(1–0) line emission together with high spatial resolution IRAS CPC maps of the 50 and 100 $\mu$m continuum. We also briefly describe recent observations of various molecular species observed in both emission ($^{13}$CO, C$^{18}$O) and absorption (HCN, HCO$^+$, CS, CN, $^{12}$CO, $^{13}$CO, C$^{18}$O) against the nuclear continuum source.

## 2 Observations

### 2.1 Millimeter spectroscopy

The millimeter line data were obtained with the Swedish-ESO Submillimeter Telescope (SEST; Booth et al. 1989) in La Silla, Chile during two observing runs in May/June 1988 and July 1989. The beamwidths (FWHM) are 45″ and 22″ at 115 GHz (the frequency of the $J = 1-0$ transition of $^{12}$CO) and 230 GHz (for the $J = 2-1$ transition) respectively. For the $^{12}$CO map and the lines observed in emission the total integration time per spectrum per position was 240 seconds. The absorption measurements required total integration times of between 2 and 10 hours depending on the observing frequency and the molecular species. Further details of the data acquisition and reduction techniques can be found in Eckart et al. (1990a,b). The $^{12}$CO map is presented in figure 1 where it is overlaid on an optical photograph of the galaxy. The spatially integrated CO luminosity is 72 K km s$^{-1}$ kpc$^2$.

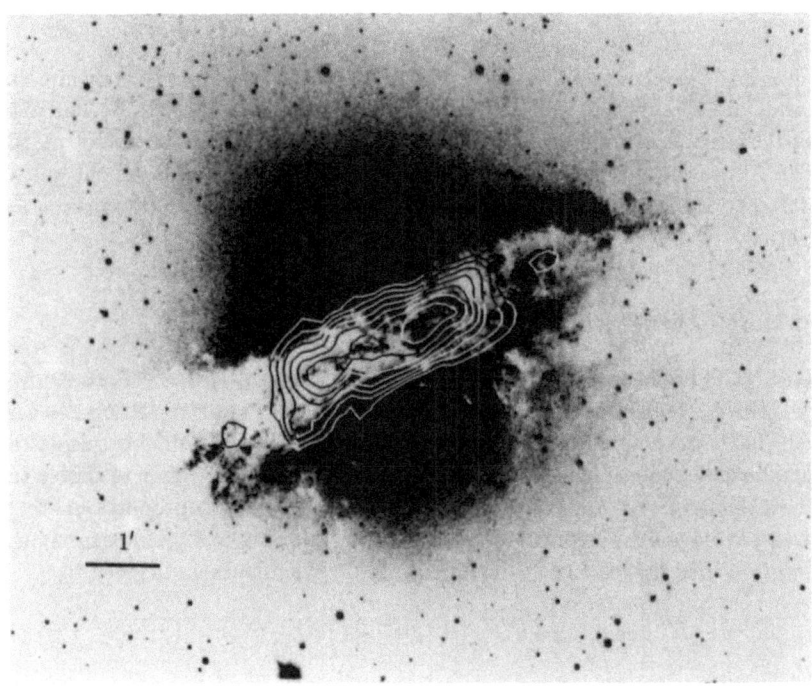

**Figure 1:** Map of the integrated $^{12}CO(1-0)$ line emission of Centaurus A superimposed on an optical image. The emission is well concentrated along the dust lane. The contour levels are 17.5, 22.5, 27.5... K km s$^{-1}$. The peak intensity is 54 K km s$^{-1}$.

## 2.2 IRAS CPC observations

The CPC instrument made up one part of the Dutch Additional Experiment (DAX) onboard IRAS and was designed for high resolution mapping in the far-IR.

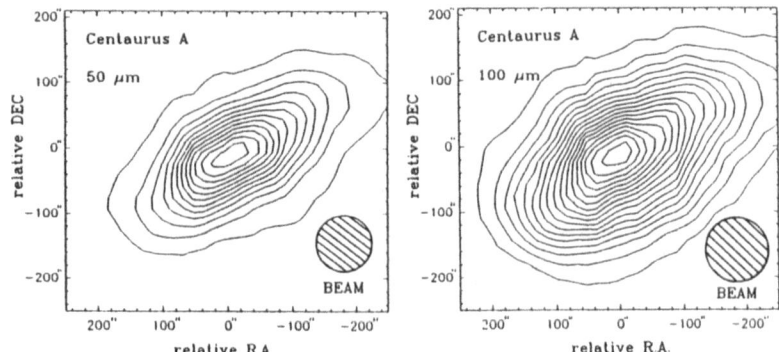

**Figure 2:** IRAS CPC maps of the dust emission in the disk of Centaurus A at (a) 50 $\mu$m and (b) 100 $\mu$m. Comparison with the extent of the gas disk in figure 1 shows that the dust emission is significantly extended. The lowest contour level and the contour interval is 8% and 6% of the peak brightness of 399 and 581 MJy sr$^{-1}$ in the 50 and 100 $\mu$m maps respectively.

The instrument had two channels centred at 50 and 100 $\mu$m with approximately circular Gaussian beams of 88″ and 100″ FWHP respectively (Wesselius et al. 1985). The data were obtained by raster scanning the beam in a 9′ ×9′ box whilst sampling at 30″ intervals. The 10$\sigma$ sensitivity of both channels was 7 Jy. The total fluxes of the galaxy at 50 and 100 $\mu$m are 194 and 435 Jy respectively. CPC maps of the far-IR emission are shown in figure 2.

# 3  Masses and the gas-to-dust ratio

The integrated $^{12}$CO luminosity of 72 K km s$^{-1}$ kpc$^2$ implies an H$_2$ mass of $\approx$3 10$^8$ M$_\odot$ when the galactic N$_{H_2}$/I$_{CO}$ conversion factor of 4 10$^{20}$ cm$^{-2}$/K km s$^{-1}$ (Solomon et al. 1987) is used. The total cool neutral gas mass is then M$_{H_2}$+M$_{HI}$=6 10$^8$ M$_\odot$, using the mass of neutral atomic determined by van Gorkom et al. (1990). In order to derive the dust mass from the far-IR data we assume that the dust emissivity is proportional to $\frac{1}{\lambda}$. This leads to an average dust colour temperature across the disk of 42 K. Assuming the typical grain properties suggested by Hildebrand (1983) the dust mass is given by

$$M_{dust} = 4.5 D^2 F_{100} \left( e^{\frac{144}{T}} - 1 \right) M_\odot \tag{1}$$

where D is the distance to the galaxy in Mpc, F$_{100}$ is the 100 $\mu$m IRAS flux in Jy, and T is the dust colour temperature. This results in a dust mass of M$_{dust}$=5.3 10$^8$ M$_\odot$, and when combined with the gas mass leads to a gas-to-dust mass ratio of $\approx$1100 for Centaurus A. This is far in excess of the canonical value ($\sim$100–200) usually quoted for the Milky Way and would appear to indicate that even with substantial star formation activity the galaxy is highly gas rich. However, it will be shown later that ratio is probably unrealistically high.

# 4  Comparison of the dust and gas distributions

The molecular gas appears to be well concentrated along the dust lane (figure 1) and the emission shows a sharp drop in intensity at a distance of about 90″ to the south-east and north-west of the centre, almost coinciding with the edges of the nuclear bulge. When deconvolved with the SEST 45″ beam, the molecular disk has an extent of 180″±10″ along and 35″±8″ perpendicular to the dust lane, which correspond to linear extents of 2.6×0.5 kpc at our assumed distance to Centaurus A of 3 Mpc. The gas disk is orientated at a position angle of 117°±5° . If we make the assumption that the disk is thin and flat, then the above dimensions are consistent with a disk observed at an inclination of 78°±3°. This estimate should be treated with some caution since it is well known from optical observations that the disk is a highly warped structure, however the derived inclination is consistent with those determined for the blue stellar component (Dufour et al. 1979) and the ring of H II regions (Graham et al. 1979).

In contrast to the relatively tight confinement of the cold gas to the disk, the CPC maps indicate that cool dust (T$_{dust}$ <50 K) is present well away from the bulk of the optical obscuration. This is most apparent in the scale heights of the two phases —the FWHM of the CO emission *along the dust lane* is 180″, whereas that of the dust emission is 220″, when account is taken of the differing beam sizes. The far-IR emission therefore arises in a more extended region. Hence it appears that the bulk of the molecular gas is confined to the disk where star formation is active whilst substantial far-IR emission is observed well away from young stars.

With these two data sets we can derive the star formation efficiency (SFE) —the ratio of the far-IR luminosity to observed molecular mass— for Centaurus A. Since the far-IR emission traces the extent of star formation (via absorption of optical and UV photons from young massive stars which are then re-radiated at longer wavelengths) and the CO luminosity is a measure of the total molecular mass, the ratio of these two quantities should be an indication of the fraction of the molecular gas in any particular region that has been destroyed in the production of young stars. We obtain $SFE_{CenA} \approx 18 \ L_\odot/M_\odot$ which compares with $\sim 1$–10 for giant molecular cloud complexes in our own galaxy and contrasts with an average value of $\sim 78$ for merging and interacting galaxies.

However, in the case of Centaurus A we can, for the first time, determine the SFE as a function of position in the galaxy. This leads to some apparently non-physical results since the differing scale heights of the far-IR and CO emission give rise to a SFE which increases steeply beyond the optical boundary of the disk. Since this can hardly indicate that star formation activity is enhanced in regions in which there is little molecular material, it must imply the presence of a large amount of cool dust which is not intimately linked with star formation —probably similar to the far-IR cirrus observed in the Galaxy.

## 5 The spatial distribution of the cirrus component

We have attempted to decompose our infrared data to distinguish between the cool dust or "cirrus" component and that more intimately associated with the bulk of the molecular gas and hence star formation (the "warm" component). This was achieved as follows:

1. we assume that the distribution of the "true" SFE, derived from far-IR emission associated only with star forming sites, does not rise towards the edge of the disk but is, to a first approximation, flat across across the dust lane.

2. we determine the minimum offsets, $O_{min}$, for the 50 and 100 $\mu$m data needed to achieve a flat distribution of the SFE.

3. In order to determine the spatial distribution of dust not intimately associated with the bulk of the molecular gas we derive:

$$F_{cirrus}(r) = F(r) - \left\langle \frac{F(r) - O_{min}}{I(r)} \right\rangle_{disk} I(r) \qquad (2)$$

where

| | |
|---|---|
| $F_{cirrus}(r)$ | Spatial distribution of the cirrus emission |
| $F(r)$ | Far-IR emission as a function of position |
| $I(r)$ | CO emission as a function of position |
| $O_{min}$ | Minimum offset to the far-IR data needed for a flat SFE |

The output at the various stages of this procedure and the derived spatial distribution of the cirrus dust component are shown in figure 3.

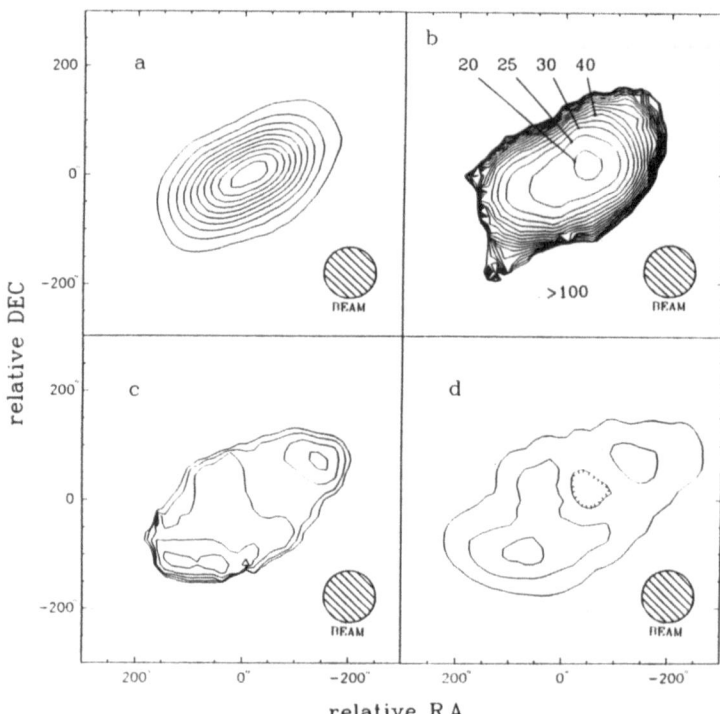

**Figure 3:** a) $^{12}$CO map convolved to a resolution of 100″. Contour levels are 2.5, 7.5,... 25 K km s$^{-1}$.

b) Straight quotient between the total far-infrared flux density map and the $^{12}$CO(1-0) line flux density map —both at a spatial resolution of 100″. The ratio rises sharply towards the edges. Contour levels are 20, 25, 30,... MJy sr$^{-1}$/km s$^{-1}$.

c) The same ratio after subtracting a constant offset, $O_{min}$, of 100 MJy sr$^{-1}$ from the far-infrared map. This offset is consistent with the contribution to the 100 $\mu$m flux density expected from the HI column density (van Gorkom et al. 1990) based on the properties of Galactic "cirrus" emission (de Vries et al. 1987). Contour levels are 5, 10, 15, ... MJy sr$^{-1}$/K km s$^{-1}$.

d) The total far-infrared map after subtracting the $^{12}$CO(1-0) map times $O_{min}$. This residual map is the best estimate of the far-infrared emission that is not intimately associated with massive star formation under the assumption that, at a spatial resolution of 100″, this component is evenly distributed and that star formation is approximately constant over the molecular disk. Contour levels are 100, 150, 200 MJy sr$^{-1}$.

Comparison of $F_{cirrus}(r)$ with the HI gas seen in emission (van Gorkom et al. 1990) shows that the distribution of these two components are well correlated, just as in the case of our own galaxy where the bulk of the cirrus emission arises from diffuse interstellar clouds. With information on these two dust components we should now revise our estimate of the total galactic dust mass:

| Dust Phase | $F_{50}/Jy$ | $F_{100}/Jy$ | T /K | Mass /$M_\odot$ |
|---|---|---|---|---|
| Warm Component | 145 | 217 | 47 | 2 $10^5$ |
| Diffuse Cirrus | 49 | 217 | 34 | 6 $10^5$ |

We must also now re-determine the total molecular mass since the $N_{H_2}/I_{CO}$ conversion factor appears to be lower, by a factor of ~8, for galactic cirrus clouds (de Vries et al. 1987). These revised masses lead to a gas-to-dust mass ratio of ~625, considerably lower than that derived earlier for a single dust temperature/distribution. This is still undoubtedly a substantial over-estimate since IRAS was insensitive to very cold dust ($T_{dust}$ <20 K) which is located within the interiors of molecular clouds away from sites of ongoing star formation and largely shielded from the galaxy's interstellar radiation field. The fraction of cold dust is probably sufficiently large within Centaurus A to bring the gas-to-dust mass ratio down to near that found for the Galaxy. A more quantitative estimate must await sensitive submillimetre continuum measurements.

# 6  Conclusions

We have obtained a fully sampled map of the $^{12}CO(J = 1 - 0)$ emission of Centaurus A which indicates that the cold molecular gas is confined to the dust lane. This contrasts with the far-IR emission which reveals that the cool dust has a significantly greater scale height than the cold gas. Using a relatively simple model, based on assumptions concerning the global character of the star formation efficiency, we have shown that the use of a single dust component for Centaurus A severely over-estimates the galaxy's gas-to-dust ratio. It is also apparent that SFEs derived for spatially unresolved galaxies, based on IRAS PSC data, may be misleading because they do not account for dust components which are not intimately associated with star formation.

Other important results of our investigation are summarised in the following points:

1. The velocity dispersion of the molecular material is of the order of 60 km s$^{-1}$ and approximately constant along the disk. At our resolution we find no evidence for strong velocity disturbances.

2. The absorption lines indicate that the properties of the giant molecular clouds in the dust lane of Centaurus A are comparable to those in the galaxy.

3. A simple one component model of the molecular gas at the centre indicates that molecular emission arises from dense, optically thick gas. The beam averaged column density is of the order of $N_{CO}$=7 $10^{17}$ cm$^{-2}$, the number density is of the order of the critical density required to populate the $J = 2$ level, and kinetic temperatures are around 10–15 K. If the number density in the disk is comparable to that at the centre the molecular disk material is cold with temperatures <10 K.

4. There are differences between the south-eastern and north-western part of the dust lane: in the SE the dust lane exhibits stronger 21cm HI emission (van Gorkom 1987), weaker $^{12}CO(1-0)$, and stronger $H_\alpha$ line (Bland et al. 1987) emission than in the NW. We conclude that, compared to the NW, star formation in the south-east of the disk is higher and that, due to the high radiation field of young massive stars, a significant fraction of the molecular material has been dissociated into atomic gas.

5. For the central absorption feature, the J=2–1 and J=1–0 line ratios for $^{12}$CO and $^{13}$CO as well as a limit on this ratio for CN, indicate that this absorption (which is at the systemic velocity of Centaurus A) occurs in very cold ($\sim$6–10 K) or subthermally excited gas.

Two more detailed reports on these findings have been submitted to the *Astrophysical Journal* (Eckart et al. 1990a,b).

# Acknowledgements

We would like to thank J. Jackson, H. Rothermel, G. Rydbeck and H. Zinnecker for helpful discussions and assistance with the observations. M.C. thanks the Royal Society and the Max Planck Gesellschaft for funding his attendance at this workshop.

# References

Bland, J., Taylor, K., Atherton, P. D.: 1987, M.N.R.A.S., **228**, 595..

Booth et al.: 1989, Astron. Astrophys., **216**, 315..

Crawford, M. K., Genzel, R., Townes, C. H., Watson, D. M.: 1985, Astrophys. J., **291**, 755..

Dufour, R. J., van den Bergh, S., Harvel, C. A., Martins, D. H., Schiffer III, F. H., Talbot Jr., R. J., Talent, D. L., Wells, D. C.: 1979, Astron. J., **84**, 284..

Ebneter, K., Balick, B.: 1983, P.A.S.P., **95**, 675.

Eckart, A., Cameron, M., Rothermel, H., Wild, W., Zinnecker, H., Rydbeck, G., Olberg, M., Wiklind, T.: 1990a, submitted to the Astrophysical Journal.

Eckart, A., Cameron, M., Genzel, R., Jackson, J., Rydbeck, G.: 1990b, submitted to the Astrophysical Journal.

van Gorkom, J. H.: 1987, in I.A.U. Symp. on *" Structure & Dynamics of Elliptical Galaxies"*, de Zeeuw, T. (ed.).

van Gorkom, J. H., van der Hulst, J. M., Haschick, A. D., Tubbs, A. D.: 1990, submitted to the Astrophys. J.

Graham, J. A.: 1979, Astrophys. J., **232**, 60..

Hildebrand, R. H.: 1983, Q. J.R.A.S., **24**, 267.

Joy, M., Lester, D. F., Harvey, P. M., Ellis, H. B., Astrophys. J., **326**, 662..

Marston, A. P., Dickens, R. J.: 1988, Astron. Astrophys., **193**, 27..

Phillips, T. G., Ellison, B. N., Keene, J. B., Leighton, R. B., Howard, R. J., Masson, C. R., Sanders, D. B., Veidt, B., Young, K.: 1987, Astrophys. J. Letters, **322**, L73..

Phillips, T. G., Sanders, D. B., Sargent, A. I.: 1988, in Proc. of the Symp. on Submillimeter & Millimeter Astronomy, held in Kona, Hawaii.

Solomon, P. M., Rivolo, A. R., Barrett, J., Yahil, A.: 1987, Astrophys. J., **319**, 730..

de Vries, H. W., Heithausen, A., Thaddeus, P.: 1987, Astrophys. J., **319**, 723..

Wesselius, P. R. et al.: 1985, IRAS-DAX Chopped Photometric Channel Explanatory Supplement, Laboratory for Space research, Groningen, the Netherlands.

# INTERACTION OF DUST PARTICLES WITH ELECTRONS AND IONS

S. Pintér[1], J. Svestka[2], E. Grün[1]
[1]*Max-Planck-Institut für Kernphysik, Heidelberg, F.R.G.*
[2]*Prague Observatory, Czechoslovakia*

## 1 Abstract

Submicron sized dust particles ($SiO_2$ and tungsten) were levitated in an electrodynamic quadrupole field and charged by ions and electrons. The electric particle surface potential is limited, when charged with ions, at field strength of approx. $1 \times 10^9$ V/m by a temperature dependent discharge mechanism. The particle interaction with 2 to 20 keV electrons always leads to positive surface potentials. The particle potential is determined by the absorption i.e. transmission and backscattering of electrons and by the secondary electrons emission. The dependences of the surface potentials and of the absorption on the electron energy were measured.

## 2 Introduction

Electric charging of dust particles by electrons, ions and UV radiation occurs in many cosmic environments (see e.g. review of Whipple, 1981). In the case of interstellar dust particles the value of their electric charge can have important consequences for instance for their destruction rate in supernova remnants shock-waves and can globally influence the overall life cycle of dust particles in galaxies (see e.g. Seab, 1987).

There are many phenomena as regards dust particles within the solar system which can be explained by their electric charging with consequent interactions with electromagnetic fields or electrostatic fragmentation (see e.g. Morfill and Grün, 1979; Fechtig et al., 1979; Grün et al., 1984 and Mendis et al., 1984).

Theoretical calculations of charging are often based on unreliable data extrapolated from the results of measurements with plane surfaces (parameters of secondary electron emission, photoemission, capture probabilities of electrons and ions etc.). The experimental laboratory work on the simulation of charging processes and the study of physical phenomena related to them (e.g. electrostatic fragmentation) promise to improve our knowledge in this field of research.

## 3 Experimental set-up

The investigation of the interaction of dust particles with monoenergetic ions or electrons requires the levitation of single dust particles and the possibility to determine

*E. Bussoletti and A. A. Vittone (eds.), Dusty Objects in the Universe, 139–146.*
© 1990 *Kluwer Academic Publishers.*

can be used to suspend single charged particles and to determine their charge to mass ratio. The suspension system consists of three conducting hyperboloids which for our set-up were machined out of aluminum with the proper contour corresponding to $r_0 = 25$ mm (see Fig. 1).The electrical field needed for the suspension of particles is generated by a voltage U-V$\cos\omega t$ applied to the middle electrode. The upper and lower pole pieces are grounded. For passage of the ion or electron beam and for illumination and observation of the center of the system, where the particle is suspended, six 7 mm - i.d. radial ports were drilled into the ring electrode. The upper and lower electrodes have 12 mm - i.d. axial holes through them for the introduction and extraction of the particles. To avoid field distorsions the 12 mm - i.d. holes are covered with grids. The electrodes are coated by a layer of black nickel in order to reduce the optical reflectivity and the emission of secondary electrons due to imping-ing electrons deflected from the beam by the electrical field inside the quadrupole. The electrode system is placed in a vacuum chamber in which the pressure can be adjusted down to about $5 \times 10^{-7}$ Torr. An expression for the charge to mass ratio Q/M of the charged particle is derived (Eq. (1)) from the equation of motion, which is a special case of Mathieu's Differential equation.

$$\frac{Q}{M} = 2\sqrt{2}\pi^2 r_0^2 \frac{f f_z}{V} \tag{1}$$

$r_0$ is the minimum radius of the ring electrode, f and V respectively are the frequency and the peak amplitude of the ring drive voltage. $f_z$ corresponds to the frequency of motion of the suspended particle in z-direction. For the derivation of Eq. (1) and more details see Wuerker, Shelton, Langmuir, 1958.

Particles are introduced into the Paul-trap from the reservoir placed above the upper electrode. While falling through the ion (electron) beam, some of them can acquire the proper initial Q/M value to be suspended. All but one of the particles are removed by manipulation of the electrical parameters U, V or $\omega$. The particle is illuminated by a HeNe laser and the scattered light can be observed with a telescope. The frequency of motion is measured by means of a photomultiplier or by varying the frequency of a sinoidal voltage applied to the upper electrode until resonance can be observed. The sizes of the suspended particles are determined by the time for the decrease of the amplitude of motion in z-direction in a gas (He) at known pressure. This method for the size determination, however, has to be calibrated. A more precise method to derive the particle size which can also be used for the calibration is to measure the particle charge and using its charge to mass ratio determined by Eq. (1). When forcing the particle out of the Paul-trap through the cylinder placed below the lower electrode the charge can be determined by measuring the influenced charge on the cylinder.

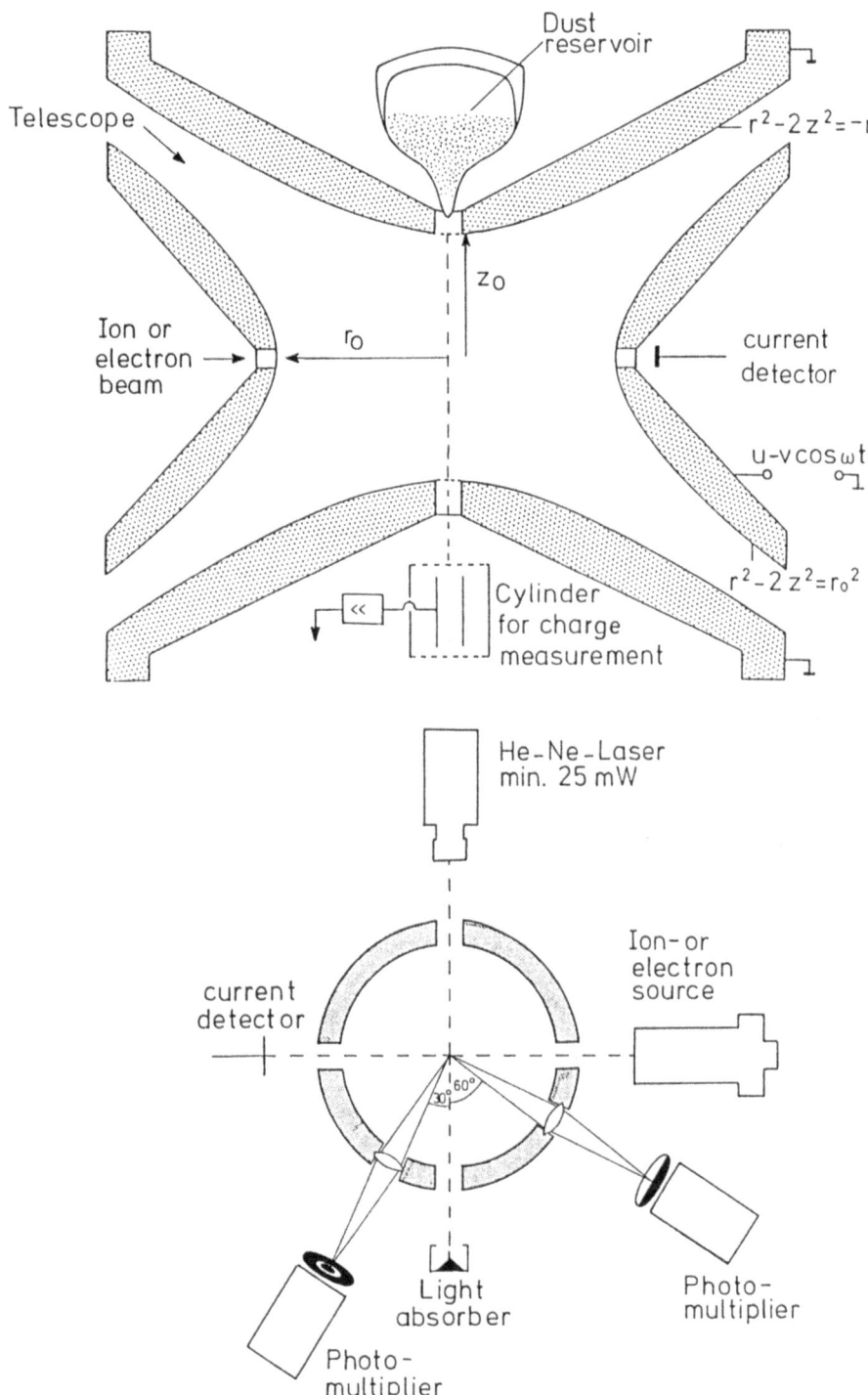

Fig. 1: Vertical (above) and horizontal (below) cross section of the suspension system.

# 4 Results and discussion

## 4.1 Interactions of dust particles with ions

The initial investigations of dust plasma interactions were started with He, Ar and $H_2$ ions of energies up to 5 keV and spherical glass particles with radii ranging from 0.25 to 3 $\mu$m. A comparison of the maximum particle surface potential attainable in the experiment with that expected theoretically from a simple Coulomb Interaction of ions with dust particles can be used to determine the disturbing influences of non ideal experimental conditions. For instance the influence of the electric field inside the quadrupole can result in a strong modification of the original energy distribution of the ions or electrons. Ionization of rest gas and production of electrons and ions on the surfaces of the quadrupole electrodes by the primary beams and the acceleration of these by the electrical field can lead to complex plasma conditions and can affect the equilibrium particle potential significantly.

To characterize all these influences quantitatively, the dependences of the equilibrium particle surface potential on the applied quadrupole voltage, vacuum pressure and the ion energy were measured. It was found for all experimental conditions that the maximum surface potential is much less than predicted by theory Eq. (2).

$$\Phi_{max.} = \text{energy of ions [Volt]} \qquad (2)$$

$\Phi_{max.}$ = maximum particle surface potential

The maximum potential attained in the experiment depends on the ion current and shows no significant dependence on the ion energy and on the different ions. It can be concluded that a negative current to the particle limits the potential. The equilibrium surface depends on the quadrupole voltage. This can be explained in terms of a reduced ion current to the particle due to a stronger deflection of the beam in higher quadrupole fields. At the equilibrium potential the net current is zero, and a decrease of the positive current to the particle thus results in a lower value for the potential.

Since at pressures lower than $1\times10^{-5}$ Torr the potential is nearly independent of the gas pressure, the factor which is limiting the maximum attainable potential cannot be ascribed to the ionization of the gas. During the charging process the dependence of the potential on time showed that the discharge mechanism starts abruptly for all particle sizes at field strength of about $8\times10^8$ V/m. It is known that for tungsten field emitter tips field ionization of water vapor adsorbed on the emitter surface starts at a field strength of approx. $2\times10^9$ V/m (Schmidt, 1964). At the onset of field ionization the ion emission current does not depend on the gas pressure. The assumption that the negative discharge current might be due to the electrons tunneling to the particle surface, when field ionization is starting, was confirmed, when the glass particles were replaced by tungsten particles and the same measurements were performed. The maximum attainable surface field strength for tungsten particles was found to be about $4\times10^9$ V/m. Furthermore, a temperature dependence was found for the value of the maximum attainable potential. The maximum surface potential was decreasing to half of its value at room temperature, while the vacuum chamber was heated up to 100 °C. The temperature dependence of the discharge current is an additional hint at the field ionization of adsorbed water vapor. The onset of field ionization at such low fields is explained by a field enhanced dissociation of neutral water molecules adsorbed in the surface layer (Beckey, 1960). An increase of the discharge current with increasing temperature can therefore be explained by the increase of the dissiciation constant.

## 4.2 Interactions of dust particles with electrons

Since silicates play a significant role in the interstellar and interplanetary space, the investigations were carried out with glass particles. The glass particles were levitated in the Paul-trap as described in the previous sections. The dust particles were charged with different electron energies for different particle sizes and with the same electron current at the entrance to the quadrupole system.

The electric particle surface potential is determined by the negative current of absorbed primary electrons and the 'positive' current due to the released secondary electrons. If the number of secondary electrons produced by an electron is higher than unity, the particle potential becomes positive. In case of positive potentials, only a fraction B of all secondaries has sufficient energy to escape from the attractive potential. The energy distribution of the secondaries can be approximated by a Maxwell-Distribution (Kollath, 1956) Eq. (3).

$$f_{SEE}\left(E_S, E_{Smax.}\right) = \frac{1}{\sqrt{2\pi}\left(2E_{Smax.}\right)^{\frac{3}{2}}}\sqrt{E_S}e^{-\frac{E_S}{2E_{Smax.}}} \qquad (3)$$

$E_S$ = Energy of secondary electrons
$E_{Smax.}$ = mean energy of secondary electrons
$E_{Smax.}$ = 2.5 eV for $SiO_2$
The fraction of the secondaries with sufficient energy to escape from the attractive particle surface potential $\Phi$ is given by Eq. (4).

$$B\left(\Phi, E_{Smax.}\right) = \int_{E_{Smin.}}^{\infty} f_{SEE}\left(E_S, E_{Smax.}\right) dE_S \qquad (4)$$

$E_{Smin.}$ = minimum energy of secondaries to overcome the potential barrier
$E_{Smin.} \equiv$ particle surface potential $\Phi$

$$B\left(\Phi, E_{Smax.}\right) = 1 - erf\left(\sqrt{\frac{E_{Smin.}}{E_{Smax.}}}\right) + \frac{2}{\pi}\sqrt{\frac{E_{Smin.}}{E_{Smax.}}}e^{-\frac{E_{Smin.}}{E_{Smax.}}} \qquad (5)$$

The current of secondary electrons escaping from the particle at potential $\Phi$ is given in Eq. (6).

$$I^+\left(\Phi\right) = I_S B\left(\Phi, E_{Smax.}\right) \qquad (6)$$

144

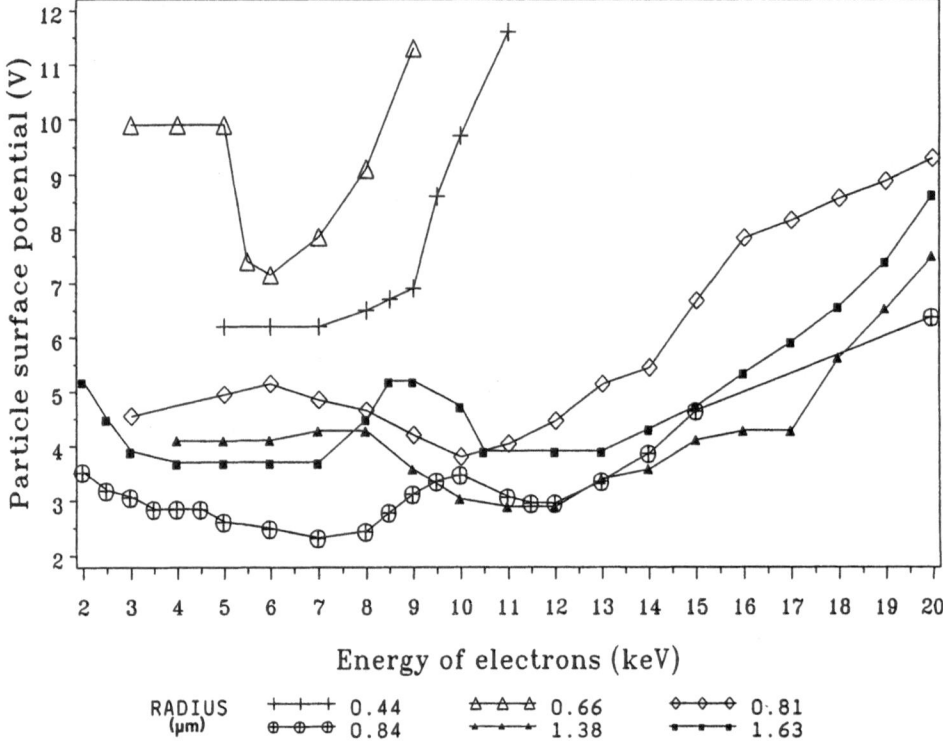

Fig. 2: Particle surface potential as function of the electron energy. For the particle with a radius of 0.66 $\mu$m a potential of 23 V was achieved with 11 keV electrons after 3 hours of charging. The errors of measurements for the potential are about 10%.

$I_S$ = total current of produced secondary electrons

When the equilibrium surface potential is reached, the net current to the particle is zero and therefore

$$I_a = I^+ (\Phi) \tag{7}$$

$$\frac{I_a}{I_S} = B (\Phi, E_{Smax.}) \tag{8}$$

$I_a$ = current of absorbed electrons

The ratio of the absorbed and the total secondary electron current can thus be obtained by measuring the equilibrium particle surface potential.

The potentials of particles with radii of 0.44, 0.66, 0.81, 0.84, 1.38 and 1.63 $\mu$m measured in our experiment are shown in Fig. 2 for 2 to 20 keV electrons. The low values of the potential i.e. the low charge to mass ratio, is restricting the measurements to particles of sizes smaller than 2 $\mu$m. The suspension of bigger ones is not possible, since the field parameters adjustable in the Paul-trap corresponding to the low Q/M values are limited. The increase of the potential with increasing electron energy above approx. 10 keV (6 keV for particles with 0.44 and 0.66 $\mu$m radii) in Fig. 2 can be explained by the decreasing absorption current due to the increasing transmission of electrons through the particle. The effect of surface contamination

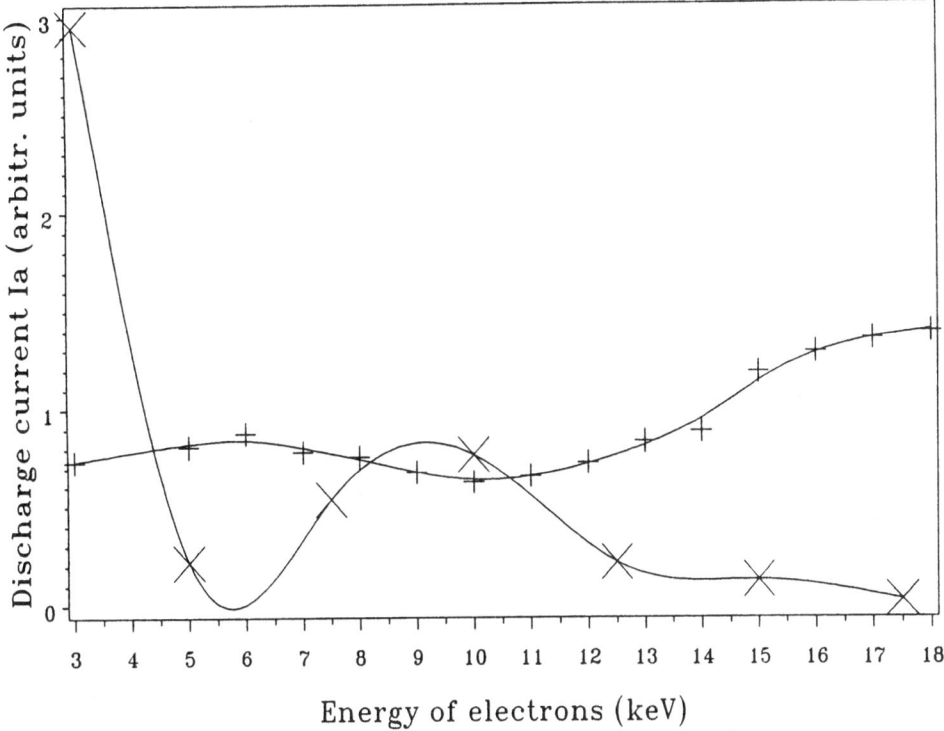

Fig. 3: Discharge current $I_a$ as function of the electron energy for the particle with a radius of 0.81 $\mu$m. The errors of measurements for $I_a$ are smaller than 10%. The dependence of the particle surface potential is shown for comparison (+).

on the particle has to be taken into account, when interpreting the values for the equilibrium potential. This effect of contamination can result in a significant enhancement of the secondary electron emission yield whereby the shape of the energy dependence of the yield remains unchanged (Kollath, 1956). The change of the potential with time for the smaller particle radii and electron energies above 6 keV in Fig. 2 is lower, the higher the potential is. This slow change is due to the small fraction of secondaries which are able to overcome the potential barrier.

The following procedure was performed in order to find out the energy dependence of the absorption: When the equilibrium potential was reached by charging with electrons, the particle was additionally charged by ions to obtain a higher potential. After obtaining a potential of about twice the value reached by charging only with electrons, the ion current was switched off and the decrease of Q/M with time was monitored until the initial equilibrium potential was reached again. This procedure was repeated with different electron energies. The discharge current is given by the slope in this diagram and corresponds to the absorbed electrons. The values for $I_a$ shown in Fig. 3 were deduced at the same particle potential in the region, where the dependence is linear. The deviation from this linearity starts when the current of secondaries escaping from the particle diminishes the negative current to the particle. It has to be mentioned that the measured discharge current $I_a$ in Fig. 3 should be corrected. The correction is due to the different primary electron flux to the particle as a result of the stronger deflection of the electrons with lower energy in the electrical quadrupole field. This, however, would result in an even higher value of $I_a$ at low electron energies. Disregarding this correction it can be concluded from the measurements that the features in the dependence of the particle potential on the electron energy can rather be ascribed to the energy dependence of the absorption than to that of the secondary electron emission.

# 5 References

Beckey, H.D. (1960): Z. Naturforschg. 15a, 822.

Fechtig, H., Grün, E.,Morfill, G. (1979): Planet. Space Sci., 27, 511.

Grün, E., Morfill, G.E. and Mendis, D.A. (1984): Planetary Rings (eds. R. Greenberg and A. Brahic; Univ. of Arizona Press), 275.

Kollath, R. (1947): Zur Energieverteilung der Sekundärelektronen. II. Ann. Physik (6) 1, 357.

Kollath, R. (1956): Sekundärelektronen-Emission fester Körper bei Bestrahlung mit Elektronen in: Handbuch der Physik Bd. 22, (ed. S. Flügge), Springer.

Mendis, D.A., Hill, I.R., Ip, W-H, Goertz, C.K. and Grün, E. (1984): Saturn (eds. T. Gehrels and M.S. Matthews, Univ. of Arizona Press), 545.

Morfill, G.E. and Grün, E. (1979): Planet. Space Sci. 27, 1269.

Schmidt, W.A. (1964): Massenspektrometrische Untersuchung der Feldionisation von Wasserdampf an Spitzen aus Wolfram, Platin und Iridium, Z. Naturforschg. 19a, 318.

Seab, C.G. (1987): Interstellar Processes (eds. D.J. Hollenbach and H.A. Thronson, Reidel, Dordrecht), 491.

Whipple, E.C. (1981): Potentials of surfaces in space, Rep. Prog. Phys., 44, 1197.

Wuerker, R.F., Shelton, H. and Langmuir, R.V. (1959): J. of appl. Phys., 30, 3, 342.

# ARE SPIRAL GALAXIES OPTICALLY THICK?

Rh. Evans, M.J. Disney, J.I. Davies, S. Phillipps
*University of Wales College, Cardiff, U.K.*

ABSTRACT. We have re-examined the classical optical evidence for the traditional view that spiral discs are optically thin and find it to be highly model dependent and unconvincing. We find that the data can be equally well fitted by discs with a physically thin but optically thick dust layer. An optically thick model is consistent with the near infra-red and IRAS observations, with the surface brightness, HI column densities, CO intensities and with Hα measurements. Optically thick galaxies could have up to 2 magnitudes of extinction in the blue, and the stirring upwards of the dust could account for the most luminous IRAS galaxies without recourse to massive bursts of star formation. If disc galaxies are optically thick then this would affect the observed mass-to-light ratio and lead to obscuration of background QSOs.

## 1 Introduction

The conventional view that spiral galaxies are optically thin is essentially based on a study made by Holmberg [11] of 119 galaxies (53 of type Sa-Sb and 66 of type Sc). He observed galaxies at different inclinations and calculated their projected surface brightness. He found a face-on correction to the luminosity, due to internal extinction, of no more than 0.7 mag. in the blue, implying that galaxies do not have large amounts of internal absorption. Similar values were found for spirals in the RC2 catalogue [21] and the RSA catalogue [19]. If galaxies are optically thin the FIR radiation of luminous IRAS galaxies cannot be due to absorption and re-emission of normal starlight, so the current view is that the radiation comes from newly formed stars embedded in giant molecular clouds (e.g. [23]). Such starbursts would not remain hidden, however, without recourse to large amounts of obscuration outside of the clouds [12,22].

In our own work on the surface brightness of galaxies we have found a deficiency of high surface brightness discs [8]. This could be naturally explained if discs are optically thick.

## 2 Optically thin or optically thick?

Figure 1 shows the Holmberg data. He defined a parameter $\Delta \overline{S}$[1] and plotted this as a function of inclination. Although we have shown our models also in terms of this

---

[1]Holmberg defined a projected surface brightness $\overline{S}$ for each galaxy as $\overline{S} = m + 5 log a$, where he assumed that $a$ was independent of inclination, and that $\overline{S}$ varied as $csc(\alpha)$. He then defined $\Delta \overline{S} =$

*E. Bussoletti and A. A. Vittone (eds.), Dusty Objects in the Universe, 147–153.*
© *1990 Kluwer Academic Publishers.*

148

parameter, in the text we have used the more conventional definition of surface brightness $I$. The middle curve is the best fit of his model to the data.

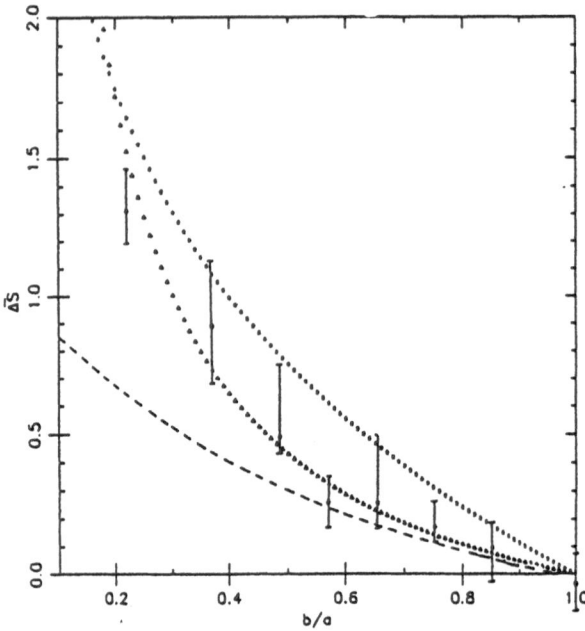

**Figure 1:** This diagram shows the projected surface brightness $\Delta\overline{S}$ plotted against inclination $b/a$. The middle curve is Holmberg's best fit to the data using a screen model (equation 1). The upper curve shows a slab model with infinite optical depth (equation 4). The lower curve shows an optically thick sandwich model with $\varsigma = 0.5$ (equation 5).

This curve actually corresponds to a screen model of a galaxy (see [6], hereafter DDP) where the obscuring dust lies between the stars and the observer (see equation 17 in DDP). Fitting this model to the data Holmberg derived a value for the average face on optical extinction of $A_0B = 0.43$.

However, if we assume a model where the dust and stars are uniformly mixed we find that the variation of surface brightness with inclination is:

$$I(i) = E^*T \quad \text{when optically thin} \quad (\tau \ll 1)$$
$$= E^*\lambda \quad \text{when optically thick} \quad (\tau \gg 1)$$

where $E^*$ is the luminosity of starlight per unit volume, $T$ is the thickness of the stars, $\lambda$ is the mean free path of visual light and $\tau = T/\lambda$ is the optical depth. So, in the optically thin case we see the whole of the galaxy, and in the optically thick case we only see one mean free path into the galaxy. The variation of $\Delta\overline{S}$ with inclination for an infinitely optically thick slab is given by the upper curve in figure 1.

The observed surface brightness of such a slab model will depend upon the surface density of stars and dust. We can obviously derive a relationship between the surface

$\overline{S}(\alpha) - \overline{S}(\pi/2)$ where $\alpha$ is $\pi/2 - i$, $i$ being the inclination to the line of sight. For an optically thin galaxy his $\overline{S}$ will be constant with inclination (whereas the more conventional measure of S.B. we have used will get brighter), and conversley for an optically thick galaxy $\overline{S}$ gets fainter (our S.B. remains constant).

density and the optical depth. If $N^*$ is the number of solar masses of material of all kinds (stars, gas and dust) per $pc^2$ in a column of depth $T$ and area $A$ then

$$\tau = N^* a^* / A \equiv N^* A^*$$

where $a^*$ is the total absorption cross-section of dust associated with $1 M_\odot$ of slab material. Using a value for $A^*$ of 0.01 (see [1,3]) we get, for different mass to light ratios Q, the curves shown in figure 2.

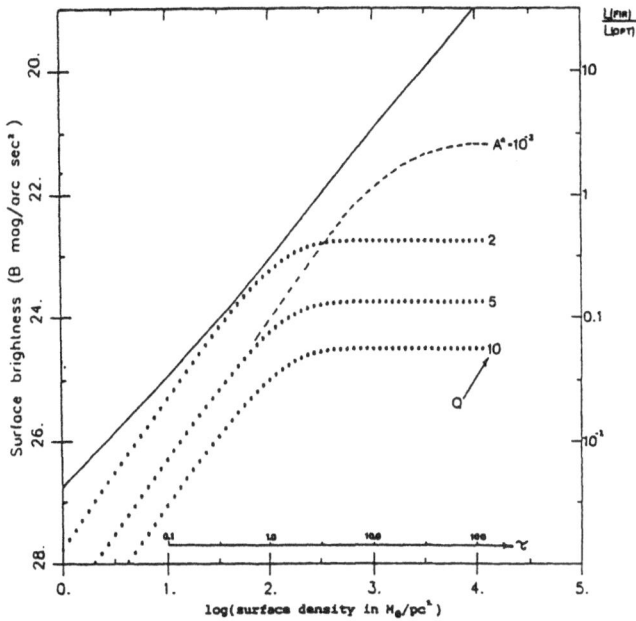

**Figure 2:** This shows the surface brightness for slabs as a function of the surface density for different $M/L$ ratios. The solid line gives the FIR/opt ratio. The curves are for an $A^*$ of 0.01 $pc^2 M_\odot^{-1}$. We also show a curve where $A^*$ has been decreased by a factor of 10.

As we can see from this figure, by the time we reach an optical depth of $\sim 1$ we can only see one $\lambda$ into the galaxy and so increasing the surface density has little effect on the measured surface brightness. Starlight from further into the disc is blocked by foreground dust. If we assume that all the starlight absorbed is eventually re-radiated in the FIR then we can write

$$L_{FIR}/L_{OPT} = I_{FIR}/I_{OPT} = (I_{BOL} - I_{OPT})/I_{OPT}$$

$I_{BOL}$ is obviously $E^* T$ (i.e. the optically thin case where we see the light from all the galaxy) and $I_{OPT}$ is given by equation 17 in DDP (we will only consider the face on case) so

$$I_{FIR}/I_{OPT} = E^* T - E^* \lambda (1 - \exp{(-\tau)})/E^* \lambda (1 - \exp{(-\tau)})$$

or, as $\tau = T/\lambda$

$$\begin{aligned}
L_{FIR}/L_{OPT} &= \tau - (1 - \exp(-\tau))/(1 - \exp(-\tau)) \\
&= \tau/2 &\text{if } \tau \ll 1 \\
&= \tau - 1 &\text{if } \tau \gg 1
\end{aligned}$$

The solid line in figure 2 shows the ratio $L_{FIR}/L_{OPT}$ and we can see that any value can be achieved by varying the optical depth $\tau$.

We can make the following comments about figure 3:

1. If disc galaxies have an exponential light profile and an extrapolated central surface brightness of 21.5 $B\mu$ [9] then some 60% of the light is contained within the 24 $B\mu$ isophote. From the figure we see that this means much of the light originates in an optically thick region.

2. The inner regions of the galaxy will have a higher optical depth than the outer regions. High surface brightness and high optical depth go together.

3. By the same argument we would expect the FIR/optical ratio to be higher towards the centre of disc galaxies (this has recently been observed, see [18]). We would expect discs with a high $L_{FIR}/L_{OPT}$ ratio to have higher than average surface brightness (cf. [17]).

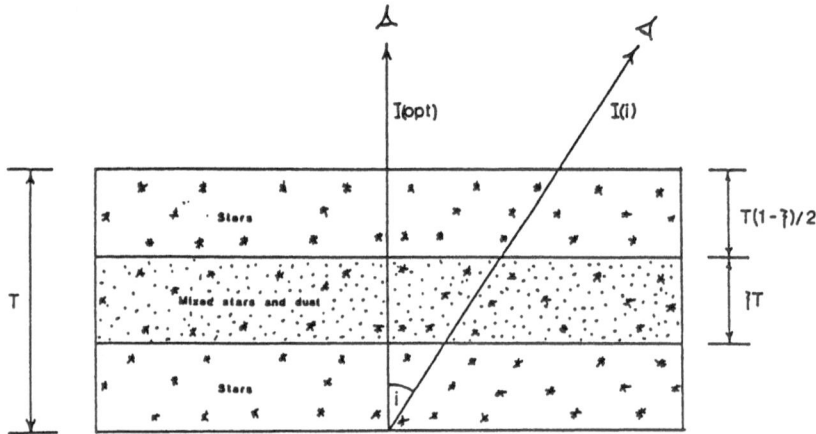

**Figure 3:** Diagram of the sandwich model for a galaxy of thickness $T$ and a layer of dust of thickness $\varsigma T$.

## 3  More sophisticated models

In a real galaxy we would expect the dust and gas to lie in a layer in the plane of the disc, with relatively unobscured stars on either side (cf. the situation in our own galaxy [4] and observations of edge on galaxies [13]). A simple model of this situation is shown in figure 3. $T$ is the total thickness of the galaxy, and $\varsigma T$ is the thickness of the layer of dust. The upper layer of unobscured stars will lead to the surface brightness varying with inclination in the same manner as for an optically thin galaxy. If the dust lies in a very thin layer in the middle of the disc then we have

$$I(i) = E^*T \sec{(i)}/2$$

as we only receive light from the upper half of the galaxy, and the SB increases as in the optically thin case. The luminosity $L(i)$ projected on the sky is independent of $i$ (since $L(i) = \pi a2I(bol)/4$). $L_{FIR}$ is also independent of inclination, and so it follows that, for an optically thick sandwich model, the ratio $L_{FIR}/L_{OPT}$ is independent of inclination. This is an important result because it shows that one cannot simply argue, as for instance [15] have done, that because the $L_{IRAS}/L_B$ ratios appear to be largely independent of $i$ spiral galaxies must be optically thin.

For a high enough value of $\tau$ (in fact $\tau > 2\varsigma/1-\varsigma$) the FIR/optical ratio is controlled almost entirely by $\varsigma$, the relative thickness of the obscuring layer. The variation of $\Delta\overline{S}$ with inclination for the sandwich model is shown in figure 1 by the lower curve. As can be seen from this figure the data is bounded below by an optically thick sandwich model and above by an optically thick slab model. Thus we can conclude from this that the variation of $\Delta\overline{S}$ with inclination tells us little about the optical depth of the galaxy.

Although the sandwich model is a fairly realistic one, we have in fact made a more realistic 'triplex' model. In this model the surface density of the stars and the dust falls off exponentially with radius, and also the stars and dust fall off exponentially in a direction perpendicular to the plane of the disc (see DDP). We have shown that a wide range of 'triplex' models (i.e. large ranges of $\tau$ and $\varsigma$ ) are consistent with Holmberg's data.

The table below shows our calculated values for the face on extinction $AB_0$ in magnitudes for the simpler models discussed above.

| Optical Depth | Screen | Slab | Sandwich | | |
|---|---|---|---|---|---|
| | | | $\varsigma = 0.75$ | $\varsigma = 0.5$ | $\varsigma = 0.25$ |
| 1 | 1.09 | 0.5 | 0.48 | 0.45 | 0.25 |
| 2 | 2.18 | 0.91 | 0.83 | 0.75 | 0.68 |
| 3 | 3.26 | 1.25 | 1.08 | 0.94 | 0.81 |
| 5 | 5.43 | 1.75 | 1.4 | 1.14 | 0.92 |
| 10 | 10.9 | 2.5 | 1.75 | 1.31 | 0.99 |
| 30 | 32.6 | 3.69 | 2.06 | 1.43 | 1.04 |
| 100 | 109 | 5 | 2.19 | 1.48 | 1.06 |

# 4 What do observations tell us?

One could show that optical studies of the variation of surface brightness with inclination of disc galaxies tell us little about the optical depth. What do observations at other wavelengths tell us?

a. Let us first look at the IRAS evidence. If galaxies contain signifiant amounts of dust then the absorbed optical light should be re-radiated in the FIR. IRAS observations show that spirals have a wide range of $L_{IRAS}/L_B$, ranging from less than 0.1 to over 100. We can, in fact, explain this whole range by pure internal extinction without recourse to starbursts if we assume suitable values of $\varsigma$ and the bolometric conversion factor $\Lambda_C$ (defined by $L_{FIR}/L_{OPT} = \Lambda_C L_{IRAS}/L_B$, see DDP). The extra extinction will imply brighter bolometric magnitudes for all galaxies. Optically selected galaxies for which $L_{IRAS}/L_B < 1$ are consistent with a wide range of $\varsigma$ and $\Lambda_C$. ELFIRS (Extremely Luminous Far Infrared Sources) can be explained by choosing values of $\varsigma$ close to 1 and large $\tau$, corresponding to a galaxy where the gas and dust have been stirred up from the plane, possibly by interaction.

b. The apparent lack of high surface brightness galaxies (see [8]) could be explained by figure 2. As the surface density increases discs become optically thick and so the surface brightness cannot rise above the actually observed values. Also IRAS failed to find any low SB galaxies which hadn't been found in the optical. Although we know of selection effects against finding low SB galaxies in the optical (see [7]), we would not expect *a priori* any strong selection effects against finding such galaxies in an infrared survey unless SB and IR luminosity were related (which it appears they are, see [17]).

c. Assuming other spiral galaxies have gas and dust ratios similar to the ratios observed in our own, the optical thickness can be inferred from the measured column densities of gas. HI profiles of spirals [2] imply column densities of $> 10^{21}$ cm$^{-2}$ within the 23–24 $B\mu$ isophote. Such column densities would imply $\tau > 1$, and so HI column densities would suggest discs to be optically thick out to the 23 or 24 $B\mu$ isophote (typically 2 scale lengths out from the centre).

Towards the centre H$_2$ begins to dominate over HI (see e.g. [24]). Since we can measure optical depths of dark clouds within the solar neighbourhood we can directly correlate optical depth and CO intensity (e.g. [5,10]). Typically observed central intensities of CO are in the range 2–40 K km s$^{-1}$ (e.g. [26,20]), eqivalent to optical depths of 1–20 in the blue ($I(CO) \approx 24$ K km s$^{-1}$ corresponding to $\tau \approx 1$). CO profiles (e.g. [25]) imply similar radial extents of the optically thick regions of galaxies as the HI data, i.e. out to the 23 or 24 $B\mu$ isophote.

d. One could explain the great majority of ELFIRS simply as heavily obscured spirals. But for the 5% which exhibit strong H$\alpha$ emission (>100 Å EW) starbursts must also be invoked. In general extra star-formation and extra extinction, possibly caused by stirring up dust in interactions, may be involved in ELFIRS.

# 5  Summary

We have not proven that disc galaxies are optically thick. We have, however, shown that the long-standing optical evidence for their being optically thin is not convincing, and that the optical data can be equally well explained by realistic optically thick models. We have also shown that the IRAS observations can be explained by an optically-thick disc hypothesis, with the stirring up of gas and dust possibly explaining the ELFIRS. Finally, observations of surface brightness and gas column densities argue in favour of optically thick discs. If the discs of galaxies are indeed optically thick then there are many obvious implications. The bolometric luminosities, colours and mass-to-light ratios will need to be reviewed. Even galaxies with low extinction may have high optical depth (if the dust has sunk to a thin layer in the plane of the disc). Such galaxies would block off much of the Universe beyond Z=3.

# References

[1] Bohlin, R.C., Savage, B.D., Drake, J.: 1978, Astrophys. J., **224**, 132.

[2] Bosma, A.: 1978, Ph.D. Thesis, Groningen.

[3] Burstein, D., Heiles, C.: 1978, Astrophys. J., **225**, 40.

[4] Cowie, L.L., Songaila, A.: 1986, Ann. Rev. Astron. Astrophys., **24**, 499.

[5] Dickman, R.L.: 1978, Astrophys. J. Suppl., **37**, 407.

[6] Disney, M.J., Phillipps, S.: 1983, M.N.R.A.S., **205**, 1253.

[7] Disney, M.J., Davies, J.I., Phillipps, S.: 1989, M.N.R.A.S., **239**, 939.

[8] Disney, M.J., Phillipps, S.: 1985, M.N.R.A.S., **216**, 53.

[9] Freeman, K.C.: 1970, Astrophys. J., **160**, 811.

[10] Frerking, M.A., Langer, W.L., Wilson, R.W.: 1982, Astrophys. J., **262**, 590.

[11] Holmberg, E.: 1958, Med. Lund Astr. Obs., Ser. 2, no. 136.

[12] de Jong, T., Brink, E.: 1987, Star Formation in Galaxies (NASA), p. 323.

[13] Kylafis, N.D., Bahcall, J.N.: 1987, Astrophys. J., **317**, 637.

[14] Leech, K.J., Lawrence, A., Rowan-Robinson, M., Walker, D., Penston, M.V.: 1988, M.N.R.A.S., **231**, 977.

[15] Lonsdale, C.J., Persson, S.E., Rice, W., Bothum, G.D.: Star Formation in Galaxies (NASA), p. 273.

[16] Moorwood, A.F.M., Veron-Cetty, M.P., Glass, I.S.: 1986, Astron. Astrophys., **160**, 39.

[17] Phillipps, S., Disney, M.J.: 1988, M.N.R.A.S., **231**, 359.

[18] Rice, W., Carol, J., Lonsdale, B., Soifer, B.T., Neugebauer, G., Kopan, E.L., Lawrence, A.L., de Jong, T., Habing, H.J.: 1988, Astrophys. J. Suppl., **68**, 91.

[19] Sandage, A., Tammann, G.A.: 1981, Revised Shapley Ames Catalog of Bright Galaxies (Carnegie Institute, Washington).

[20] Stark, A.A., Knapp, G.R., Bally, J., Wilson, R.W., Penzias, A.A., Rowe, H.E.: 1986, Astrophys. J., **310**, 660.

[21] de Vaucouleurs, G., de Vaucouleurs, A., Corwin, H.G.: 1976, Second Reference Catalogue of Bright Galaxies (University of Texas, Austin).

[22] Whitworth, A.P.: 1988, private communication.

[23] Wynn-Williams, C.G.: 1987, Star Formation in Galaxies (NASA), p. 125.

[24] Wyse, R.F.M.: 1986, Astrophys. J., **311**, L41.

[25] Young, J.S.,Tacconi, L., Scoville, N.Z.: 1983, Astrophys. J., **269**, 136.

[26] Young, J.S., Schloerb, F.P., Kenny, J.D., Lord, S.D.: 1986, Astrophys. J., **310**, 660.

# On the possible interaction between magnetic field and HII region in AB Aur

**G.G. Corciulo**[1], **A. Bianchini**[2], **F. Strafella**[3], **A.A. Vittone**[1]

[1] Osservatorio Astronomico di Capodimonte, Napoli

[2] Osservatorio Astronomico di Padova

[3] Università degli Studi di Lecce

## ABSTRACT

The Ae Herbig's star, AB Aur, shows alternatively at different times, a Beals Type III and Type II P Cygni profile, in $H_\alpha$ and $H_\beta$ lines. We investigate the possibility that such behaviour is produced by a dipolar magnetic field which forces the inner parts of the expanding envelope (stellar wind) to corotate with the star out to a few stellar radii.

An explanatory model is proposed for the corotation region (the so called magnetosphere) including HII region. Since the magnetosphere has a non-spherical symmetry, if magnetic and rotation axes are misaligned, the magnetosphere axis undergoes a precessional motion around the rotation axis, showing to the observer different corotation radii and consequently different radial velocity dispersion in the hydrogen lines. It is proposed that when the projected velocity of the corotating portion of the envelope is larger than the velocity of the stellar wind in the line-forming region, we observe an emission component in the blue side of the absorption feature (Beals Type III profile). Furthermore, the combined effects of the oblique rotator geometry and the mass loss produced by the centrifugal force give rise to the changing Type II P Cyg profile at $H_\alpha$ and $H_\beta$ lines.

## Introduction

The Herbig Ae/Be stars are high mass (ranging from 3 to 5 solar masses) early type objects still in the pre-main sequence (PMS) stage of evolution (Herbig

*E. Bussoletti and A. A. Vittone (eds.), Dusty Objects in the Universe, 155–163.*
© 1990 *Kluwer Academic Publishers.*

1960; Strom et al. 1972; Cohen and Kuhi 1979). In the evolution of young objects toward the main sequence, these stars will populate the spectral range B-A, while the T Tauri stars are thought to be the low main-sequence counterparts.

Moreover, since Herbig's and T Tauri stars are young objects, they are surrounded by a fairly extensive envelope of gas (Taurus-Auriga complex, Orion etc.) whose interaction with the stellar source affects both the line and continuum spectrum.

The original list of 26 Ae/Be stars compiled by Herbig in his pioneering work, has been extended to 57 objects by Finkenzeller and Mundt (1984). These authors have obtained high resolution profiles for $H_\alpha$ lines and have divided these stars into three subclasses: in the first the stars show in the $H_\alpha$ line two emission peaks; the stars belonging to the second, show one emission peak in the $H_\alpha$ line and finally, in the third, we find those showing a P Cyg profile at $H_\alpha$.

One of the brightest stars of the Finkenzeller and Mundt sample, AB Aur (A0Ve), belongs to the subclass of stars showing P Cyg profiles at $H_\alpha$ (Felenbok et al. 1983; Garrison and Anderson 1977; Catala et al. 1986a, hereafter CCFP; Catala and Kunasz 1987). Furthermore, this object shows rapid variability (on the scale of hours) in the shape of spectral lines (Praderie et al. 1986 and references cited herein).

In this contribution we focus our attention on the P Cyg profile variations of the $H_\alpha$ and $H_\beta$ lines in AB Aur, attempting an interpretation in terms of interaction of the ionized envelope with the stellar magnetic field. On the other hand, the possible presence of a magnetic field on this object has been supported recently by Güdel et al. (1989) through the detection of non-thermal emission in the radio range.

## Observations

The spectroscopic observations have been carried out at the 182 cm telescope located at Cima Ekar (Asiago) equipped with an echelle spectrograph providing a dispersion of 10 Å/mm at $H_\alpha$. In fig. 1a,b,c and fig. 2a,b,c we present the sequence of line profiles observed at $H_\alpha$ and $H_\beta$ during three observing nights. Inspection of these figures reveals a short term variability of the blue side of the lines: the P Cygni profile goes from a Beal's Type III (fig. 1a) (Beals 1951) to a Type II (fig. 1e) in approximately 48 hours. The spikes visible on the line wings in some spectrum, are due to impurities on the CCD detector.

Many interpretations are found in the current literature to explain the small emission component on the blue edge of $H_\alpha$ line (CCFP) and the correspondent "shoulder" at $H_\beta$ (fig. 2a). According to Mihalas and Conti (1980), the

Beal's Type III profile may be interpreted as a pure P Cyg profile on which is superimposed a second peak due to a *large radial velocity dispersion*. More precisely, the velocity dispersion is a consequence of the interaction between a stellar magnetic field and the ionized region that is forced to corotate with the same angular velocity. This model is proposed to explain the filling-in of the P Cyg absorption and the consequent appearance of a *blueshifted emission*, but it does not account for the observed variability.

**The model**

To obtain a preliminary description for the inner regions of the AB Aur envelope in the stationary case and assuming an aligned rotator, we consider a dipolar magnetic field given by

$$B(r,\vartheta) = \frac{B_*(4 - 3sin^2\vartheta)^{\frac{1}{2}}r_*^3}{r^3} \tag{1}$$

where $B_*$ is the magnetic field at the equator on the star, $r_*$ is the stellar radius and $r, \vartheta$ are polar coordinates ($\vartheta$ colatitude). The simplest criterion to be fulfilled for the gas to corotate is

$$\frac{1}{2}\varrho(r,\vartheta)v^2(r,\vartheta) \leq \frac{B^2(r,\vartheta)}{8\pi} \tag{2}$$

where $v(r,\vartheta)$ is the velocity of the gas along a field line (poloidal velocity). The equation (2) establishes the region where the magnetic energy is larger than the thermal one. To estimate the extent of the corotating region (Alfvèn radius), we adopt the calculation scheme given by Mestel (1968).
The continuity equation in a dipolar field can be written as

$$\frac{\varrho(r,\vartheta)v(r,\vartheta)}{\varrho(r_*,\vartheta)v(r_*,\vartheta)} = \frac{B(r,\vartheta)}{B(r_*,\vartheta)} \tag{3}$$

On the other hand, the conservation of the energy per unit mass along a stream-line in a non-rotating frame is described by the equation (Bernoulli's equation)

$$\frac{1}{2}v^2(r,\vartheta) + \frac{1}{2}\Omega^2d^2 + a^2log\varrho(r,\vartheta) - \frac{GM}{r} - \Omega d^2 = constant \tag{4}$$

in which $\Omega$ is the angular velocity of the star, $a$ the isothermal sound speed and $d$ the distance of the point $(r,\vartheta)$ from the rotation axis. The use of equations (1),(2),(3) and (4) with appropriate boundary conditions allows us to derive velocity and density fields and consequently the extent of the corotating region.

This has been done numerically and in fig. 3 we present the results obtained for appropriate values of AB Aur parameters ($\Omega = 5.5 \times 10^{-5}$ rad/s; $a = 10$ Km/s; $r_* = 3R_\odot$; $M = 2.5 M_\odot$; $B_* = 80$ G).

For AB Aur, the magnetosphere coincides approximately with HII region. In fact, equating the 0 - 912 Å ionizing photon number and the recombination number per second, we can obtain the dimension of the hydrogen photoionized envelope. This calculation provides, for AB Aur HII region, $r_{HII} \sim 2r_*$, assuming as typical value of terminal wind velocity $v_\infty = 300$ Km/s, and a mass loss rate $\dot{M} = 10^{-8} M_\odot$/yr. Then, out to the Alfvèn radius, the HII region corotates with the star, while expansion dominates in the outer regions. Therefore, if the maximum rotation velocity is larger than the maximum expansion velocity (stellar wind) in the line-forming zone, a small blueshifted emission is observed (Beals Type III profile).

But, is it possible to explain, through the same model, both the Beals Type III and Type II P Cyg profile at $H_\alpha$ and the correspondent behaviour at $H_\beta$ line? Many interpretations exist about this problem (CCFP; Catala and Kunasz 1987) which are principally founded on the variation of the wind velocity law. Since our spectroscopic observations of AB Aur seem to indicate a sort of cyclically changing P Cyg profiles, we propose that both the blueshifted emission and absorption component are the effects of mass losing of an *oblique magnetic rotator*. In fact, if the magnetic dipole is inclined to the rotation axis, the magnetosphere axis revolves round the rotation axis (fig. 4), showing to the observer different corotation radii and consequently different radial velocity dispersion at the $H_\alpha$ line profile (fig. 5). In this case, mass loss should occur from those areas of the magnetosphere which are farthest from the rotation axis (where the centrifugal force is maximum). For any position angle of the rotator, the material will be ejected into space along the tangent to the circular trajectory of the mass losing areas. The result is the formation of two revolving spirals. The different observed P Cyg profiles could then be the effect of the various intersections of these spirals with the line of sight.

## Conclusions

Since our interpretation is based on the precessional motion of the magnetic axis around the rotation axis, it is crucial to prove a periodicity in $H_\alpha$ and $H_\beta$ profile variations. This is the main goal of next observation runs at the 182 cm telescope at Asiago. It is remarkable, however, to consider that the time-scale of variations in the hydrogen lines is comparable with the estimated rotation period of the star ($\sim 32$ hours at the photosphere; Catala et al. 1986b).

# References

Beals, C.S.: 1951, *Publ. Dominion Astrophys. Obs.* **9**, 1

Catala, C., Czarny, J., Felenbok, P., Praderie, F.: 1986a, *Astron. Astrophys.* **154**, 103

Catala, C., Felenbok, P., Czarny, J., Tavalera, A., Boesgaard, A.M.:1986b, *Astrophys. J.* **308**, 791

Catala, C., Kunasz, P.B.: 1987, *Astron. Astrophys.* **174**, 158

Cohen, M., Kuhi, L.V.: 1979, *Astrophys. J. Suppl.* **41**, 743

Felenbok, P., Praderie, F., Tavalera, A.:1983, *Astron. Astrophys.* **128**, 74

Finkenzeller, U., Mundt, R.:1984, *Astron. Astrophys. Suppl.* **55**, 109

Garrison, L.M., Anderson, C.M.:1977, *Astrophys. J.* **218**, 438

Güdel, M., Benz, A.O., Catala, C., Praderie, F.:1989, *Astron. Astrophys.* **217**, L9

Herbig, G.H.:1960, *Astrophys. J. Suppl.* **4**, 337

Mestel, L.:1968, *Monthly Notices Roy. Astron. Soc.* **138**, 359

Mihalas, D., Conti, P.:1980, *Astrophys. J.* **235**, 515

Praderie, F., Simon, T., Catala, C., Boesgaard, A.M.: 1986, *Astrophys. J.* **303**, 311

Strom, S.E., Strom, K.M., Yost, J., Carrasco, L., Grasdalen, G.L.:1972, *Astrophys. J.* **173**, 353

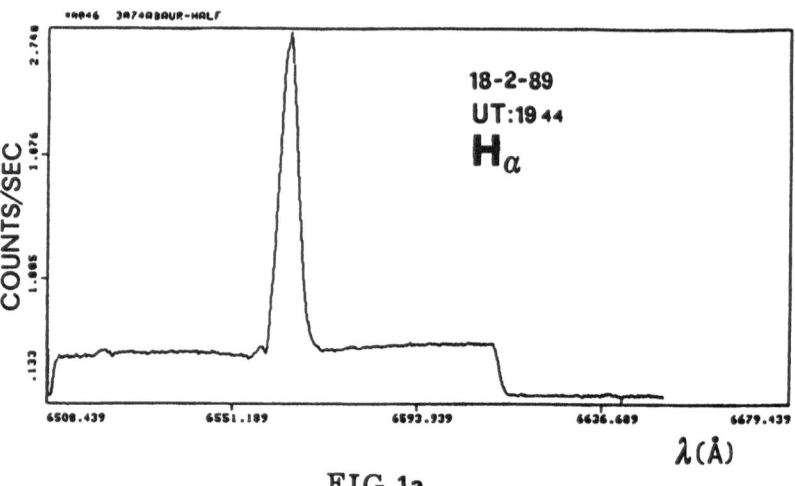

FIG. 1a

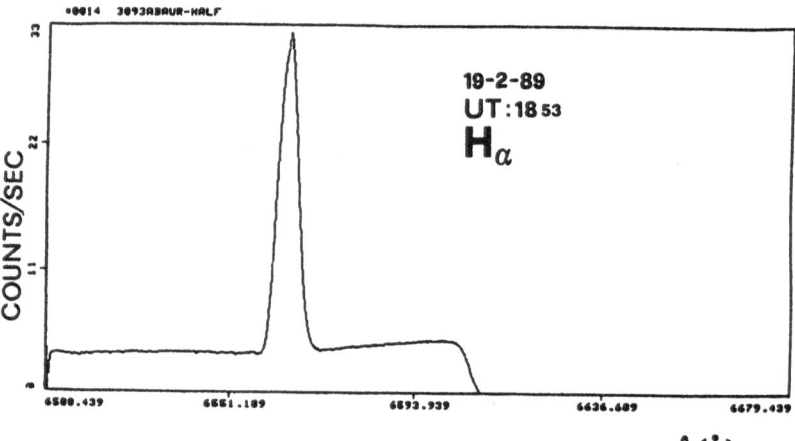

FIG. 1b

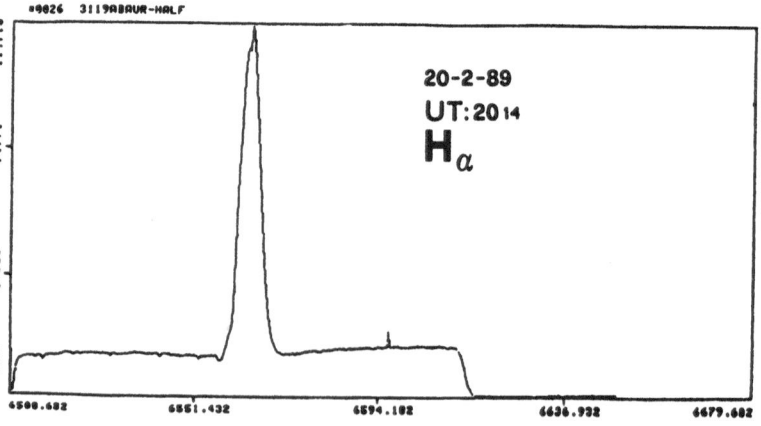

FIG. 1c

161

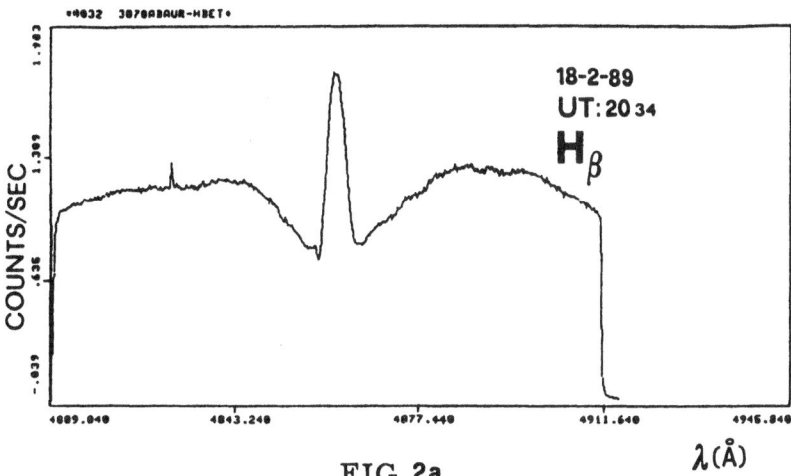

FIG. 2a

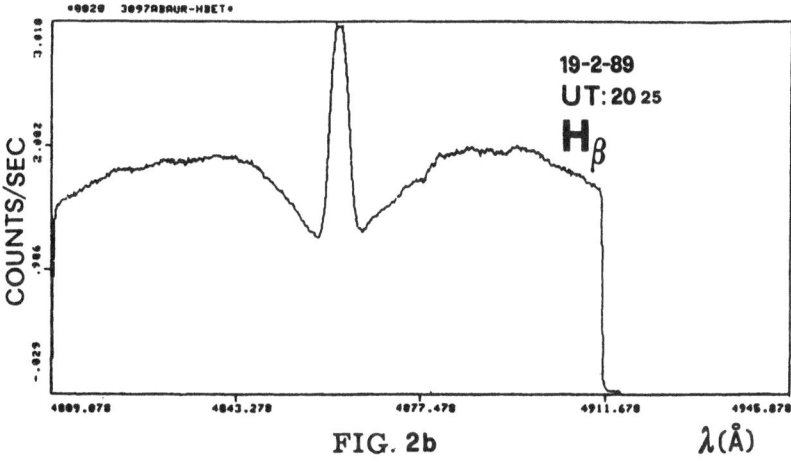

FIG. 2b

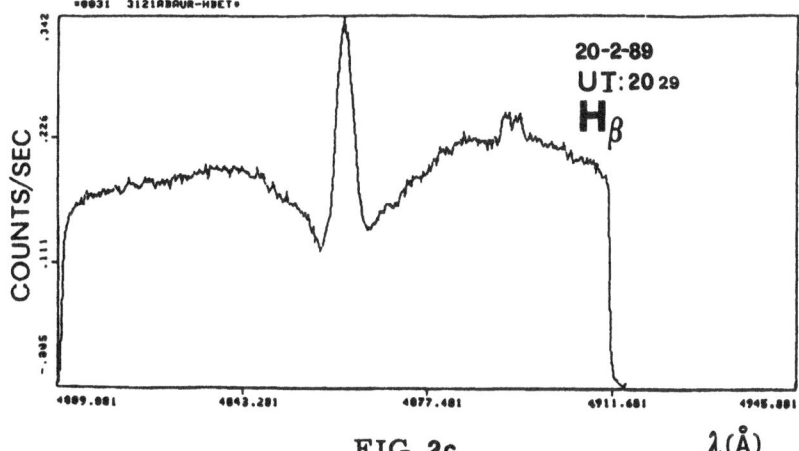

FIG. 2c

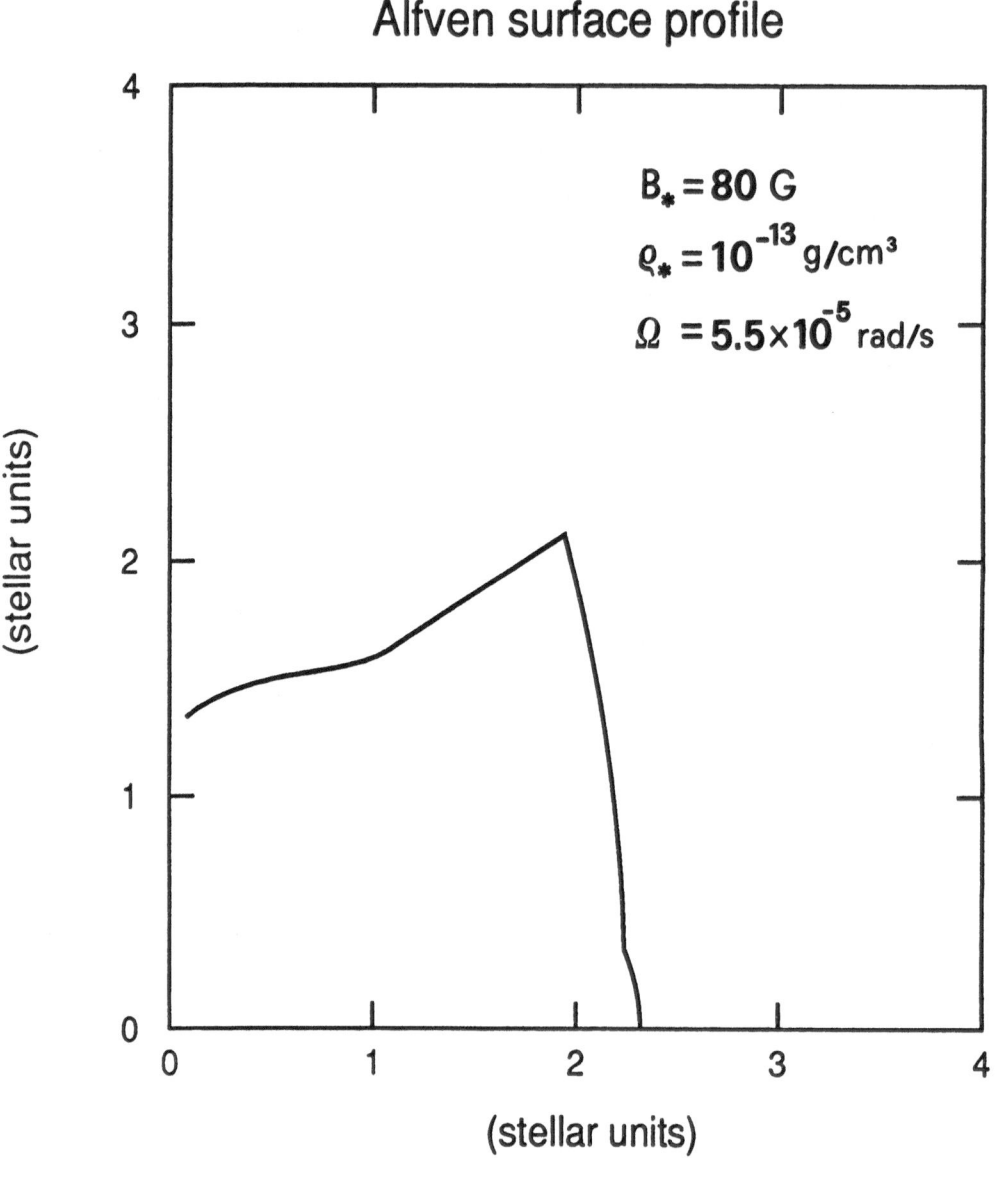

Alfven surface profile

$B_* = 80$ G
$\varrho_* = 10^{-13}$ g/cm³
$\Omega = 5.5 \times 10^{-5}$ rad/s

(stellar units)

(stellar units)

FIG. 3

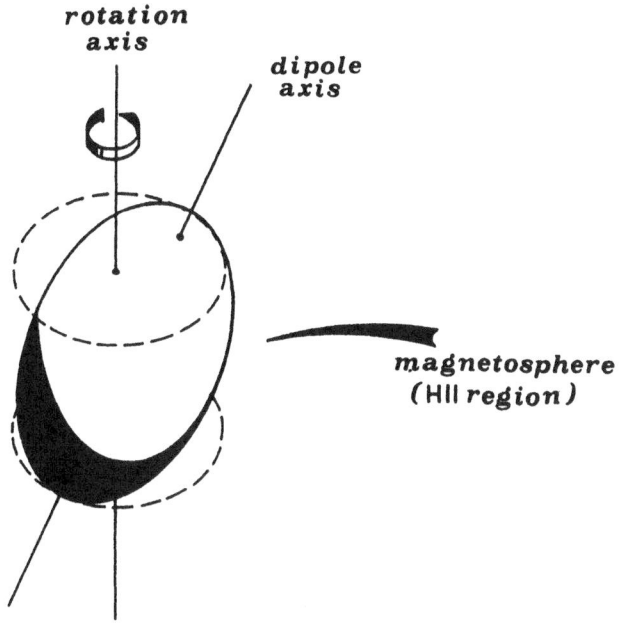

rotation
axis
dipole
axis
magnetosphere
(HII region)

FIG. 4

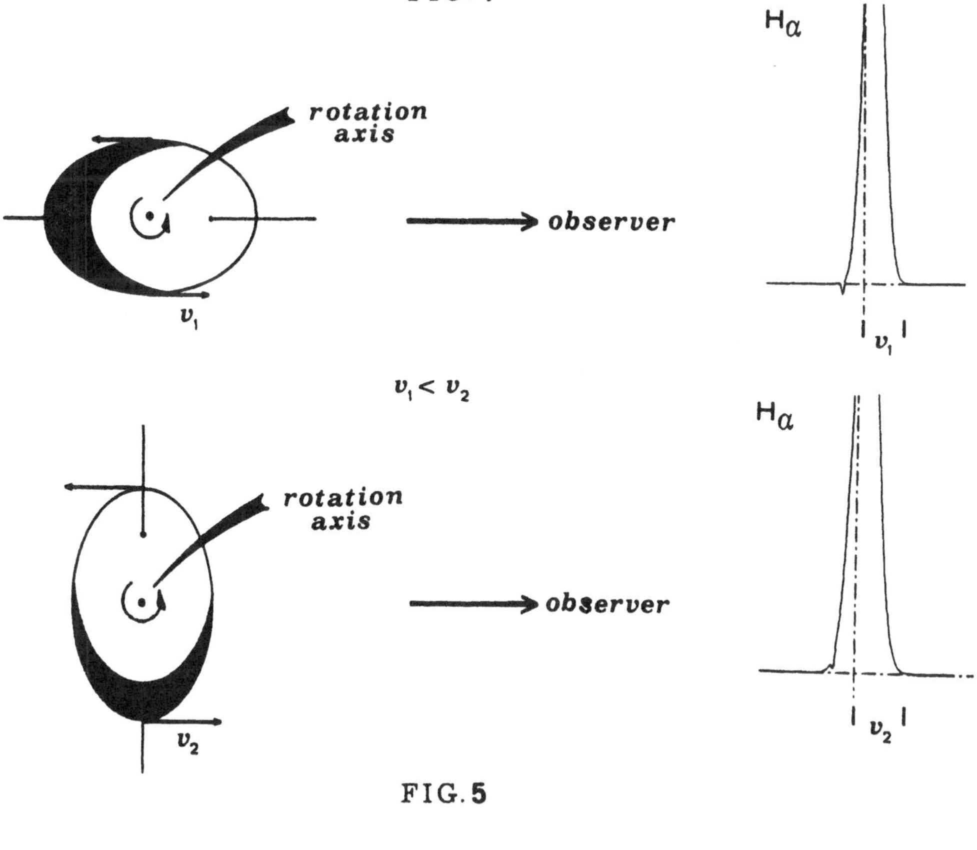

rotation
axis
observer
$v_1$
Hα
$v_1$
$v_1 < v_2$
Hα
rotation
axis
observer
$v_2$
$v_2$

FIG. 5

# A MODEL OF THE COMETARY DUST TRAILS DETECTED BY *IRAS*

M. FULLE
*Osservatorio Astronomico*
*Via Tiepolo 11*
*I-34131 Trieste*
*Italy*

ABSTRACT. Dust tail analysis has provided detailed information on the loss mass and number rates, the ejection velocities and the size distribution of the meteoroids ejected by comets P/Encke and P/D'Arrest along their orbits (Fulle, 1989b). In this paper we consider these results to model the dust trail detected by IRAS along the orbit of comet P/Encke (Sykes, 1988) and to show preliminary results about the dependence of the trail shape on the number of revolutions of the meteoroids.

## 1. Introduction

One of the major discoveries of the *Infrared Astronomical Satellite (IRAS)* was of cometary dust trails covering a large sector of the parent body orbits (Sykes, 1988). The shape and the luminosity of these thin structures depend on the dust production of the parent comets, which is properly described by the dust loss rate along the orbit, the ejection velocity from the inner coma (depending on time and on the grain size) and by the time-dependent size distribution of the meteoroids which are injected into the trail. However, it is extremely difficult to infer these quantities from the trail data themselves, because the trail is composed of the overlapping of several populations of meteoroids, one for each comet revolution around the Sun.

On the other hand, dust tail analysis supplies all the necessary information about the physical quantities which properly describe the dust environment of a comet. In this paper we consider the results of the dust tail analysis of comet P/Encke (Fulle, 1989b) to reconstruct its dust trail. The comparison between our synthetic dust trail and the observed one will give further constraints on the Encke's dust environment determined by means of dust tail analysis.

## 2. The Model

The physical properties of the dust produced by P/Encke during its passage of 1977 (Fig.1, from Fulle, 1989b) were deduced by the analysis of the dust tail data obtained by Sekanina and Schuster (1978). In Fig.2 we show the comparison between the observed isophotes and those computed by means of the results shown in Fig.1. For this tail reconstruction we considered only the dust ejected during the revolution of 1977. In order to build up the dust trail, we simply consider the ejection of meteoroids characterized by the physical properties shown in Fig.1 during n revolutions before 1977. Therefore, the trail reconstructions depend on three free parameters, the size dependence of the dust ejection velocity u and the anisotropy parameter w (see Table 1) typical of dust tail interpretation, and the new free parameter n, the revolutions before that of 1977.

165

*E. Bussoletti and A. A. Vittone (eds.), Dusty Objects in the Universe, 165–171.*

The numerical method used to analyse dust tails (Fulle, 1989a) is able to directly build-up the dust trails because it considers the rigorous motion of each ejected meteoroids, and therefore determines the major semiaxis  a, the eccentricity  e  and the perihelion orientation of the orbit of each grain.  The only difference with respect to dust tail analysis is that on long times the Poynting - Robertson drag becomes efficient, and this implies a slow variation of the orbital parameters:

$$\frac{da}{dt} = - C \frac{1-\mu}{a} (2 + 3 e^2) (1 - e^2)^{-3/2} \qquad (1)$$

$$\frac{de}{dt} = - \frac{5}{2} C \frac{1-\mu}{a^2} e (1 - e^2)^{-1/2} \qquad (2)$$

where  $C = 6.24\ 10^{-4}\ AU^2\ years^{-1}$  and  $1-\mu$  is the solar radiation pressure force acting on the grain expressed in solar gravity force units (Burns et al., 1979).  For the sizes and the time intervals (shorter than 100 years) considered here, the Poynting-Robertson corrections are negligible, but they will become necessary for larger values of the revolution number  n.

## 3.  Results

In Fig.3 we show a very wide filed view of the Encke's trail as built-up by the first values of  n.  The luminosity levels of the trail are much lower than those of the tail built-up at n = 0, which appears as a overexposed object at the center of the image.  This figure cannot give information on the trail width, because it is always much smaller than the size of the cells sampling the image.

In Figs. 4 and 5 we show a sector of the trail close to the comet, which appears also in this image as an overexposed object at the center of the field.  Obviously the luminosity of the trail increases with  n, because the meteoroids number in the trail is directly proportional to  n.  However we point out that the luminosity increases also with  w, because isotropic ejections give trails twice brighter than anisotropic ones.  This fact cannot depend on the number of sample grains on a single shell  $\mathbf{N_s}$, because the trail luminosity levels are normalized with respect to  $\mathbf{N_s}$.  Similarly, the trail luminosity dependence on the ejection anisotropy cannot be due to differences of the dust loss rates, because isotropic ejections are related to loss rates lower than anisotropic ejections (Fig.1).  Therefore we are led to conclude that isotropic ejections should really inject at least twice more meteoroids into the trails than anisotropic ones.

The trail width is quite independent of the dust ejection anisotropy.  As in the case of dust tails, the decrease of the shell size for increasing anisotropies is probably balanced by the increase of the dust velocity.  On the contrary, the trail width depends on the revolution number, increasing from  $\approx 10^4$ km  for  n = 4  to  $\approx 2\ 10^4$ km  for  n = 32, at  $S_{10} = 1$  and  $10^6$ km  behind the comet.  Finally, we point out that the short and wide sunward spike which was observed in the antitail of P/Encke cannot be explained by the trail, because the trail luminosity level for  n = 32  is  $\approx 100$  times lower than that observed for the spike, thus confirming the interpretation in terms of a Neck-Line Structure (Fulle, 1989b).

## 4. Conclusions

The luminosity of the trail shows a significant dependence on the anisotropy of dust ejection and on the number of revolutions during which the comet continues to produce meteoroids injected into the trail. Obviously the luminosity increases with the revolution number. A less expected result is that the luminosity decreases when anisotropy increases. Since the meteoroids of the inner trail should be the largest ones, which were ejected mainly before perihelion, this fact shows that a significant contribution to the trail is given by the meteoroids ejected in the direction opposite to the comet motion. The significant dependence of the trail luminosity on the free parameters suggests that trail data might not give very strong constraints to the dust environment of the parent comet. Further analyses will take into account increasing revolution numbers and thermal properties of the meteoroids, so that comparisons with the IRAS data will become possible.

*Acknowledgements* The calculations were performed on the Apollo computers of Astronet Trieste center. The diagrams were generated using Astronet AGL standard graphics.

## References

BURNS, J.A., P.L. LAMY, AND S. SOTER 1979. Radiation forces on small particles in the Solar System. *Icarus* **40**, 1 - 48

FULLE, M. 1989a. Evaluation of cometary dust parameters from numerical simulations: comparison with analytical approach and role of anisotropic emissions. *Astron.Astrophys.* **217**, 283 - 297

FULLE, M. 1989b. Meteoroids from short period comets. *Astron.Astrophys.* (in press)

SEKANINA, Z., AND H.E. SCHUSTER 1978. Dust from Periodic Comet Encke: Large Grains in Short Supply. *Astron.Astrophys.* **68**, 429 - 435

SYKES, M.V. 1988. IRAS observations of extended zodiacal structures. *Astrophys.J.* **334**, L55 - L58

**Table 1.** Parameters of the model of the Encke's dust tail. $u = \partial \log v(t, d) / \partial \log d$. w, half width of the dust ejection cone: isotropic ejection (half width of $\pi$), hemispherical ejections (half width of $\pi/2$), and strongly anisotropic ejections (half width of $\pi/4$). $\mathbf{N}_s$, $\mathbf{N}_\mu$, $\mathbf{N}_t$, dust samples on a dust shell, in the modified size and in time. $N_t$, $N_\mu$, samples of the solution $F(t, 1-\mu)$ in time and in the modified size. $N_M$, $N_N$, samples of the source image in the M and N directions. Trials, number of test functions $v(t)$. $\mathbf{M}$, total mass of ejected meteoroids ($10^{12}$ grams for $Ap(\alpha) = 0.03$). S, Symbol in Fig. 1.

| u | w | $\mathbf{N}_s$ | $\mathbf{N}_\mu$ | $\mathbf{N}_t$ | $N_t$ | $N_\mu$ | $N_M$ | $N_N$ | Trials | $\mathbf{M}$ | S |
|------|------|------|-----|-----|----|----|----|----|----|-----|---|
| -1/6 | 180° | 2578 | 100 | 225 | 15 | 10 | 30 | 30 | 59 | 4.2 | O |
| -1/6 | 90°  | 1285 | 100 | 225 | 15 | 10 | 30 | 30 | 33 | 4.2 | □ |
| -1/6 | 45°  | 382  | 100 | 225 | 15 | 10 | 30 | 30 | 63 | 7.0 | Δ |
| -1/4 | 180° | 2578 | 100 | 225 | 15 | 10 | 30 | 30 | 34 | 4.3 | + |
| -1/4 | 90°  | 1285 | 100 | 225 | 15 | 10 | 30 | 30 | 32 | 4.0 | × |
| -1/4 | 45°  | 382  | 100 | 225 | 15 | 10 | 30 | 30 | 61 | 3.9 | * |

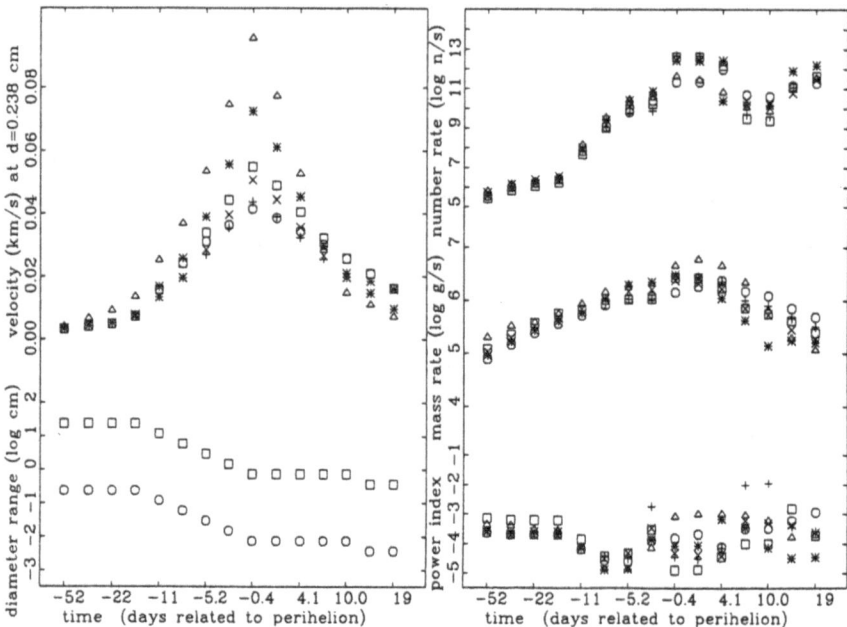

**Figure 1:** Dust environment of Comet P/Encke assuming the albedo $Ap(\alpha) = 0.03$: the dust loss rates (depending inversely on $Ap(\alpha)$), the dust ejection velocity, the power index of the time-dependent size distribution and the diameter interval to which all the solutions are related. The symbols are related to Table 1. The time sampling steps correspond to true anomaly steps of $15°$.

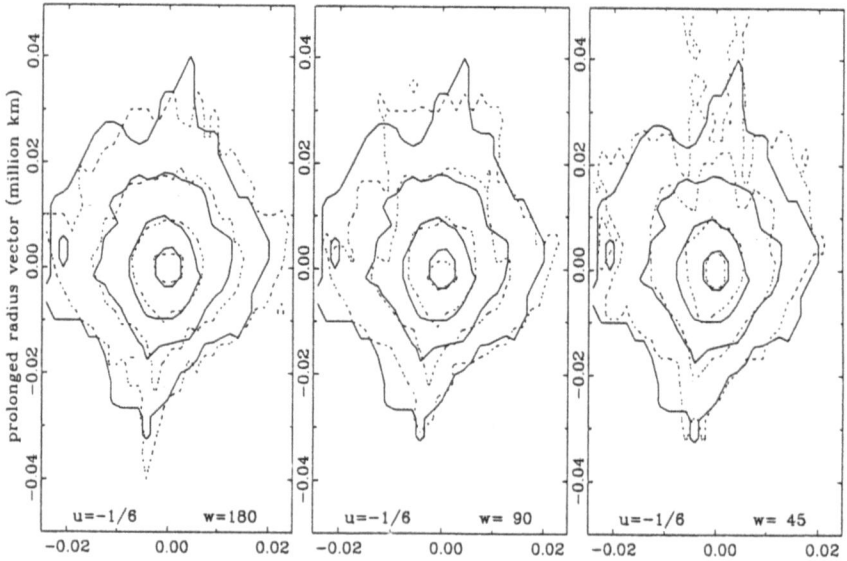

**Figure 2:** The dust tail of Comet P/Encke on 1977, Sep 12. Continuous lines: observed isophotes (from the data of Sekanina and Schuster, 1978) for the values $S_{10} = 390, 610, 1200, 3500$ ($S_{10}$ is the number of 10 R-magnitude stars per square degree). Dashed lines: computed isophotes. w is the anisotropy parameter. $u = \partial \log v(t,d)/\partial \log d$ (see Table 1). $n = 0$.

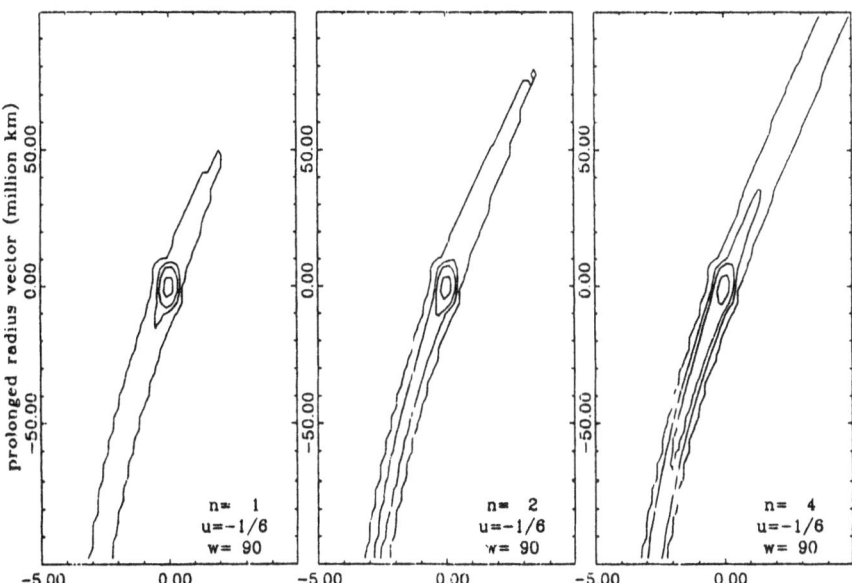

**Figure 3:** Isophotes of the dust trail of comet P/Encke for the levels $S_{10} = 0.001, 0.02, 0.03, 0.07$ for the same observations time of Fig. 2. The image is centered on the comet nucleus. The horizontal scale is enlarged of a factor 10. $u = \partial \log v(t,d)/\partial \log d$. $w$ is the anisotropy parameter. $n$ is the number of revolutions before that of 1977. $N_e = 49, N_\mu = 100, N_t = 225, N_t = 15, N_\mu = 10, N_N = N_M = 50$ (the definition of these parameters is given in Table 1).

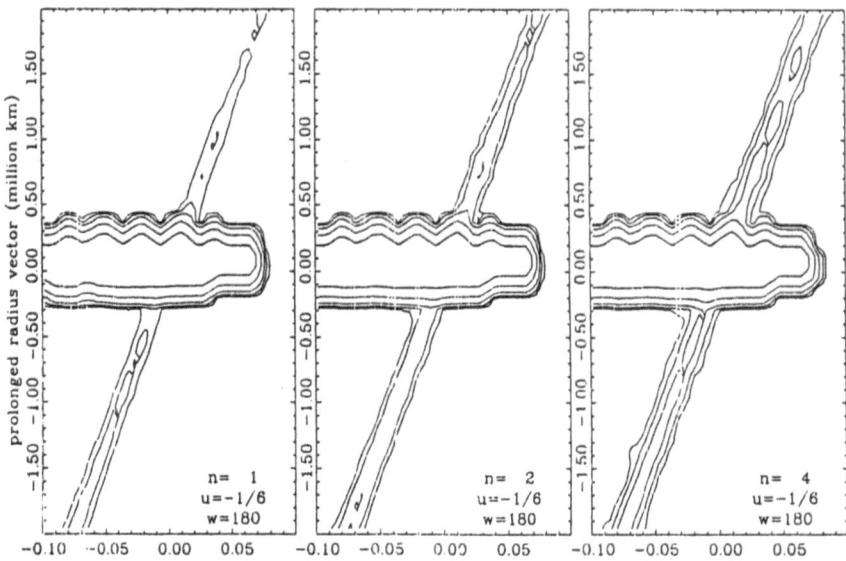

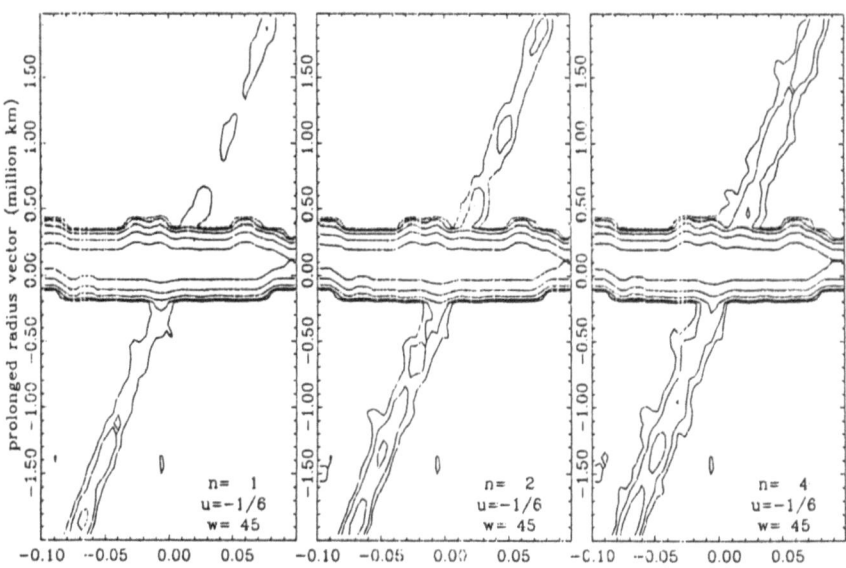

**Figure 4:** Isophotes of the dust trail of comet P/Encke for the levels $S_{10} = 0.3, 0.5, 1, 2, 3.5$ for the same observations time of Fig. 2. The image is centered on the comet nucleus. The horizontal scale is enlarged of a factor 10. $u = \partial \log v(t,d)/\partial \log d$. w is the anisotropy parameter. n is the number of revolutions before that of 1977. $N_s = 104$ (upper images) and $N_s = 18$ (lower images), $N_\mu = 100, N_t = 225, N_i = 15, N_p = 10, N_N = N_M = 50$.

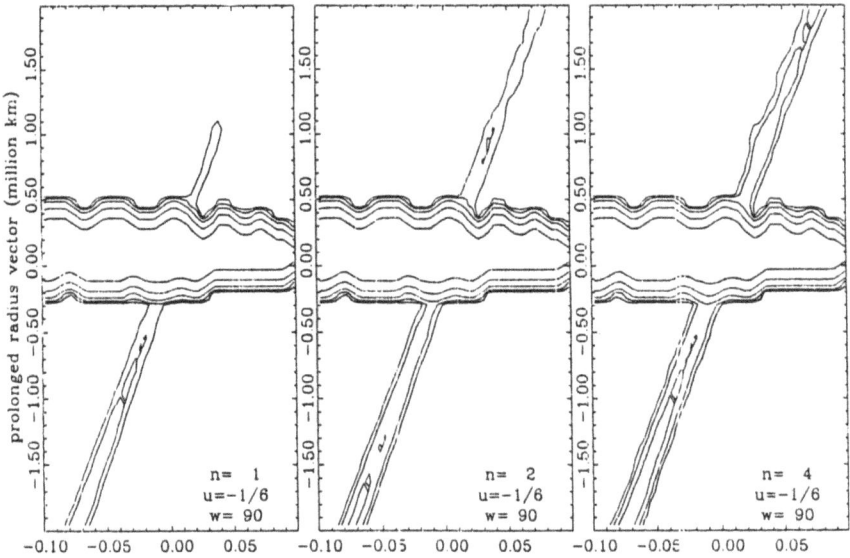

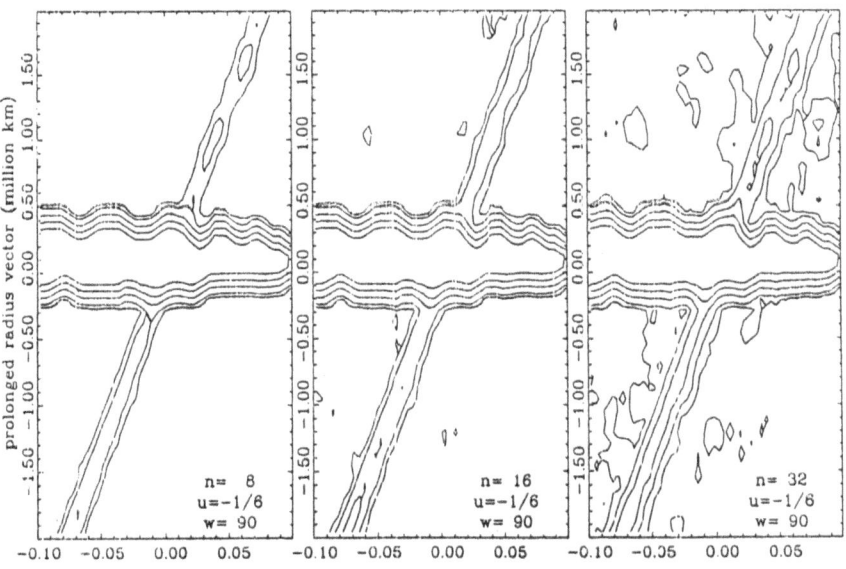

**Figure 5:** Isophotes of the dust trail of comet P/Encke for the levels $S_{10} = 0.3$, 0.5, 1, 2, 3.5 (upper images) and $S_{10} = 0.5$, 1, 2, 3, 4 (lower images) for the same observations time of Fig. 2. The image is centered on the comet nucleus. The horizontal scale is enlarged of a factor 10. w is the anisotropy parameter. $u = \partial \log v(t,d)/\partial \log d$. n is the number of revolutions before that of 1977. $N_s = 49$, $N_\mu = 100$, $N_t = 225$, $N_t = 15$, $N_\mu = 10$, $N_N = N_M = 50$.

# THE DUST TAIL OF COMET BRADFIELD 1987XXIX

M. FULLE
Osservatorio Astronomico
Via Tiepolo 11
I-34131 Trieste
Italy

G. CREMONESE
Osservatorio Astronomico
Vicolo dell'Osservatorio 5
I-35122 Padova
Italy

A. CIMATTI
Dipartimento di Astronomia
Universita' di Bologna
I-40100 Bologna
Italy

ABSTRACT. In this paper the reduction and the analysis of two images comcerning the dust tail of Comet Bradfield 1987XXIX are discussed. The plates (CRT7 + W16) were obtained by means of a 20/25/50 cm Schmidt camera. The adopted emulsion and filter combination (pass band 520 - 570 nm) approximates the standard V photometric system, thus dropping most of the plasma emissions, being sensitive only to neutral emissions in the inner coma due to $C_2$. We use SAO and GSPC-I calibration stars to measure the sky brightness, which resulted $V = 20.0 \pm 0.3$ mag arcsec$^{-2}$. To these data we apply the inverse numerical method (Fulle, 1989) which gives information about the dust loss rates, the dust ejection velocity from the inner coma and the size distribution. We consider grains of diameters between 1 μm and 1 cm ejected during the time interval $-280 < t < +40$ (days related to perihelion). The total mass ejected during this time interval is $(4 \pm 1) \, 10^{14}$ g (for an albedo $Ap(\alpha) = 0.03$), higher than that of Comet Kohoutek 1973XII, and the mass of meteoroids injected into bound orbits is $(2 \pm 1) \, 10^{14}$ g. The power index of the time-averaged size distribution is $-3.2 \pm 0.2$.

## 1. Introduction

Comets are probably the main source of interplanetary dust. The majority of such dust should come from short period comets, but a significant fraction of such dust may be supplied by other comets, as required by collisional models of the interplanetary cloud (Leinert et al., 1983), which have shown that only an unbound source can explain the radial dependence of the cloud density. However some bright comets (e.g. Arend-Roland 1957III, Kohoutek 1973XII) are characterized by hyperbolic orbits, so that their contribution of meteoroids in bound orbits is much lower than that from short period comets. On the contrary long period comets (e.g. Bennett 1970II, Bradfield 1987XXIX) are characterized by orbital eccentrities low enough to inject a large mass into bound orbits, in particular when the size distribution of the released dust has a power index higher than -4. Therefore they are the best candidates to supply the unbound fraction of interplanetary dust. In this paper we analyse the dust environment of Comet Bradfield 1987XXIX in order to check if the large contribution of meteoroids released by Comet Bennett 1970II (Fulle, 1987) is a common characteristic of long-period comets.

## 2. Data Reduction

The plates were obtained by means of a 20/25/50 cm Schmidt camera (technical data in Table 1) and were digitized by means of the PDS of the Padova Observatory adopting a square scanning window of 50 μm². The adopted emulsion and filter combination (CRT7 + W16, pass band 520 - 570 nm) approximates the standard V photometric system.

E. Bussoletti and A. A. Vittone (eds.), Dusty Objects in the Universe, 173–179.

The photographic densities were linearized into intensity by means of the related calibration wedges. To perform the absolute calibration of the images, we selected calibration stars from the SAO and the GSPC-I (Lasker, Sturch et al., 1988) catalogues, which were digitized by means of a square scanning window of 50 $\mu m^2$ and linearized by means of the same calibration wedges used for the comet images. For each star of visual magnitude V, we measured the integrated intensity $I_A$ over a sky area $A_{sky}$ (arcsec$^2$) covering the whole star trail. Then we measured the integrated intensity of sky background $I_B$ over a same area near to the star trail. Therefore the integrated star intensity $I_S$ is given by the subtraction of $I_B$ from $I_A$ (Table 2). The sky background surface light intensity $S_{10B}$ expressed in number of 10 V-magnitude stars per square degree is then given by

$$S_{10B} = 1.296 \frac{I_B}{A_{sky} I_S} 10^{11 - 0.4 V} \qquad (1)$$

## 3. Results

In Table 3 we show the parameters of the applications of the inverse numerical approach to dust tail interpretation (Fulle, 1989), which directly supplies the dust loss rates and the time-dependent size distribution. As free parameters, we consider two size dependences of the dust ejection velocity and three different dust ejection anisotropies by means of ejections on cones with the symmetry axis pointing to the Sun. For each parameter combination, we tested the number of trial velocity functions v(t) given in the table, and for each trial velocity function we tested different weights regularizing the solutions of our inverse ill-posed problem. We adopted the smallest regularizing weight which allowed to avoid large instabilities of the solutions.

To test the accuracy of our solutions and the stability of our constrained inverse problem, we reconstructed the isophote fields used as input by means of the solutions themselves (Figs. 1 and 2). In order to best fit the reconstructed images to the observed ones it was necessary to introduce normalization factors (Table 1) which differ from the unity by a factor close to the relative errors of the background surface light intensities. Only anisotropic dust ejections allow to best fit the antitail, (isotropic ones give a too long antitail), whereas the split tail of C/1987XXIX is automatically well interpreted by all parameter combinations, without the necessity of ad-hoc mechanisms to obtain it. The adopted filter plate combination should drop the strongest ion emissions, but is sensitive to the $C_2$ bands, which may have seriously contaminated the innermost isophote, although the strongest $C_2$ emissions have wavelengths shorter than 520 nm, out of our passband.

The solutions concerning the dust ejection velocities, the range of diameters of the considered sample grains, the dust loss rates and the power indexes of the time-dependent size distributions are shown in Fig.3. In order to uniformly sample the images, it was necessary to consider time-dependent size intervals. We confirm the strong dependence of the dust velocity on the ejection anisotropies which was observed for comets C/1962III, C/1973XII and P/Encke (Fulle, 1989, 1990). The dust loss rates, which were computed adopting the albedo $Ap(\alpha) = 0.03$ ($\alpha \approx 55°$, Hanner and Newburn, 1989, phase angle $\alpha$ in Table 1), do not show relevant differences. The slow increase of the dust number loss rate is mostly due to the decreasing size interval which was considered.

The mass loss rate is about constant close to perihelion (for $-60 < t < +10$ days related to perihelion), similarly to the behaviour of C/1973XII. The power index of the size distribution shows small variations. During two weeks around perihelion, its value is close to -4, otherwise it is higher than -4, and this implies the release of very large grains. This fact is confirmed by the quite flat time averaged size distribution (Fig. 4), characterized by the power index of $-3.2 \pm 0.2$.

The amount of mass of meteoroids injected into bound orbits reaches the 50% of the total released dust mass (Table 3), and such a value is so high that a Comet Bradfield every 5 years supplies the same amount of dust given by all short period comets (Fulle, 1990), thus confirming the large contribution already found for Comet Bennett 1970II (Fulle, 1987). This result mainly depends on the high luminosity of the dust tail of C/1987XXIX, that is on the sky brightness discussed in Section 2. We point out that if such sky brightness should be higher (lower mag arcsec$^{-2}$), the tail surface light intensity, and therefore the mass loss rate, would furtherly increase. On the contrary, if we admit that our site was characterized by a quite improbable sky brightness fainter of 1 V-magnitude ($V_{sky} \approx 21$ magnitude arcsec$^{-2}$), the mass loss rates would decrease of a factor $\approx 3$.

*Acknowledgements* The calculations were performed on the Apollo computers of Astronet Trieste center. The diagrams were generated using Astronet AGL standard graphics.

## References

FULLE, M. 1987. Meteoroids from Comet Bennett 1970II. *Astron.Astrophys*. **183**, 392 - 396

FULLE, M. 1989. Evaluation of cometary dust parameters from numerical simulations: comparison with analytical approach and role of anisotropic emissions. *Astron.Astrophys*. **217**, 283 - 297

FULLE, M. 1990. Meteoroids from short period comets. *Astron.Astrophys*. (in press)

HANNER M.S., AND NEWBURN R.L.. 1989. Infrared photometry of Comet Wilson (1986l) at two epochs. *Astron.J*. **97**, 254 - 261

LASKER, B.M., C.R. STURCH, C. LOPEZ, A.D. MALLAMA, S.F. MCLAUGHLIN, J.L. RUSSELL, W.Z. WISNIEWSKI, B.A. GILLESPIE, H. JENKNER, E.D. SICILIANO, D. KENNY, J.H. BAUMERT, A.M. GOLDBERG, G.W. HENRY, E. KEMPER AND M.J. SIEGEL 1988. The Guide Star Photometric Catalog. I. *Astrophys.J.Suppl.Ser*. **68**, 1 - 90

LEINERT, C., ROSER, S., AND BUITRAGO, J.: 1983. How to mantain the spatial distribution of interplanetary dust. *Astron.Astrophys*. **118**, 345 - 357

**Table 1.** Geometry of the observations. Plate, serial number of the photographic plate. UT, time of midexposure, Dec 1987. r, $\Delta$, Sun-Comet and Earth-Comet distances (AU). $\alpha$, Phase angle. Exp., exposure time (minutes). f, normalization factor of the reconstructed images. $S_{10B}$, sky background surface light intensity in number of 10 V-magnitude stars per square degree. $V_{sky}$, sky background V-magnitude arcsec$^{-2}$.

| Plate | UT | r | $\Delta$ | $\alpha$ | Exp. | Emulsion | f | $S_{10B}$ | $V_{sky}$ |
|-------|--------|------|------|------|------|-------------|-----|----------------|----------------|
| 2012 | 20.792 | 1.17 | 0.85 | 56° | 30 | CRT7 + W16 | 1.0 | $1410 \pm 230$ | $19.9 \pm 0.2$ |
| 2312 | 23.733 | 1.20 | 0.87 | 54° | 30 | CRT7 + W16 | 1.2 | $1350 \pm 370$ | $20.0 \pm 0.3$ |

**Table 2.** Plate calibration. Plate, serial number of the photographic plate. $A_{sky}$, sky area covering the star trails and the sky background (arcsec$^2$). Star, calibration star of V magnitude. $I_S$ and $I_B$, light intensity (integrated over $A_{sky}$) of the star and of the background (arbitrary units). $S_{10B}$, sky background surface light intensity expressed in number of 10 V-magnitude stars per square degree.

| Plate | $A_{sky}$ | Star | V | $I_S$ | $I_B$ | $S_{10B}$ |
|-------|-----------|------|---|-------|-------|-----------|
| 2012 | 47600 | SAO 90304 | 8.5 | 599 | 799 | 1446 |
|      | 47600 | SAO 90243 | 8.6 | 602 | 1185 | 1297 |
|      | 40800 | SAO 90282 | 8.6 | 573 | 659 | 1326 |
|      | 47600 | SAO 90386 | 8.8 | 340 | 762 | 1843 |
|      | 47600 | SAO 90368 | 8.9 | 388 | 810 | 1565 |
|      | 40800 | GSPC-I P399A | 9.13 | 339 | 656 | 1370 |
|      | 27200 | GSPC-I P399B | 10.27 | 163 | 462 | 1053 |
| 2312 | 34000 | SAO 90598 | 8.4 | 554 | 321 | 964 |
|      | 34000 | SAO 90588 | 8.4 | 406 | 374 | 1533 |
|      | 34000 | SAO 90622 | 8.7 | 364 | 371 | 1286 |
|      | 34000 | SAO 90611 | 9.3 | 151 | 409 | 1967 |
|      | 34000 | GSPC-I P400A | 9.31 | 229 | 318 | 999 |

**Table 3.** Parameters of the solutions. $u = \partial \log v(t, d) / \partial \log d$. w, half width of the dust ejection cone: isotropic ejection (half width of $\pi$), hemispherical ejections (half width of $\pi/2$), and strongly anisotropic ejections (half width of $\pi/4$). $\mathbf{N}_s$, $\mathbf{N}_\mu$, $\mathbf{N}_t$, dust samples on a dust shell, in the modified size and in time. $N_t$, $N_\mu$, samples of the solution in time and in the modified size. $N_M$, $N_N$, samples of the $N_k$ source images in the M and N directions. T, number of test functions $v(t)$. $\mathbf{M}$, total ejected dust mass ($10^{14}$ grams) for $Ap(\alpha) = 0.03$. $\mathbf{M}_b$, total mass of meteoroids injected into bound orbits ($10^{14}$ grams) for $Ap(\alpha) = 0.03$. S, Symbol in Figs. 3 and 4.

| u | w | $\mathbf{N}_s$ | $\mathbf{N}_\mu$ | $\mathbf{N}_t$ | $N_t$ | $N_\mu$ | $N_k$ | $N_M$ | $N_N$ | T | $\mathbf{M}$ | $\mathbf{M}_b$ | S |
|---|---|------|------|------|------|------|------|------|------|------|------|------|------|
| -1/6 | 180° | 2578 | 100 | 180 | 20 | 10 | 2 | 30 | 30 | 10 | 4.3 | 2.3 | O |
| -1/6 | 90°  | 1285 | 100 | 180 | 20 | 10 | 2 | 30 | 30 | 15 | 3.6 | 1.9 | □ |
| -1/6 | 45°  | 382  | 100 | 180 | 20 | 10 | 2 | 30 | 30 | 36 | 4.0 | 1.9 | Δ |
| -1/4 | 180° | 2578 | 100 | 180 | 20 | 10 | 2 | 30 | 30 | 24 | 4.2 | 2.1 | + |
| -1/4 | 90°  | 1285 | 100 | 180 | 20 | 10 | 2 | 30 | 30 | 15 | 5.5 | 3.7 | × |
| -1/4 | 45°  | 382  | 100 | 180 | 20 | 10 | 2 | 30 | 30 | 17 | 3.2 | 1.2 | * |

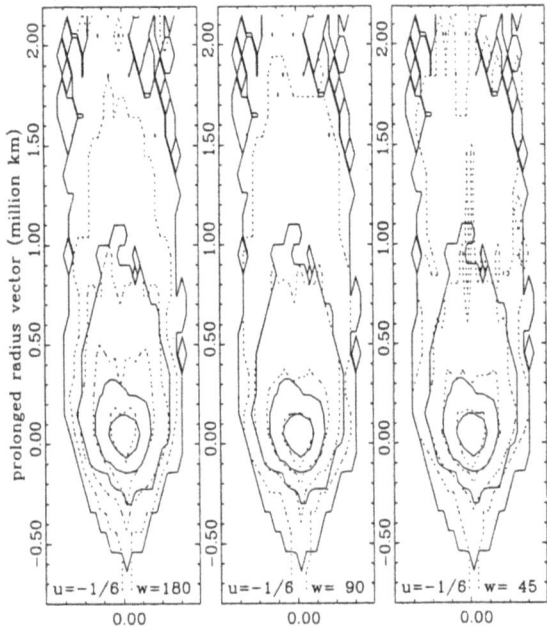

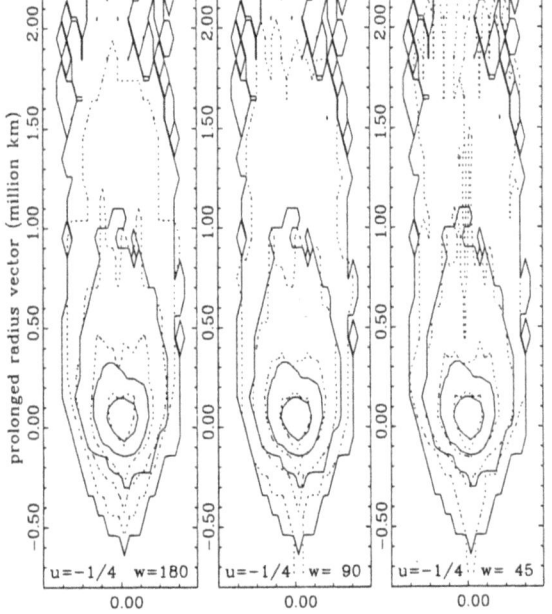

**Figure 1:** Isophotes of the dust trail from image 2012 for the levels $S_{10}$ = 315, 675, 1425, 3825. The prolonged radius vector is opposite to the Sun direction. Continuous lines: observed isophotes. Dashed lines: computed isophotes. $u = \partial \log v(t,d)/\partial \log d$. w is the anisotropy parameter.

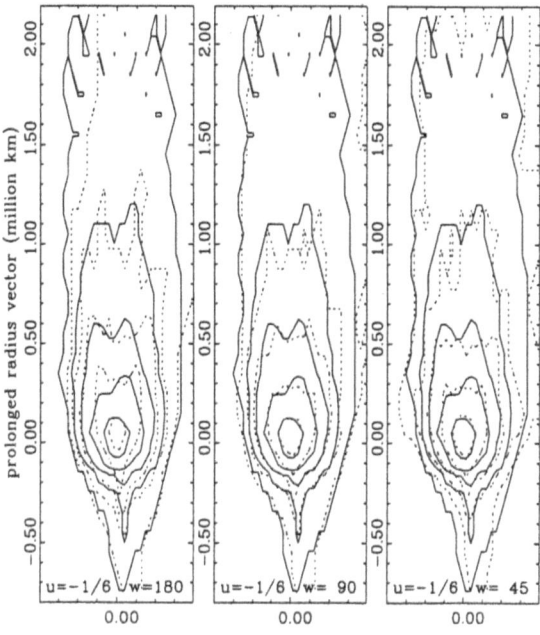

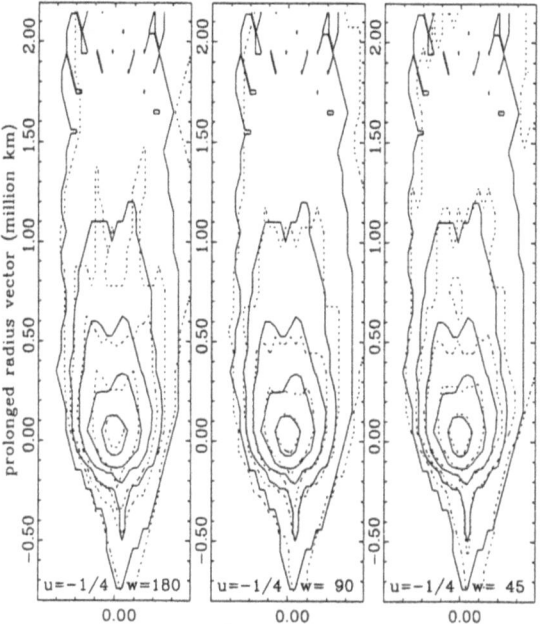

**Figure 2:** Isophotes of the dust trail from image 2312 for the levels $S_{10} =$ 315, 675, 1125, 2325, 6075. The prolonged radius vector is opposite to the Sun direction. Continuous lines: observed isophotes. Dashed lines: computed isophotes. $u = \partial \log v(t,d)/\partial \log d$. w is the anisotropy parameter.

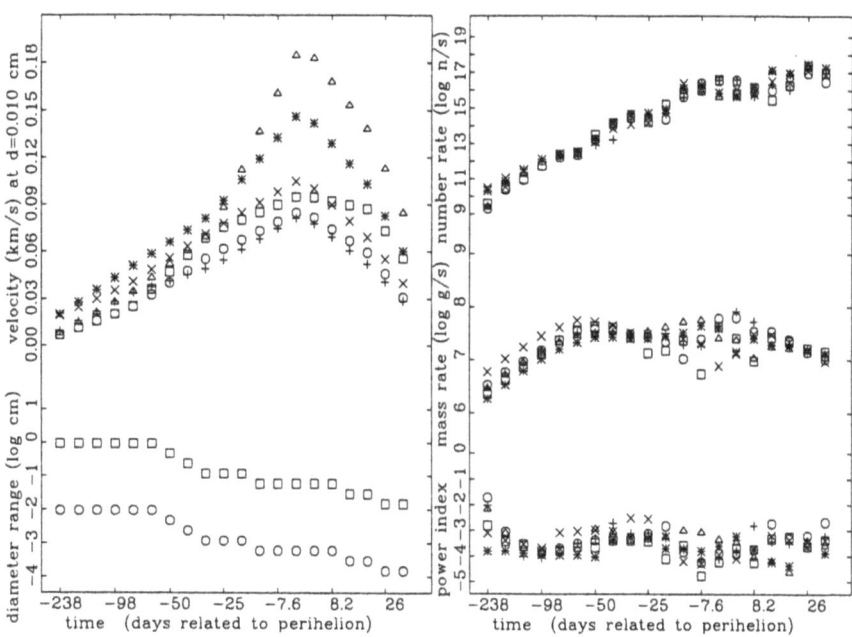

**Figure 3:** Dust environment of Comet Bradfield 1987XXIX assuming the albedo Ap($\alpha$) = 0.03: the dust loss rates (depending inversely on Ap($\alpha$)), the dust ejection velocity, the power index of the time-dependent size distribution and the diameter inteval to which all the solutions are related. The symbols are related to Table 3. The time sampling steps correspond to anomaly steps of 9°.

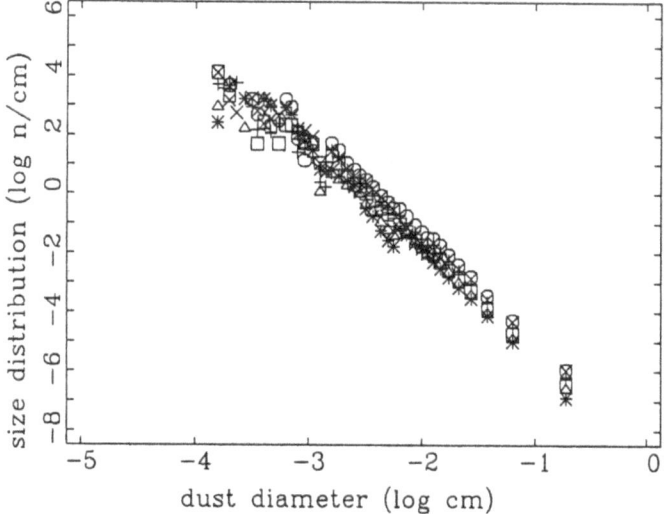

**Figure 4:** Time averaged size distribution assuming a dust bulk density of 1 g cm⁻³ (the diameter values d depend inversely on the assumed dust bulk density). The power index of the distribution of C/1987XXIX is - 3.2±0.2. The symbols are related to Table 3.

# THE DUST PROPERTIES OF HIGH-MASS STAR-FORMING REGIONS

M. Barsony and J. Najita
*Department of Astronomy*
*University of California*
*Berkeley, CA 94720*
*USA*

ABSTRACT Using the IRAS database, we have constructed radial intensity profiles of nearby, high-mass, star-forming regions. The model calculations we use to try to fit these profiles require four (4) fixed parameters and four (4) free parameters. The 100 $\mu$m profiles can be well, but non-uniquely, fit. However, only one model is consistent with other, independently determined source properties, such as total mass and total 60 $\mu$m and 100 $\mu$m fluxes. In general, the best-fit parameters consistent with the 100 $\mu$m profiles yield models of the 60 $\mu$m, 25 $\mu$m, and 12 $\mu$m emission which *do not* fit the data.

## 1. Introduction

The goal of this project was to determine the radial density, temperature, and grain size distribution of dust in galactic, high-mass, star-forming regions using the IRAS database. The sources to be modelled were chosen because they are bright, symmetric, and dominated by a central source. The data to be modelled were derived from the IRAS Coadd Intensity Grids (Barsony, 1989). After suitable background subtraction, each source was fitted by ellipses at logarithmic steps of the radial coordinate. The resultant azimuthally averaged profiles of the intensity distribution at each IRAS band are presented in Figure 1 for two representative sources, S106 and LkH$\alpha$ 101.

## 2. The Models

### 2.1 CALCULATIONS

For our modelling, we assumed spherical symmetry. The four (4) fixed input parameters in the models are: $r_0$, a fiducial radius, $r_{max}$, the maximum radial extent of the source, $p_{min}$, the minimum projected distance from the cloud center, and $\kappa_\nu$, the mass absorption coefficient at frequency $\nu$, averaged over the relevant IRAS bandpass. We have adopted $\kappa_\nu = 3.4, 3.5, 0.84$, and $0.29$ cm$^2$ gm$^{-1}$ at 12 $\mu$m, 25 $\mu$m, 60 $\mu$m, and 100 $\mu$m, respectively (Draine and Lee 1984). For LkH$\alpha$101, $r_{max} = 4.1$ pc. We set $r_0$, the fiducial radius at which $T_0$ and $\rho_0$ are specified, to be at the cloud edge. The minimum "impact parameter," $p_{min}$, was set to $3.6 \times 10^4$ AU. This is the smallest radius at which the dust should be undisturbed by the immediate circumstellar envrironment of this particular pre-main-sequence star (Barsony *et al.* 1990).

There are four (4) input parameters which we varied to produce a best fit to the 100 $\mu$m data. These are: $\rho_0$ and $T_0$, the density and temperature at the fiducial radius, and the two power-law exponents, $p$ and $q$, which specify the radial dependences of the cloud's density and temperature structure via

*E. Bussoletti and A. A. Vittone (eds.), Dusty Objects in the Universe, 181–187.*
© 1990 *Kluwer Academic Publishers.*

$$\rho = \rho_0 (r/r_0)^{-p}$$
$$T_d = T_0 (r/r_0)^{-q}.$$

We then compute the surface brightness profiles at the four IRAS frequencies by evaluating the integral:

$$\int_{-s_{max}}^{+s_{max}} B_\nu(T_d(r)) \, exp(-\tau(p,s)) \kappa_\nu \, \rho(r) \, ds,$$

where

$B_\nu$ = the blackbody function at frequency $\nu$

$p$ = projected distance from center of cloud to point on sky

$s$ = coordinate along line of sight ranging from $-s_{max} \le s \le +s_{max}$ along a given line of sight

$r = (p^2 + s^2)^{1/2}$

$T_d(r)$ = the dust temperature at r

$\tau(p,s)$ = the optical depth from front edge of the cloud to the point in the cloud $(p,s)$

$\kappa_\nu$ = the opacity at frequency $\nu$

$\rho(r)$ = the density at radius r

Models were computed for $p = 0$ and $2$, corresponding to constant and isothermal sphere density distributions, respectively. Models were computed with $q = 1/2, 2/5, 1/3$, corresponding to dust emissivity laws scaling with frequency to the 0, 1, and 2 powers, respectively. To see this latter relationship of the dust emissivity law to the radial dust temperature distribution, assume that $\kappa_\nu \propto \nu^\beta$. Then, the Planck-averaged absorption coefficient is given by:

$$\kappa_P = \frac{\int_0^\infty \kappa_\nu B_\nu \, d\nu}{\int_0^\infty B_\nu \, d\nu}$$

Recalling that

$$B_\nu \propto \nu^3/(e^{h\nu/kT} - 1),$$

substituting

$$x = \frac{h\nu}{kT},$$

and evaluating the above integral, gives the result that $\kappa_P \propto T^\beta$. For dust at an equilibrium temperature, $T_d$, at a distance $r$ from the central source requires that

$$\kappa_P T_d^4 \propto \frac{L_*}{4\pi r^2}.$$

Therefore, for $\kappa_P \propto T^\beta$,

$$T_d^{4+\beta} \propto \frac{L_*}{r^2},$$

or,

$$T_d \propto r^{\frac{-2}{4+\beta}}.$$

## 2.2 RESULTS

Results of this modelling for the 100 $\mu$m specific intensity variation with projected distance for LkH$\alpha$101 are shown in Figure 2. The solid curve represents the IRAS data. The models in the left column all have an inverse square law radial density dependence, whereas the models in the right column all have a spatially constant density. The top panels show the variation with the power law of the radial dust temperature distributions, for q=1/2, 1/3, and 2/5. The middle panels show variations with $T_0$, the temperature at the cloud edge. The bottom panels show variations with $n_0$, the gas and dust density at the outer cloud boundary.

Of the above series of models, four best-fit models to the 100 $\mu$m emission of the LkH$\alpha$101 cloud were chosen for display in the top panel of Figure 3. The four sets of input parameters, $T_0$, $n_0$, q, and p used to generate these "best-fit" models are tabulated in Table 1. These same input parameters were then used to calculate models of the 60 $\mu$m emission. The resulting 60 $\mu$m models are plotted in the bottom panel of Figure 3. Models of the 25 $\mu$m and 12 $\mu$m emission for these same 100 $\mu$m "best-fit" input parameters are not plotted, because the resultant model points fall *several orders of magnitude* below the actual data.

Table 1

| 100 $\mu$m Best $-$ fit | p | q | $T_o$ | $n_o$ |
|---|---|---|---|---|
| Model 1 | 2 | 0.4 | 15 | 50 |
| Model 2 | 2 | 0.3 | 16 | 60 |
| Model 3 | 2 | 0.5 | 15 | 40 |
| Model 4 | 0 | 0.5 | 15 | 200 |

In order to choose among these models, we have performed several consistency checks. These include determinations of the predicted cloud mass, the total integrated fluxes at 60 $\mu$m and 100 $\mu$m, and the $\tau$ at the minimum projected distance, for each of the four contending models. The results are shown in Table 2 below.

Table 2

| | Cloud Mass $M_\odot$ | Integrated Flux (Jy) 100 $\mu$m | Integrated Flux (Jy) 60 $\mu$m | $\tau_{100\mu m}(p_{min})$ |
|---|---|---|---|---|
| Model 1 | 2777 | $2.68\times10^5$ | $3.10\times10^5$ | 0.051 |
| Model 2 | 3332 | $2.26\times10^5$ | $1.95\times10^5$ | 0.061 |
| Model 3 | 2221 | $4.78\times10^5$ | $9.11\times10^5$ | 0.041 |
| Model 4 | 966 | $3.91\times10^4$ | $3.77\times10^4$ | 0.019 |
| Data | 425* | $1.17\times10^4$ | $7.42\times10^3$ | opt. thin |

* This value was derived using a mean column density of $10 \times 10^{20}$ atoms cm$^{-2}$ (Christie *et al.* 1982) over a circular area of radius $1.27 \times 10^{19}$ cm.

## 3. Conclusions

Note that of the four best-fit models to the radial distribution of the 100 $\mu$m emission, only one, Model 4, is consistent with the total cloud mass (determined independently) and with the integrated 100 $\mu$m and 60 $\mu$m fluxes (assuming an 800 pc distance to LkH$\alpha$ 101). We may therefore safely conclude that over spatial scales of 0.2 pc–4.1 pc, the density distribution of the LkHα 101 cloud is constant, as opposed to centrally-peaked. Furthermore, the dust temperature distribution seems to vary as $r^{-1/2}$.

We have *not* succeeded in modelling the observed 12 $\mu$m and 25 $\mu$m emission, however, and even the 60 $\mu$m models tend to fall below the 60 $\mu$m data points (see Figure 3). This tells us that the data cannot be fit by a single radial dust temperature distribution. At each radius, dust exists at many different "temperatures."

One of the major discoveries of IRAS was the unexpectedly large amount of 12 $\mu$m and 25 $\mu$m emission from the interstellar medium. This emission is far in excess of that expected from models of dust heating by ambient radiation fields. Our attempts at modelling the spatial distribution of IRAS emission in Galactic high-mass star-forming regions dramatically confirm the need for an alternate explanation for the source of this "excess" infrared radiation at the shorter IRAS wavelengths.

The most widely-held theory for the origin of this emission is the non-equilibrium heating of Very Small Grains (VSG's) with radii in the 3–10 Årange (Draine and Anderson 1985). This theory is not wholly satisfactory. In particular, the power-law of the adopted grain-size distribution, as well as the upper and lower bounds of the VSG radii, are varied to fit the two IRAS data points (the integrated fluxes at 12 $\mu$m and 25 $\mu$m) for different sources (Draine and Anderson 1985, Dwek 1986, Wieland *et al.* 1986).

Even so, one model which could reproduce the 25 $\mu$m/100 $\mu$m and 60 $\mu$m/100 $\mu$m ratios for L255, predicted a value of the 12 $\mu$m/100 $\mu$m ratio for this same source which was a factor of 1.5 too low (Draine and Anderson 1985). This led the authors to speculate that something other than the stellar radiation field could be playing an important role in the heating of small grains.

In conclusion, we still have a long way to go to understand the physical properties of interstellar grains: their heating and cooling mechanisms, their optical properties in the infrared, and their destruction mechanisms.

### References

Barsony, M. 1989, *Ph.D. Thesis*, California Institute of Technology.

Barsony, M., Scoville, N.Z., Schombert, J.M., and Claussen, M.J. 1990, *Ap. J.*, submitted.

Christie, R.A., McCutcheon, W.H., and C.P. Chan 1982, in *Regions of Recent Star Formation*, ed. R.S. Roger and P.E. Dewdney, D. Reidel Publishing Co., pp. 343–348.

Draine, B.T., and Anderson, N. 1985, *Ap.J.*, **292**, 494.

Draine, B.T. and Lee, H.M. 1984, *Ap.J.*, **285**, 89.

Dwek, E. 1986, *Ap.J.*, **302**, 363.

Wieland, J.L., Blitz, L., Dwek, E., Hauser, M.G., Magnani, L., and Rickard, L.J. 1986, *Ap.J.Letts.*, **306**, L101.

### Acknowledgements

We would like to thank Prof. Frank Shu and Dr. Steve Ruden for helpful discussions and suggestions. This research was supported by NASA's Astrophysics Data Program Grant No. NAG5-1243.

Radial Distributions of IRAS 100 $\mu$m Emission

in Two High-Mass Star-Forming Regions

Figure 1

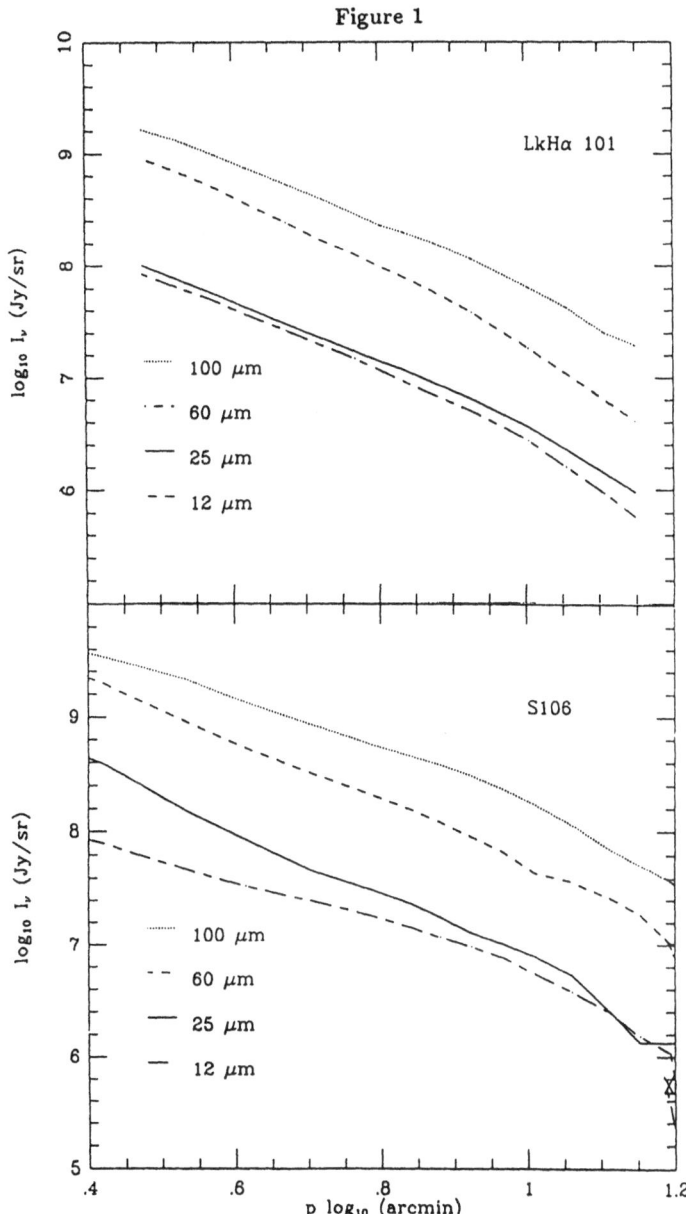

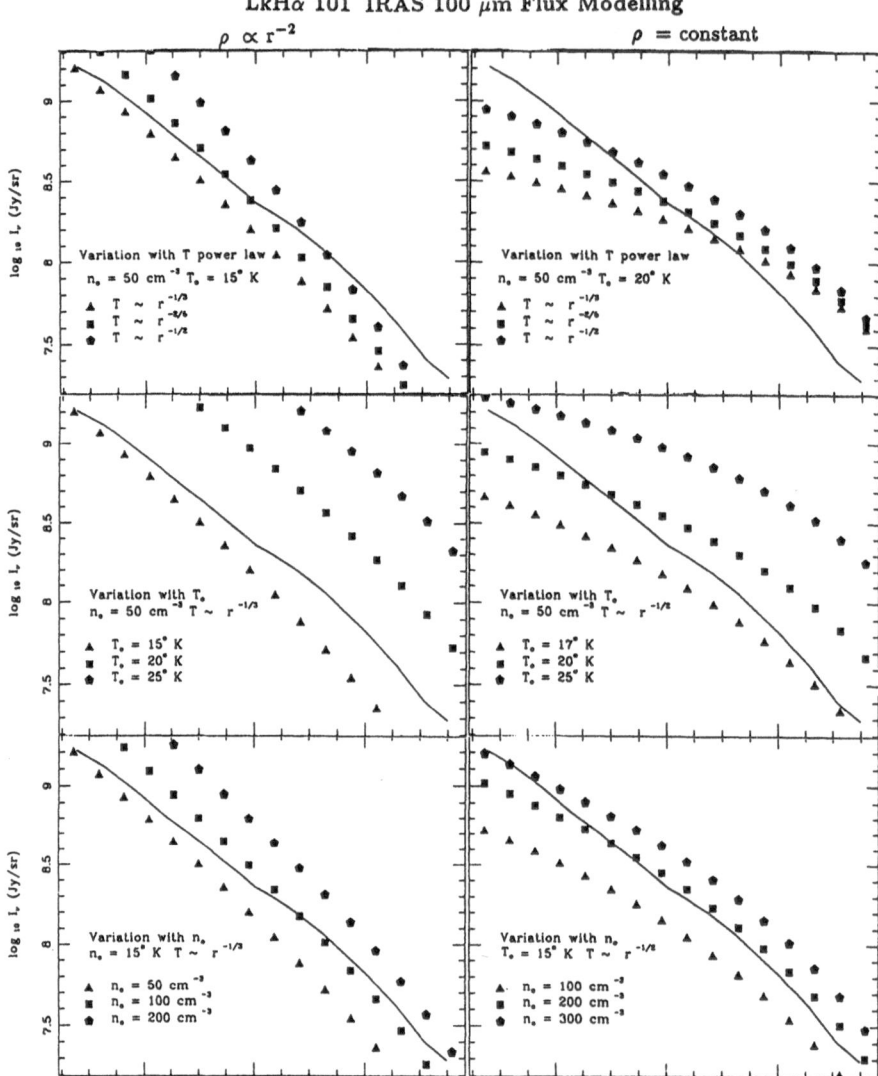

Figure 2

LkHα 101

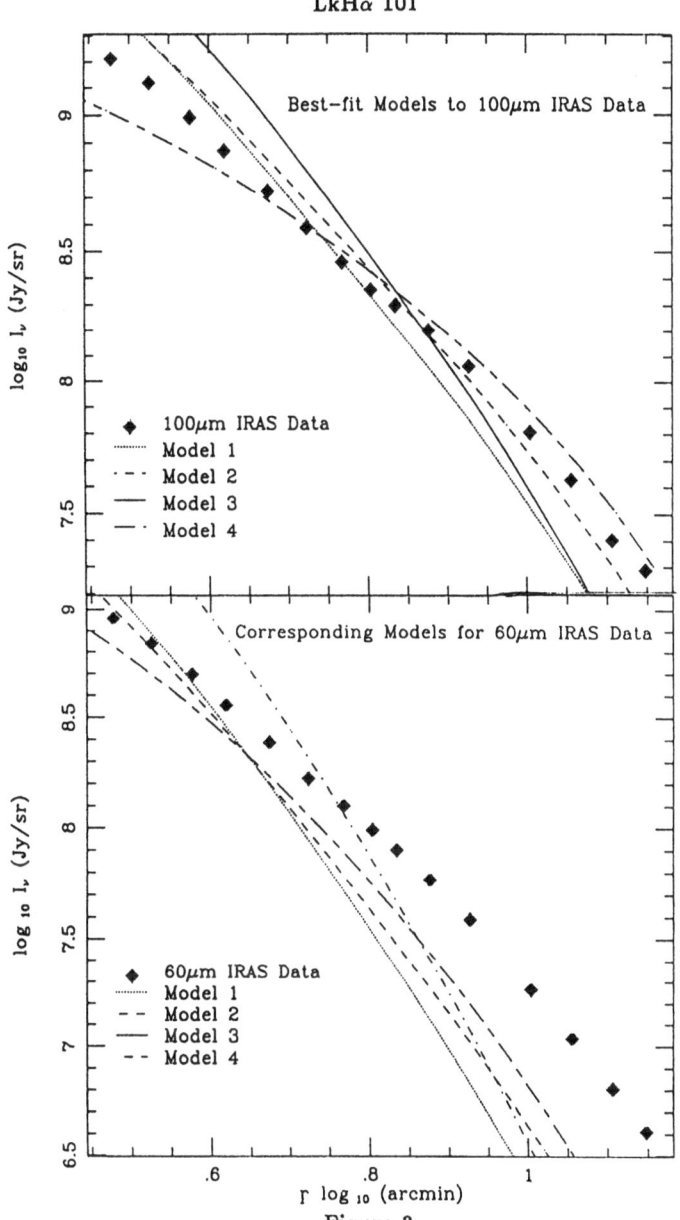

Figure 3

# ON THE ENERGETIC COMPONENT OF FU ORI WIND

M. Di Martino[1], L. Mantegazza[2], V. Pirronello[3]
[1]Osservatorio Astronomico di Torino, Pino Torinese, Italy
[2]Dipartimento di Fisica, Università di Pavia, Pavia, Italy
[3]Dipartimento di Fisica, Università della Calabria, Rende, Italy

ABSTRACT. Observations of short term fluctuations in the Johnson U-band of the star FU Orionis are presented. A preliminary deduction of the flux of the most energetic component of wind particles from such a star is also given.

## 1 Introduction

In 1976 Kuan observed very short term fluctuations (with typical time scales of the order of minutes) in the ultraviolet band in a sample of T Tauri stars. These fluctuations were interpreted as the result of the superposition of a continuous flaring activity in the stars upper atmosphere. From his data Kuan derived that the power spectra, in the frequency range typical of the observed variations, followed a power law characterized by an exponent of about "–5/3".

In 1981 Worden et al. applied the same method to a new set of data obtained from a sample of five T Tauri stars, and essentially confirmed the Kuan results. Starting from this result they were able to estimate the total flare optical energy and from this figure under suitable hypotheses the flux of energetic protons emitted, on the average, from these stars.

After these no more studies were performed on the very short term variations of T Tauri stars, despite the potential interest that they could have to investigate both the properties of these objects and those of the circumstellar matter that is present around them. Here we present some observations of short term fluctuations in the light emission of FU Orionis, an object related to T Tauri stars. Assuming that such fluctuations are originated by an atmospheric flare activity in the star (an assumption that can be however questioned) we will give a preliminary estimate of the flare energetics and of the flux of fast wind particles.

## 2 The observations

Broad-band photometric observations of FU Orionis were carried out at the European Southern Observatory (La Silla, Chile) by the 1 meter ESO reflector, equipped with a single-channel pulse counting photometer employing a RCA 31034A Quantacon photomultiplier, refrigerated with a Peltier cooling unit, and UBV filters. The data acquisition was handled by a HP 21MX computer that could write the data to a magnetic tape, to

189

E. Bussoletti and A. A. Vittone (eds.), Dusty Objects in the Universe, 189–195.
© 1990 Kluwer Academic Publishers.

a printer and to a real time display unit. Our observations were performed, through a 1.15 mm diaphragm corresponding in the sky to about 15.54 arcsec, in the high speed mode consisting of sequential 10 seconds integrations, with the aim to obtain about three data points per minute. Fu Ori was observed during two nights in the U band and for one night in the V band, with durations up to about 1.5 hours for each run. The V band observations were performed in a later run to check if the variability observed in the U band was also present at longer wavelengths. Sky readings were made during the observations at approximately 5 minute intervals, the ratio of star to sky counts was about 10.

Before and after the observation of FU Orionis, groups of standard stars, taken from the E-regions (Graham, 1982), were observed at different air masses in order to determine the standard UBV magnitudes of the star and the extinction coefficients of the night. To perform these reductions the ESO "Snopy" program was used. Counts of each nightly run have been transformed into relative intensity after that the correction for the atmospheric extinction had been performed. As unity we adopted the average intensity in each night.

Figure 1 shows the data obtained during the two observational runs in the U band. The lightcurve in the V band (that is not shown here) is however characterized by a complete flatness if compared with the U ones, where approximate 3% brightness variations on time scale of some tens of minutes are present.

## 3  Data analysis

In order to deduce the total energy released in the phenomena that are responsible for the short term fluctuations in the lightcurves (perhaps flares) we have Fourier analyzed our photometric observations. Our data may be, in fact, represented by the standard integral

$$I(t) = \int_{-\infty}^{+\infty} \tilde{I}(f)e^{-2\pi i f t}\, df$$

and $I(f)$ is computed using the inverse Fourier transform

$$I(f) = \int_{-\infty}^{+\infty} I(t)e^{2\pi i f t}\, dt$$

and being a complex function it may be written as a real part (the amplitude $A(f)$) times a complex phase $e^{i\vartheta(f)}$, i.e.:

$$\tilde{I}(f) = A(f)e^{i\vartheta(f)}$$

The so called power spectrum $P(f)$, given by

$$P(f) = \left|\tilde{I}(f)\right|^2 = A(f)^2$$

is often used to have information on the amplitude.

The observational data were then analyzed following the method used by Kuan (1976). Since our data are not exactly uniformly spaced and moreover they contain some small gaps (those corresponding to the sky measurements) it is not possible to analyze them in the frequency domain by means of the Fast Fourier Transform (FFT) because the unequal spacing destroys the orthogonality between sine and cosine terms. The analytical technique we have adopted is the least-squares power spectrum technique proposed by Vanicek (1971) (see also Antonello et al., 1986), which is essentially equivalent to the Scargle's periodogram (Scargle, 1982). This technique is suitable for the analysis of unequally-spaced data, moreover with this power spectra it is possible to compute the false-alarm probability, i.e. the probability that a given spectral peak of height "z" is due to random noise, which is currently under investigation.

The spectra of the two U data nights are quite alike (figs. 2 and 3), showing the presence of dominant power at low frequencies, however, while for the data taken on March the $18^{th}$ this low-frequency domain is statistically highly significant, for the data taken on March the $19^{th}$ it is only marginally significant: this fact, that reflects the appearance of the lightcurve, testifies that in FU Orionis the activity on very short time scales is not uniform. On the other end the power spectrum of V band data does not present any significant low frequency feature, showing that this kind of activity mainly concerns the shorter wavelengths.

Following Worden et al. (1981) we represent our U spectrum as the result of the sum of a frequency-dependent signal and a noise contribution of the type:

$$P(f) = \alpha f^{-\beta} + \gamma$$

where "$\alpha$" is a normalization constant and "$\gamma$" is the noise contribution. To do that we first subtracted the estimated noise power and then we made a least-squares fit on the spectrum for the low frequency region where the signal power is dominant. We got

$$\beta \simeq 0.412$$

We can see that the estimated "$\beta$" value for FU Ori is considerably different from that obtained by Worden and co workers (1981) for the T Tauri stars they observed.

## 4 Energetics of events and fast particle fluxes

If we still go on assuming that this short term variability arises from the superposition of many solar like flares events we can try to estimate the total flare energy by integrating over the amplitudes. The flare energy relative to the total U-band flux is then given by:

$$E = [A(0)]^{-1} \int_{f_1}^{f_2} A(f) \, df$$

where $A(0)$ is the average U light intensity (the spectral amplitude at the zero frequency) and "$f_1$" and "$f_2$" are the frequency integration limits. In this computation the critical factor is just the choice of such integration limits. For the low-frequency one "$f_1$" we have adopted that corresponding to a time-scale about equal to our observing time: 100 min. This choice is not a very serious limitation because if we change this value by a factor two (i.e. from 100 to 50 min) the relative flare energy changes of 0.0008 only. The choice of the high-frequency limit could however present a more serious problem. In this preliminary approach we have restricted our evaluation to the so-called "Nyquist limit", the frequency value over which we should not obtain any useful spectral information (relative to an approximate spacing of about 20 seconds in our data).

In such a case the relative flare energy then becomes

$$E \simeq 0.023 L_U$$

i.e. the resulting energy of the flare is about 0.023 times the ultraviolet luminosity $L_U$. However this should be considered as an upper limit, in fact, another plausible choice of the high frequency integration limit could have been the value where the logarithmic spectrum becomes flat.

Adopting the U magnitude determined averaging the values obtained in different nights, assuming the distance of the star to be of about 500 parsecs as given by Herbig (1977) and applying the correction for interstellar extinction taken from Mendoza (1968) we obtain

$$L_U(star) \simeq 7.8 * 10^{34} \quad \text{ergs/sec.}$$

In order to estimate the total flare optical luminosity Worden et al. (1981) assumed that the flare events of T Tauri stars are similar to those occurring on UV Ceti ones, adopted the relations among flare luminosities in different bands derived by Lacy et al. (1976) and defining $L_{opt} = L_U + L_B + L_V$ obtained

$$L_{opt} \simeq 2.4 L_U \quad \text{(flare)}$$

In the case of FU Ori we certainly cannot assume a complete similarity with the flares of UV Ceti type stars because of the difference in the power spectrum exponent. It is then quite questionable to use the last relation to get the optical luminosity of FU Orionis flares without a careful analysis; by the way we will use for the moment the same expression because our observations of this star give indications that flare activity, if it is the agent responsible of the short term fluctuations, mainly concerns the U band and suggests us that the error in the coefficient should not be much higher than 100%.

Using the previous analysis we obtain for the total flare optical luminosity

$$L_{opt} \simeq 4.3 \times 10^{33} \quad \text{ergs/s.}$$

If really flares are responsible for the observed U band fluctuations one can imagine that together with the electromagnetic radiation also particle radiation is emitted. This would then represent one of the very few cases in which it is possible to infer the flux of particles permeating the surrounding environment. To perform such an estimate we will assume that: the ratio between the power as optical radiation is in the case of FU Ori equal to the solar one, as it was used by Worden et al. (1981) for T Tauri stars, and that "on the average" the shape of the energy spectrum (at least from the energy value we are interested in, i.e. 10 MeV) of the accelerated ions is the same to that of the flares of these stars. Then assuming in particular as a spectrum of energetic particles that one obtained by Van Hollebeke et al. (1975) we get at most a flux "$\Phi$" of ions more energetic than 10 MeV at a distance from the star of about 1 A.U. of

$$\Phi(E > 10 \text{ MeV}) \simeq 5.5 \times 10^{10} \quad \text{H}^+ \text{ cm}^{-2} \text{ s}^{-1}$$

This is an upper limit to the flux, but remains an enormous value!

# 5   Conclusions

We have carried out observations of short term fluctuations in the Johnson's U band of the very special star FU Orionis. We have then analyzed them by means of well known Fourier algorithms deducing the the total optical power released on the average by the phenomena that are responsible of for such fluctuations. Finally accepting the interpretation that these fluctuations are due to a continuous flaring activity occurring in the atmosphere of the star, following Worden et al. (1981), we have obtained a preliminary evaluation of the flux of energetic (E>10 MeV) particles belonging to its wind.

This estimate of the particle flux, or better the final estimate of it could then be used to investigate possible effects induced in the solid component or in the gas that is present in the nebula around the star. Among these effects we can mention here:

a) the stability against sputtering (Brown et al., 1982; Pirronello and D'Arrigo, 1985; Pirronello, 1987) of icy or icy covered grains, that are known (Cohen, 1976) to exist in the nebula of T Tauri stars, the stars in which the FU Orionis flare up might occur, perhaps in a recursive way (Herbig, 1977);

b) the possible understanding of the occurrence of some isotopic anomalies encountered in the solar system that are at the moment explained as due to the explosion of supernovae.

## References

Antonello, E., Mantegazza, L., Poretti, E.: 1986, Astron. Astrophys. **159**, 269.

Brown, W.L., Augustyniak, W.M., Simmons, E., Marcantonio, K.J., Lanzerotti, L.J., Johnson, R.E., Boring, J.W., Reimann, C.T., Foti, G., Pirronello, V.: 1982, Nucl. Instr. Meth. 198, 1.

Cohen, M.: 1975, M.N.R.A.S. **173**, 279.

Graham, J.A.: 1982, P.A.S.P. **94**, 244.

Herbig, G.H.: 1977, Astrophys. J. **214**, 747.

Kuan, P.: 1976, Astrophys. J. **210**, 129.

Lacy, C.H., Moffet, T.J., Evans,D.S.: 1976, Astrophys.J. Suppl. **30**, 85.

Mendoza, V.E.E.: 1968, Astrophys. J. **151**, 977.

Pirronello, V., D'Arrigo, C.: 1985, "Particle Bombardment of Minor Bodies during Eruptive Phases of the Early Solar Evolution" in "Asteroids, Comet, Meteors", Lagerkvist C.-I. et al eds., p. 235, Uppsala, Sweden.

Pirronello, V.: 1987, Nucl. Instr. Meth. B19/20, 959.

Scargle, J.D.: 1982, Astrophys. J. **263**, 835.

Vanicek, P.: 1970, Astrophys. Space Sci. **12**, 10.

Worden, S.P., Schneeberger, T.J., Kuhn, J.R., Africano, J.L.: 1981, Astrophys. J., **244**, 520.

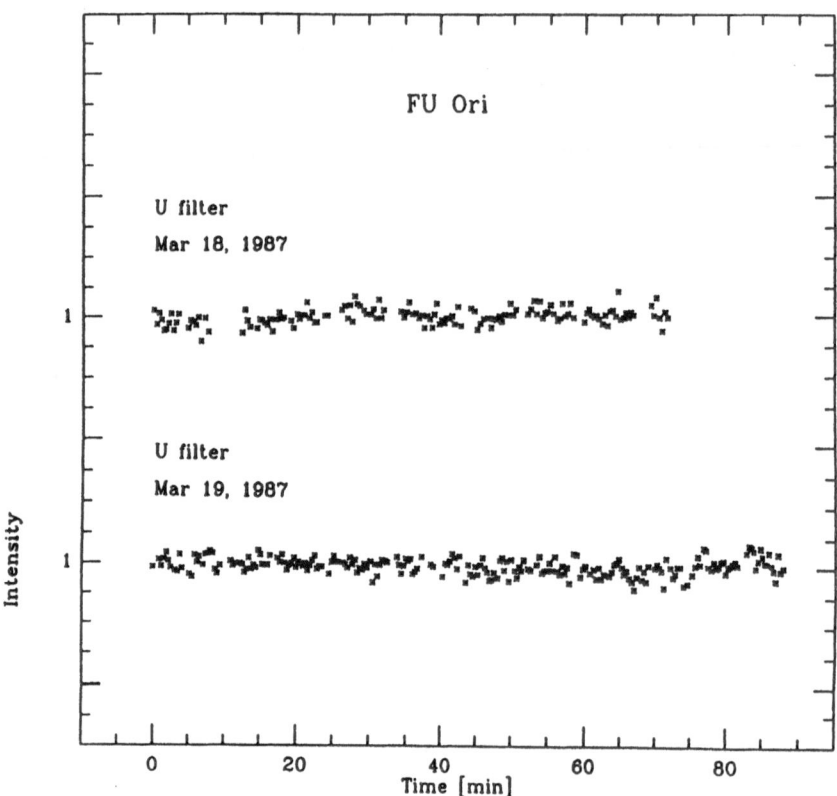

**Figure 1:** Intensity fluctuations as a function of time in the U band.

FU ORI

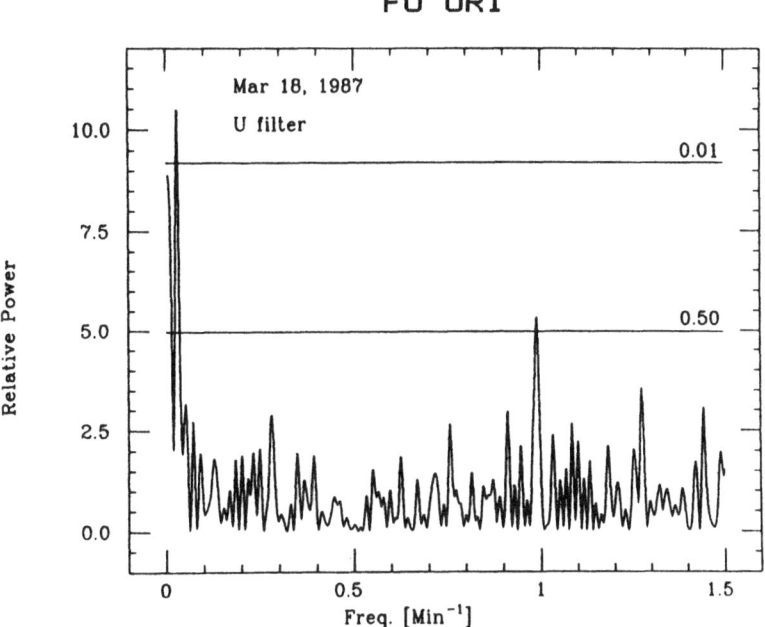

**Figure 2:** Fourier spectrum of the first night.

FU ORI

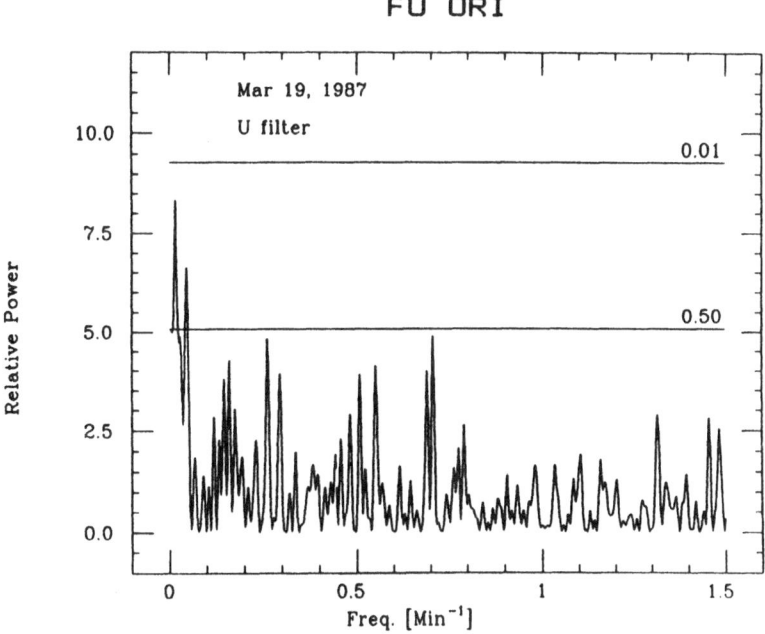

**Figure 3:** Fourier spectrum of the second night.

# MODELS FOR INFRARED EMISSION FROM GALAXIES

Michael Rowan-Robinson

*Astronomy Unit, Queen Mary College*
*Mile End Rd, London E1 4NS*

ABSTRACT.    Current ideas about the nature of interstellar dust in galaxies are reviewed, with a strong emphasis on the nature of the very small grain component needed to explain the mid-infrared diffuse emission and unidentified infrared features. Models for the infrared spectra of galaxies are reviewed and the evidence that most of the radiation in star-forming regions is being absorbed by a high visible-uv optical depth of dust is summarised.   The evidence for destruction of very small grains in regions of high radiation intensity is discussed.

A new model for interstellar grains in galaxies is presented, based on a revised version of the model of Rowan-Robinson (1986) and is compared to observed far infrared colour-colour diagrams and to far infrared spectra of galaxies which have been mapped at 800 μm by Hughes et al (1989).   Work on far infrared and submillimetre mapping of galaxies is reviewed, as also is recent work on infrared emission from ellipticals and lenticulars.   The determination of dust mass in galaxies is briefly discussed.

## 1. INTRODUCTION

My task in reviewing interstellar dust in galaxies is greatly simplified by the appearance of several excellent review articles on this area during the past year or so. Although each covers only a specific aspect of the subject, together they comprise a good introduction to our current knowledge.

A comprehensive review of infrared emission from our Galaxy, with much historical background, has been given by Cox and Mezger (1989).   They emphasize that the results from IRAS have led to a major reappraisal of estimates of the fraction of the infrared emission from our Galaxy which comes from interstellar dust illuminated by the interstellar radiation field, as opposed to regions of massive star formation.   The latter are now believed to contribute only about 10% of the total infrared emission from the Galaxy. Boulanger and Perault (1988) have given an authoritative discussion of the infrared emission observed by IRAS from the different components of diffuse emission from our Galaxy,  and the correlations between them, which must be

197

*E. Bussoletti and A. A. Vittone (eds.), Dusty Objects in the Universe, 197–226.*

the starting point for any analysis of the interstellar dust in normal galaxies. A general review of the IRAS view of the extragalactic sky has been given by Soifer et al (1987). Telesco (1988) has reviewed enhanced star formation and infrared emission in the centres of galaxies, with a strong emphasis on imaging and spectroscopic data derived from ground-based studies. Helou (1988) has reviewed the far infrared emission from Galactic and extragalactic dust seen by IRAS, emphasizing the similarity in the range of far infrared colours seen in external galaxies and in reflection nebulae in our Galaxy. Roche (1988) has given an interesting summary of the results from near and middle infrared spectroscopy of galaxies. Puget and Leger (1989) have given a very thorough review of the evidence for small grains and large aromatic molecules in the interstellar medium of our own and other galaxies. Finally Draine (1988a) has reviewed interstellar extinction in the infrared.

In this review I shall concentrate on those areas where major controversy exists and where significant progress may be expected in the next few years. The topics I have selected are grain models, first attempts to explain the infrared spectra of IRAS galaxies, the destruction of the very small grain component, a new picture of interstellar dust in galaxies, results from far infrared and submillimetre mapping of galaxies, determination of the dust mass in galaxies, and dust in ellipticals and lenticulars.

## 2. GRAIN MODELS

Classical grain models consisting of silicate and carbon grains of radius 0.01-0.1 $\mu$, for example those of Mathis et al (1977), Hong and Greenberg(1980), Draine and Lee (1984), Rowan-Robinson (1986), Tielens and Allamandola (1987), are successful in accounting for the observed visible and ultraviolet extinction curve and the emission longward of 60 $\mu$. However there is not yet a concensus on the grain properties longward of 300 $\mu$, as emphasized by Draine (1988a). I will discuss this further in section 5 below. The observations which the classical grain model definitely can not account for are (i) excess diffuse emission from the Milky Way at 2-20 $\mu$ (Price 1981, Boulanger et al 1985), (ii) 2-20 $\mu$ emission from reflection nebulae with colour temperature approximately independent of distance from the star (Sellgren 1984) and (iii) the broad features at 3.3, 6.2, 7.7, 8.6 and 11.3 $\mu$ seen ubiquitously in emission (Gillett et al 1973).

Current models for these three phenomena all involve the non-equilibrium response of very small particles to absorption of an ultraviolet photon (Greenberg 1968, Duley 1973, Allen and Robinson 1975, Purcell 1976, Andriesse 1978, Sellgren 1984, Draine and Anderson 1985). The main contenders are:
(A) Polycyclic aromatic hydrocarbons (PAH), which can be thought of as hydrogenated graphite platelets consisting of about 50 atoms (Platt 1956, Donn 1968, Leger and Puget 1984, Allamandola et al 1985, Puget and Leger 1989). To account for the full range of observed phenomena, Puget and Leger (1989) have to include also a very small carbonaceous grain (VSG) component with radii in the range 0.0015-0.01 $\mu$. Fig 1a shows how some particular examples of PAHs can give at least

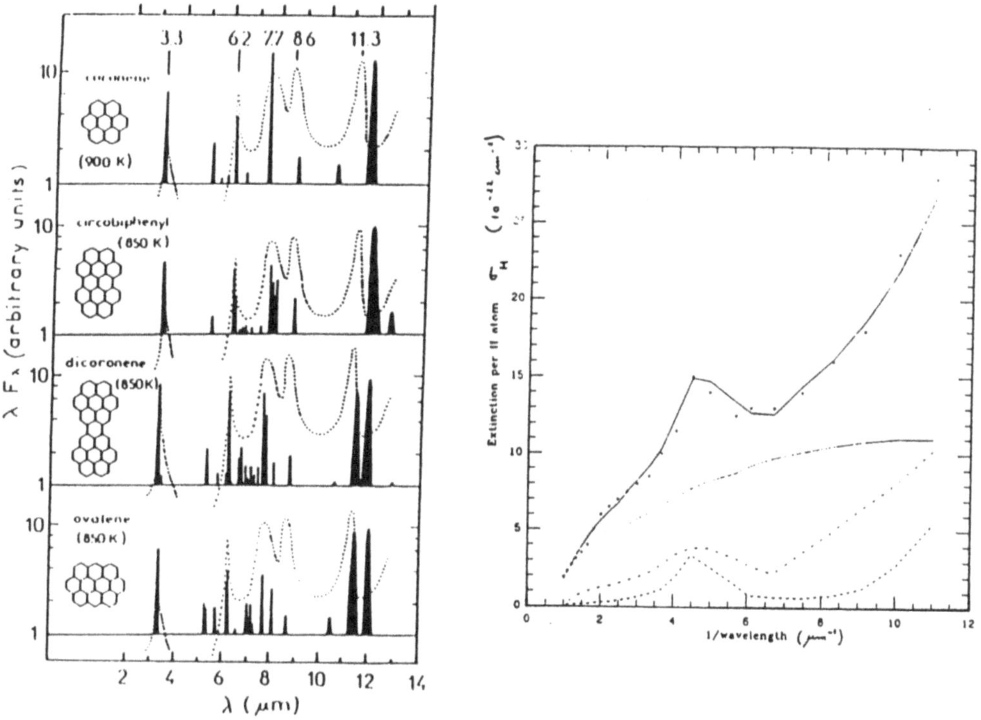

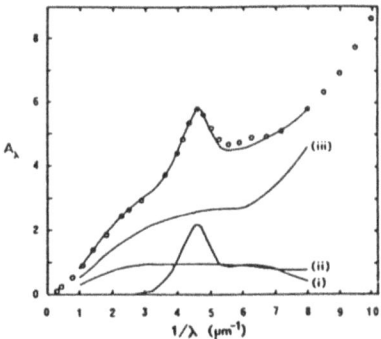

Fig 1: (a) Emission spectra of several PAHs calculated from their laboratory absorption spectra, compared with observations of the reflection nebula NGC2023. (b) Fit to the interstellar extinction curve.(Puget and Leger 1989)

Fig 2: Fit by Jones et al (1987) to the interstellar extinction curve in the visible and ultraviolet.

qualitative agreement in the wavelengths of (most of) the 3-12 μ broad-band features (Puget and Leger 1989). Fig 1b shows Puget and Leget's fit to the interstellar extinction curve in the visible and ultraviolet.

(B)  Hydrogenated amorphous carbon (HAC), which can be thought of as poorly connected PAH islands in a larger structure (Duley and Williams 1981, 1988a,b, Duley 1987, Jones et al 1987, Williams 1989). They attribute the 0.22 μ feature to small silicate particles. Fig 2 shows their fit to the visible and ultraviolet interstellar extinction curve. Broad-band emission in the 0.6-0.9 μm region is attributed to luminescence from a diamond-like component in the HAC (Duley and Williams 1988b).

(C)  Quenched carbonaceous composite (QCC) has been proposed by Sakato et al (1983,1984). This material is made in the laboratory in a process intended to simulate the expanding atmospheres of carbon stars.

(D)  Amorphous aggregates of small particles of silicates, amorphous carbon and graphite (Mathis and Whiffen 1989). These authors show that the optical properties of an aggregate can be significantly different from a simple sum of the ingredients in the aggregate. Fig 3a illustrates the appearance of the Mathis and Whiffen composite grains, Fig 3b shows their fit to the visible and ultraviolet extinction curve and Fig 3c shows the properties of their grains at 1-1000 μ.

It is clear that in the aggregate models (B-D), the very small grain component must retain its thermodynamic identity in order to explain the phenomena (i-iii) above. From the point of view of understanding infrared emission from dust, it may therefore be academic whether the very small grain component is integrated into a larger structure or not, since this integration must be so weak as to leave the specific properties of the component intact. Puget and Leger (1989) in fact query whether aggregate grain models can localize the energy of an incident photon for the several seconds required for infrared emission.

Draine (1988b) has reviewed the variety of models which have been put forward specifically to explain the 0.2175 μ feature. The models which he considers are graphite, nongraphitic carbonaceous solids, $OH^-$ on small silicate grains, PAH, small MgO or CaO particles, dessicated microorganisms, radiation-damaged $SiO_2$, charge transfer on Si, Fe or Mg, and finally the absorption edge in silicate grains. He concludes that only two are consistent with all the available observations, graphite or $OH^-$ on small silicate grains, and he notes that the latter hypothesis is less well developed than the graphite hypothesis.

In section 5 below I shall try to pull together some of these ingredients into a simple but comprehensive picture for interstellar dust in galaxies.

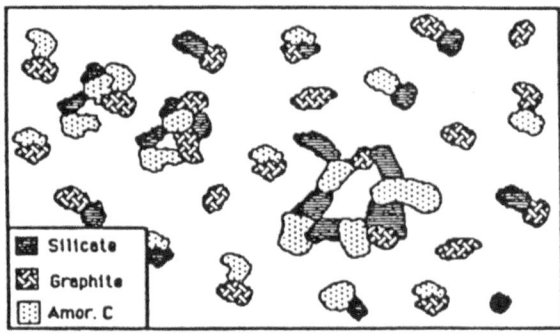

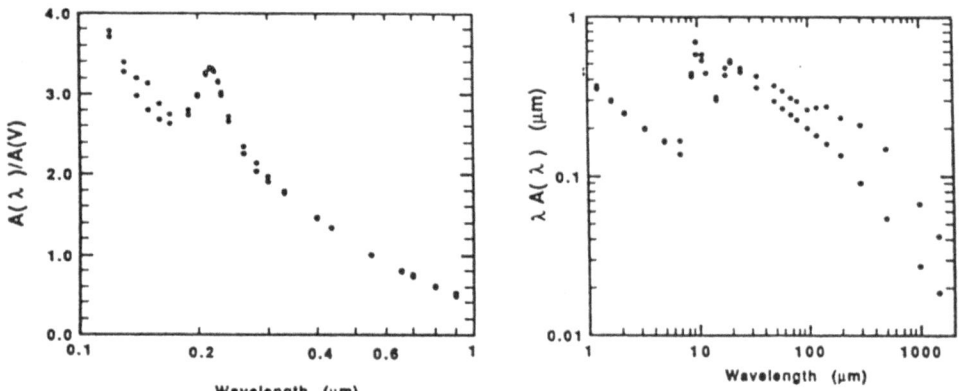

Fig 3: (a) Schematic picture of the Mathis & Whiffen (1989) grain model.
(b)  Their fit to the ultraviolet and visual interstellar extinction curve.
(c) The same for infrared wavelengths. (filled circles:calculated, open circles:
observations)

## 3. FIRST ATTEMPTS TO EXPLAIN THE INFRARED SPECTRA OF IRAS GALAXIES

Models for IRAS galaxy spectra have been reviewed by Rowan-Robinson (1987 a,b) and Helou (1988). The first model proposed was a simple 2-component model consisting of warm (50 K) dust in molecular clouds/HII regions and cool (20 K) dust in the interstellar medium heated by the interstellar radiation field (de Jong et al 1984). This model has been developed further by de Jong and Brink (1987) and has been criticized by Eales and Devereux (1989). The model is rather similar to that proposed by Cox and Mezger over a number of years (see Cox and Mezger 1989).

Helou (1986) proposed an extension of this model in which the warm component becomes a one-parameter family, with the heating intensity as the parameter. As the intensity increases from that found in the solar neighbourhood to the much higher value found in star-forming regions, the dust temperature increases from 20 to 50 K. More recently, Helou (1988) emphasizes the similar range of IRAS colours found in galaxies and in Galactic sources. Fig 4a shows $\log\{$ S(60)/S(100)$\}$ versus $\log\{$ S(12)/S(25)$\}$ for IRAS galaxies and Fig 4b shows the same diagram for Galactic sources. The sequence of colours found in the reflection nebulosity surrounding $\xi$ Per by Boulanger et al (1988) with increasing distance from the star is also shown. This appears to be telling support for Helou's hypothesis that the variation of colour is simply due to variation of the heating intensity experienced by the grains.

Rowan-Robinson and Crawford (1986,1989) have also used the analogy with Galactic sources to derive a rather different model for IRAS galaxy spectra. They propose that the galaxy spectra are a mixture of three components, the general disc emission of the galaxy consisting of reradiation of the interstellar radiation field absorbed by interstellar grains (Fig 5a), a component present in Seyferts peaking at 25 $\mu$m due to dust in the narrow-line region, and a starburst component with a spectrum similar to that for Galactic compact HII regions. Their models for the latter (Crawford and Rowan-Robinson 1987) are optically thick at visible and ultraviolet wavelengths, with $A_V \approx 20$ (they are optically thin in the far infrared, of course). Fig 5b compares their starburst model spectrum with the Telesco et al (1984) spectrum of the NGC1068 starburst component and with the average spectrum for Galactic compact HII regions/regions of massive star formation derived by Rowan-Robinson (1979). Confirmation of the fact that most of the massive star formation in galaxy starbursts takes place at high visible-uv optical depth comes from a comparison of the 60 $\mu$ luminosity of a large sample of IRAS galaxies with their H$\alpha$ luminosity (Leech et al 1988, Fig 5c). Ratios of these luminosities range from 200-4000, compared with 30-100 for the nearby normal galaxies studied by Persson and Helou (1987). The H$\alpha$/H$\beta$ ratios for these IRAS galaxies indicate values for $A_V$ of only a few, so the bulk of the far infrared radiation must come from stars whose visible light is heavily extinguished, while the H$\alpha$ radiation must come from near the surface of the star-forming volume (Leech et al 1989). Further evidence for high visual extinction comes from the Brackett-alpha and -gamma observations of Kawara et al (1989) for a sample of starburst galaxies. From these they infer values for $A_V$ in the range 7-

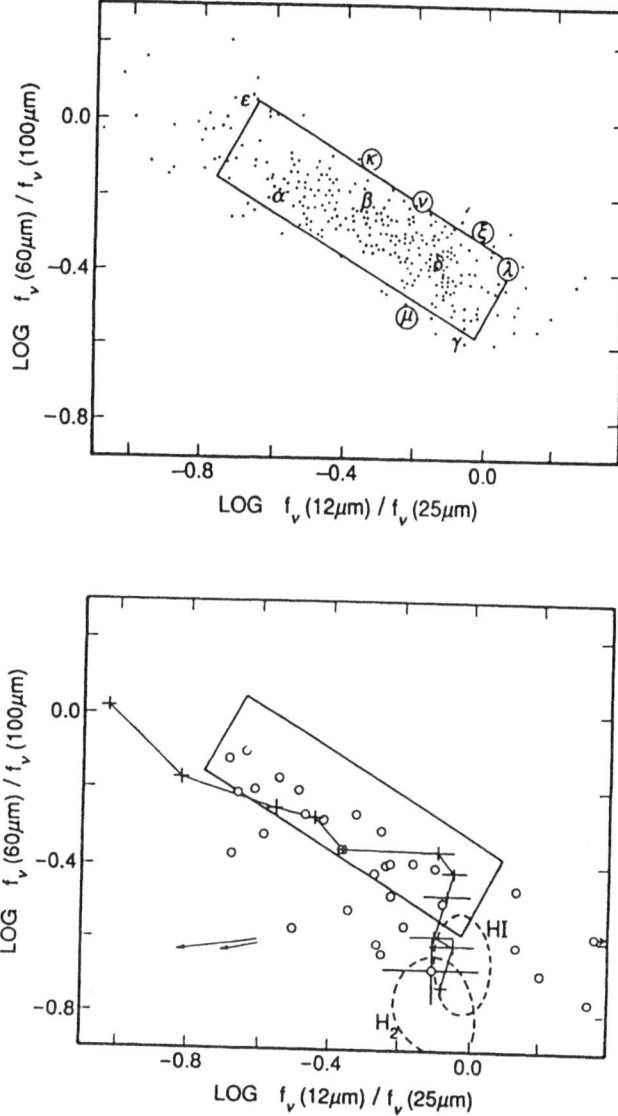

Fig 4:   IRAS colour-colour diagrams for (a) galaxies, (b) Galactic star-forming regions (Helou 1988).  In (b) the crosses denote data for ξ Per.

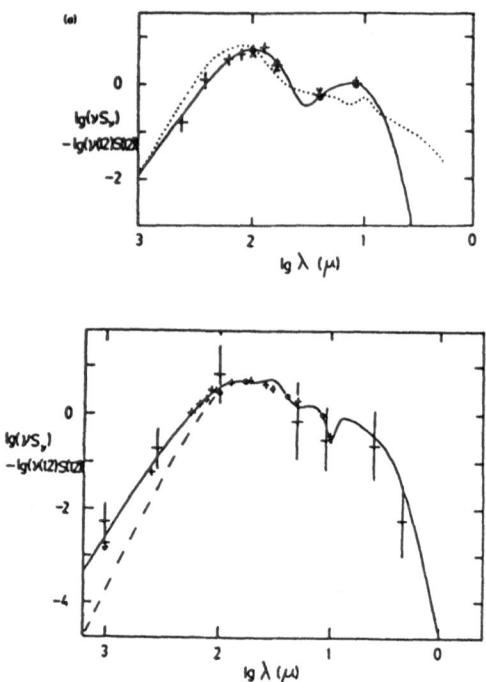

Fig 5: Models by Rowan-Robinson & Crawford (1989) for (a) the cirrus and (b) the starburst components in galaxy spectra. The broken curve in Fig (b) shows the effect of changing the wavelength at which the grain absorption efficiency steepens to 80 μm.

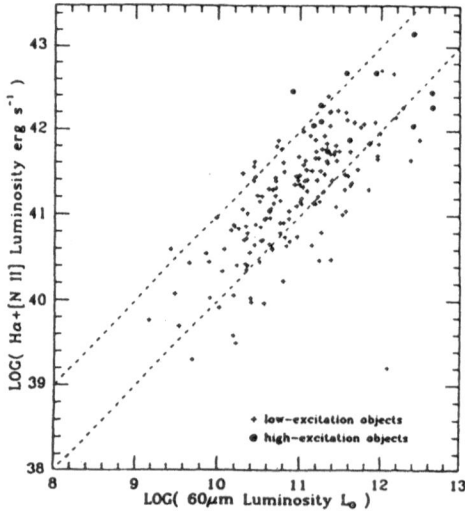

Fig 5c: H-alpha luminosity versus infrared luminosity for sample of IRAS galaxies (Leech et al 1988). The broken lines correspond to L(60μm)/L(H-alpha) = 400 and 4000.

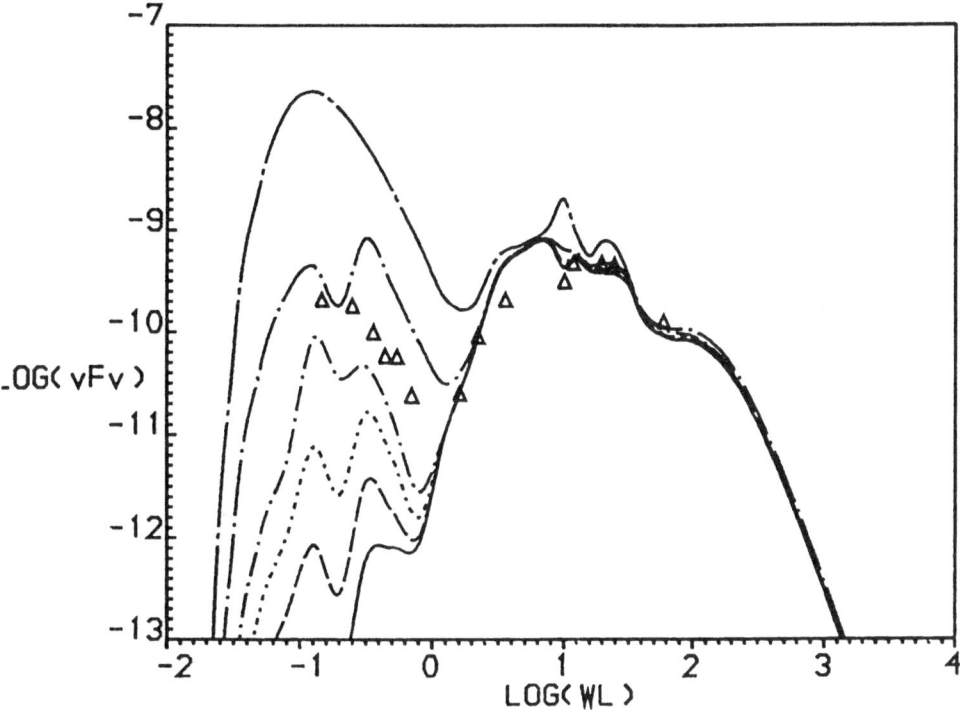

Fig 6: Sequence of flared disc models for the narrow-line region of NGC4151, as a function of the viewing angle, from face-on (top) to edge-on (bottom) (Efstathiou & Rowan-Robinson 1989)

33. These values are in agreement with those inferred from the depth of the 10 μm silicate feature in these galaxies.

Once we are dealing with dust clouds with $A_V \gg 1$, then the illumination geometry becomes of critical importance for models of the infrared spectra. Evolved HII regions in our Galaxy show strong deviations from spherical geometry, often displaying a blister geometry, although it is possible that for young compact HII regions spherical symmetry is a reasonable approximation (Rowan-Robinson 1982, Crawford and Rowan-Robinson 1987). Efstathiou and Rowan-Robinson (1989) have developed an accurate radiative transfer code for axially symmetric dust clouds. Fig 6 illustrates the crucial importance of the aspect angle when viewing a non spherically-symmetric system. Leisawitz (1989) has also studied the role of non-spherical geometry in star-forming regions.

An improved model for IRAS galaxy spectra, which is essentially a fusion of the approaches of Helou and of Rowan-Robinson and Crawford, will be described in section 5.

### 4. THE DESTRUCTION OF VERY SMALL GRAINS

In the past two years several lines of evidence have begun to point towards the destruction of very small grains in regions of very high uv radiation intensity. The most direct evidence comes from infrared spectroscopy. Roche (1988) and Desert and Dennefeld (1988) have shown that the broad 3-12 μ features attributed to very small grains are absent in the spectra of many Seyfert galaxies (Fig 7a). Destruction of very small grains is also presumably the reason that Rowan-Robinson and Crawford (1989) found that the disc component was very weak or absent in many Seyferts (Fig 7b).

Reasonably direct evidence for the destruction of very small grains in a high radiation intensity comes from the decline in the ratio of S(12)/S(100) near hot stars. Ryter et al (1987) showed this effect for σ Sco and Boulanger et al (1988) showed it for ξ Per.

Telesco et al (1989) argue that a similar effect is seen in the centre of M82. Fig 8a shows the increase in S(25)/S(12) with increasing uv intensity found by Telesco et al for M82 superposed on the curve derived from Boulanger et al's observations of ξ Per. However the spectrum of the emission from outside the nucleus of M82 (and of the integrated emission from the galaxy) is very similar to that for the NGC1068 starburst, and for compact Galactic HII regions, shown in Fig 5b, and one would normally assume that the bulk of this emission arises in regions where the visible and ultraviolet optical depth is >> 1. The 10 μm emission from such a cloud does not arise from very small grains. Fig 8b shows the integrated spectrum of M82 compared to the optically thick starburst model of Rowan-Robinson and Crawford (1989): the agreement is good. Also shown is the shape of the spectrum of the central region of

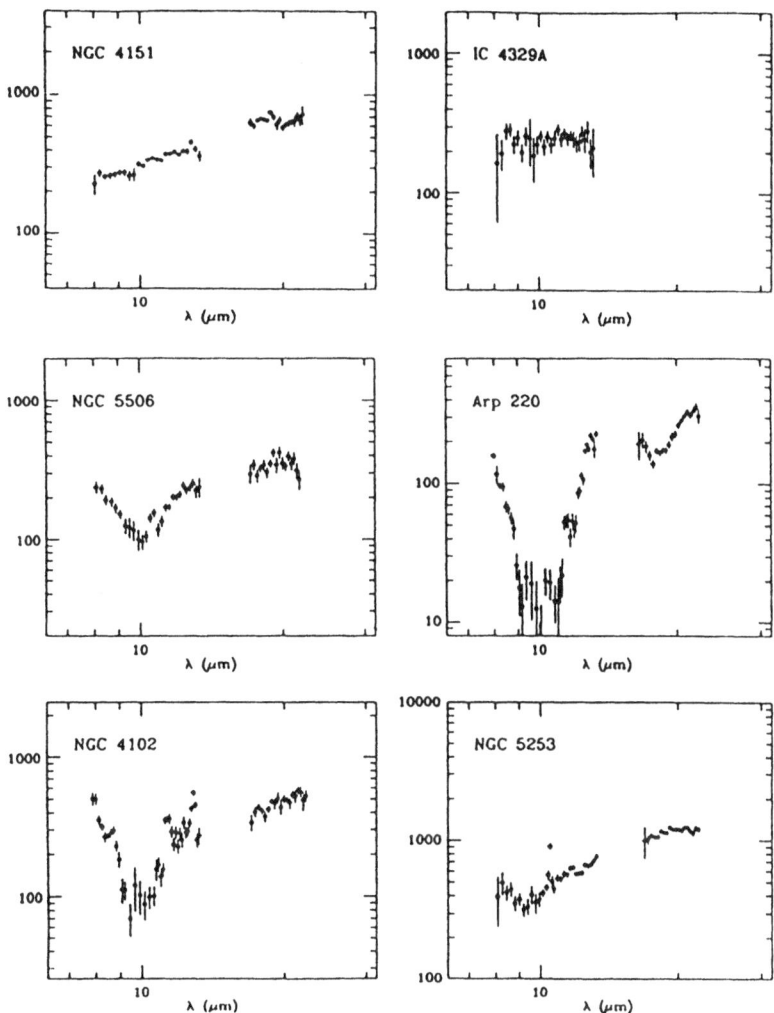

Fig 7: (a)   8-13μm  and  17-22μm  spectra  of  six  galaxy  nuclei.    Note  that  the unidentified  ir  features  are  completely  absent  from  the  Seyferts    NGC4151  and IC4329A.

208

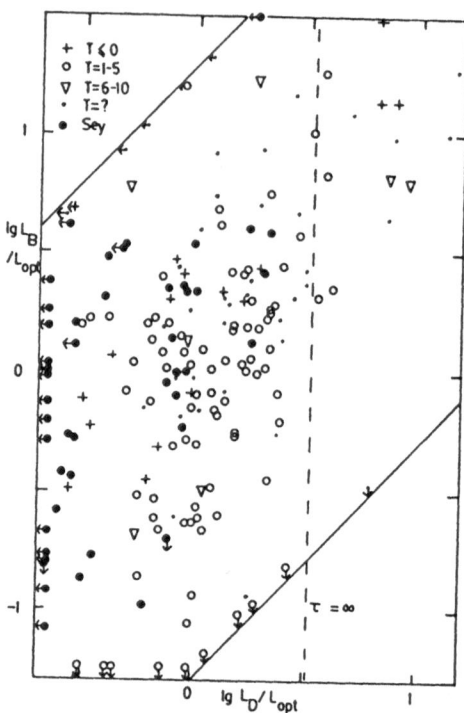

Fig 7(b)  Ratio of infrared luminosity in starburst component to optical luminosity, versus ratio of infrared luminosity in cirrus component to optical luminosity for IRAS galaxies (Rowan-Robinson & Crawford 1989).  The Seyferts (filled circles) are deficient in the cirrus component.

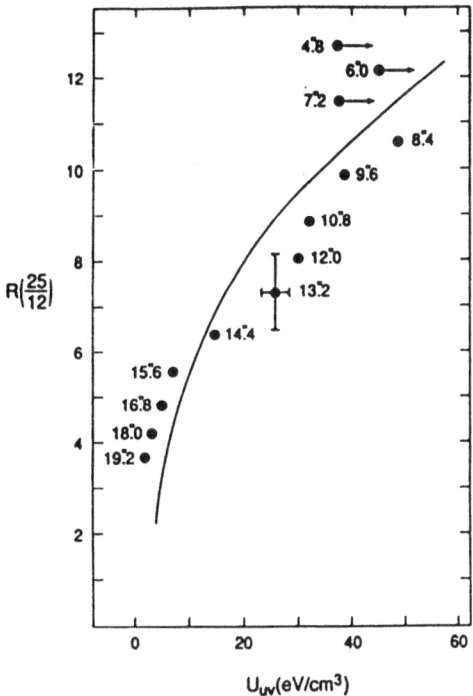

Fig 8a: Variation of 25/12 μm colour ratio with intensity of radiation field in centre of M82 (filled circles, Telesco et al 1989)) compared with relation found in ξ Per by Boulanger et al (1988).

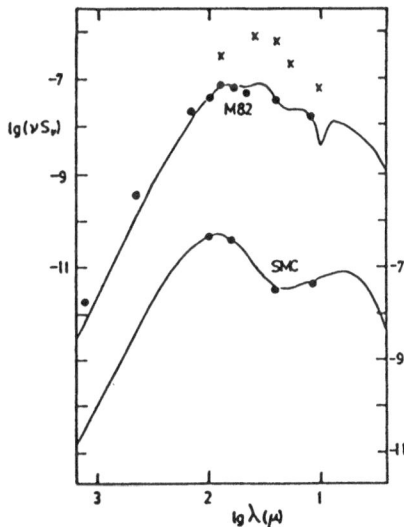

Fig 8b. Top: Integrated spectrum of M82 (filled circles) compared with starburst model (data from Telesco 1988, Smith et al 1989a). The crosses (arbitrary vertical scale) show the relative shape of the spectrum of the core of M82. Bottom: Integrated spectrum of the SMC compared with cirrus model (X=30) in which abundance of 5 Å grains has been reduced by 2/3rds.

M82, derived from the colours measured by Telesco et al (1989). The change in spectrum towards the centre of M82 is essentially a shift of the emission peak from 80 μm to 60 μm, presumably due to the increase in intensity of the radiation from the starburst towards to nucleus. It seems unlikely that we are seeing emission from optically thin dust ( the ratio of Brackett-alpha to -gamma gives a value for $A_v$ of 14 for M82 (Kawara et al 1989) ) and hence the analogy with ξ Per appears to be spurious.

Similarly unconvincing evidence comes from the far infrared colours of galaxies (Pajot et al 1986, Gosh & Drapatz 1987, Helou 1988). Here again the problem is confusion with the role of the optically thick starburst component, for which, in the model of Rowan-Robinson and Crawford (1989), $S(12)/S(60) = 0.04$ , but radiative transfer effects in normal 0.01-0.1 μm dust rather than small grain depletion is the cause. Fig 4b above showed Helou's (1988) compilation of the IRAS colours of compact Galactic HII regions and of galaxies superposed on the range of colours seen in ξ Per by Boulanger et al (1988). The agreement is good, but in my view this is fortuitous in the case of Galactic HII regions and galaxies dominated by starbursts since in most cases the optical depth in these sources is high and the analogy with ξ Per therefore of doubtful significance. If the 60/25 μm colour ratio, ignored by Helou, is also considered, the agreement with ξ Per is less impressive. However the case of the Small Magellanic Cloud (Schwering 1988) is convincing because the spectrum of this galaxy does indeed look like cirrus in which the smallest grain component is depleted (see Fig 8b).

In an interesting development, Leene and Cox (1987) have found that the 0.22 μ feature is also suppressed in regions of high radiation intensity, which suggests that this feature is associated with the very small carbonaceous grains responsible for the broad features and diffuse emission at 2-20 μ .

5. TOWARDS A NEW PICTURE OF INTERSTELLAR DUST IN GALAXIES

If we concentrate first on the 'cirrus' component in galaxies, the reradiation by interstellar dust of the energy absorbed from the interstellar radiation field, then it is clear that a satisfactory model involves a number of ingredients. Firstly a multiple (or aggregate ?) grain model is required to account for the interstellar extinction curve and it must incorporate very small grains and/or PAH. Secondly the model must allow for the fact that there is a range of heating intensities within galaxies and from galaxy to galaxy. For our Galaxy and a few other nearby galaxies we may hope to study how the observed spectrum varies with heating intensity. For more distant galaxies for which we have only the integrated spectrum we have to make do, for the moment, with a characteristic heating intensity. Let me define $X = I/I_{isrf}$ , where I is the intensity in the region under consideration and $I_{isrf}$ is the intensity of the interstellar radiation field in the solar neighbourhood, which I assume to be as characterised by Mathis et al (1983). Finally we may have to allow for the fact that for $X >$ some critical value, the very small grain component starts to be destroyed.

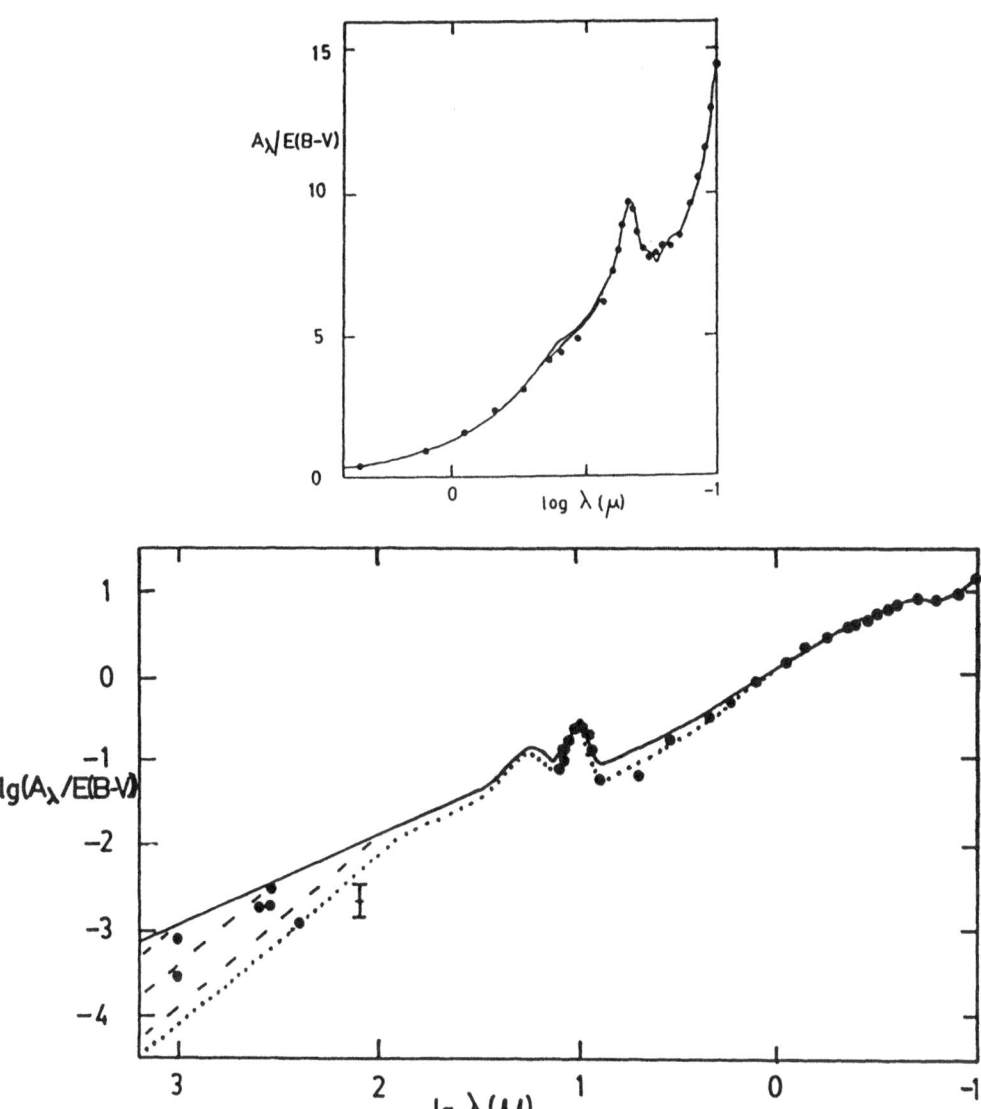

Fig 9: Fit to interstellar extinction curve for model described in section 5
(a) at visible and ultraviolet wavelengths (upper curve: Rowan-Robinson 1986,
lower curve: revised model (differs only near 0.4 μm) (b) in the infrared (solid and
broken curves: Rowan-Robinson 1986, dotted curve: revised model). References for
observations are given in Rowan-Robinson (1986).

Models which satisfy the first two of these requirements were presented by Draine and Anderson (1985). Bernard and Desert (1989) have given some details of work which satisfies all three requirements. Here I give some results from an extension of my earlier interstellar grain model (Rowan-Robinson 1986), which is intended to be the simplest possible model that fits all the present observational data. The model retains the 6 grain types of the earlier work, with some modifications: (i) 0.1 μ amorphous carbon grains, their optical properties derived from circumstellar dust shells around carbon stars. The absorption efficiency of these has been reduced by a factor of 1.5 at wavelengths > 0.4 μ to improve the fit to the interstellar extinction curve at 5-9 μ, while retaining the same total extinction at wavelengths < 1 μm. This also has the effect of increasing the visible and ultraviolet albedo to a more acceptable value of 0.7. (ii) 0.1 μ amorphous silicate grains, their optical properties derived from circumstellar dust shells around M stars. (It is worth noting that 50% of the mass of carbon and 80% of the mass of silicon in interstellar grains is in the form of these larger amorphous grains. We see them being manufactured in situ. We know that this is where the bulk of interstellar grains were last made.) (iii) 0.03 μ graphite grains, (iv) 0.03 μ silicate grains, (v) 0.01 μ graphite grains, (vi) 0.01 μ silicate grains, all four types with properties as given by Draine and Lee (1984). These components are required to explain the interstellar extinction curve in the ultraviolet and the 0.22 μ feature. The main difference from the earlier model is that the mass in 0.01 μ graphite grains is now redistributed between 0.01 μ grains, 0.002 μ (20 Å) grains and 0.0005 μ (5 Å) grains. The absorption and scattering properties of these latter two species are assumed to be the same as the 0.01 μ grains at wavelengths > 0.1 μ, but because they are so small they will not be in equilibrium with the incident radiation field. Instead we have to assume that they have a certain probability $p(T)$ $dT$ of having a temperature between $T$ and $T + dT$. The emission spectrum from these grains then has to be calculated from

$$I_\nu = \int Q_\nu \ B_\nu(T) \ p(T) \ dT \qquad\qquad (1)$$

The calculation of $p(T)$ is a complex matter but has been carried out by Draine and Anderson (1985) for the grain properties of Draine and Lee (1984) adopted here (see also Guhathakurta and Draine 1989). Their results can be approximated analytically as

$$p(T) = k \ T^{-b} \quad \text{for } T_1 \leq T \leq T_2, \qquad\qquad (2)$$

where $b = 2.75$, and $k = 6.68$, $T_1 = 2.7$ K, $T_2 = 500$ K, for the a= 20 Å grains, and $k = 0.168$, $T_1 = 2.7$ K, $T_2 = 80$ K, for the a = 5 Å grains. Here I am assuming that the very small grains emit the bulk of their radiation as a continuum. Roche (1988) estimates that galaxies emit 1% of their energy in the form of unidentified features and as only 10% of the energy of galaxies is emitted at 2-20 μ, we can infer that only about 10% of the radiation from very small grains emerges as the unidentified features. It will be relatively simple to incorporate these features into the calculation in future.

213

## TABLE 1: PARAMETERS FOR GRAIN MODEL

| type | $B_0$ | abundance % cosmic | | grain temperatures | | |
|---|---|---|---|---|---|---|
| | | $(x\ 10^{-4})$ | | X=1 | X=10 | X=500 |
| 0.1μ amor Si | 1.36 | 0.257 | 77 | 14.5 | 21.4 | 42.6 |
| 0.03μ Si | 1.10 | 0.063 | 19 | 16.8 | 24.5 | 47.0 |
| 0.01μ Si | 0.52 | 0.010 | 3 | 17.5 | 25.4 | 48.5 |
| 0.1μ amor C | 0.45 | 1.02 | 20 | 15.3 | 22.6 | 46.0 |
| 0.03μ C | 0.65 | 0.45 | 9 | 19.3 | 28.7 | 60.3 |
| 0.01μ C | 1.0 | 0.24 | 4.7 | 19.7 | 29.0 | 61.7 |
| 0.002μ C | 0.8 | 0.039 | 0.8 | | | |
| 0.0005μ C | 0.8 | 0.010 | 0.2 | | | |

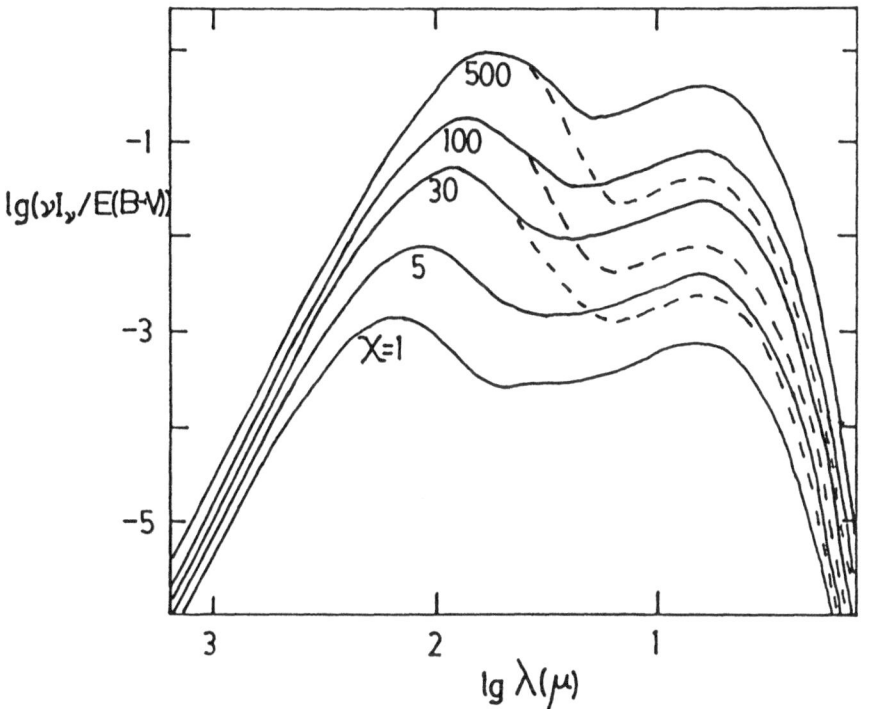

Fig 10: Predicted infrared emissivity of interstellar grains (ergs/cm$^2$/s/mag) as a function of the intensity of the radiation field. For X ≥30, the effect of a 90% destruction of the smallest grains (5 Å) is shown as broken curves.

In the earlier calculation (Rowan-Robinson 1986), I considered values of the wavelength at which the absrprtion efficiency of the 0.1 μ grains steepened from $Q_\nu$ $\propto \nu$ to $Q_\nu \propto \nu^2$ of 100, 316 and 1000 μ. In the present model I take this wavelength to be 80 μ, which is still consistent with the IRAS data for circumstellar dust shells (with the possible exception of IRC+10216, Rowan-Robinson et al 1986) and gives an acceptable fit to the data for high latitude dust clouds in our Galaxy. The possibility that $Q_\nu \propto \nu$ to a wavelength significantly larger than 100 μm is now completely ruled out by observations in our Galaxy and other galaxies. Fig 9a shows the fit to the interstellar extinction curve at visible and ultraviolet wavelengths. Fig 9b shows the overall fit at 0.1 -1000 μ. Fig 10 shows the predicted emission spectra for interstellar dust in the infrared for a range of heating intensities (the temperatures of the different grain components are given in Table 1). For $X \geq 30$, the effect of 90% depletion of the 5 Å grains is also illustrated. Fig 11a shows the predicted emissivity for grains immersed in the local interstellar radiation field compared with obervations of high latitude clouds. The agreement with observations is excellent both in the shape of the spectrum and in the absolute value of the emissivity. The 12-100 μm emissivity of the isolated cloud observed by Herter et al (1989) also agrees with that predicted in Fig 11a.

Fig 11b shows the corresponding prediction and observations for the central regions of the Galaxy (l <30°), where the intensity of the radiation field corresponds to $X = 5$: the fit is also satisfactory. Fig 12 shows the predicted IRAS colour-colour diagrams for $X = 1$-500 (colour-corrected as in Appendix A of Rowan-Robinson and Crawford 1989) compared with observations. Figs 12a and b show the data for the unresolved IRAS galaxies studied by Rowan-Robinson and Crawford (1989). Figs 12c and d show data for resolved IRAS galaxies mapped by Rice et al(1988) and Young et al (1989). Galaxies with log{S(60)/S(25)} $\leq 0.5$ need the additional Seyfert component peaking at 25 μm. For the resolved galaxies, most of which can be explained as pure disc (cirrus) emission, it can be seen that a range of heating intensities are present, from $X = 1$ for NGC205 to $X = 30$ for M33. The vast majority of the galaxies whose colours are shown in Figs 12 can be understood as a mixture of Seyfert (S) + starburst (B) + one of the cirrus models (curved lines). There are 3 classes of exception to this. (a) Two galaxies, NGC1569 and Arp 220 appear to lie on the locus of a highly extinguished starburst model. (b) Several galaxies, notably the SMC, lie to the right of the cirrus curve in the 25-60-100 μm colour-colour diagram and above and to the right of the cirrus curve in the 12-25-60 μm diagram, consistent with the effect of destruction of very small grains at high heating intensities. However not all galaxies with high heating intensity show evidence for very small grain destruction. M33, which like the SMC has a spectrum consistent with $X = 30$, appears to have a normal abundance of very small grains. It is also worth noting that there appear to be no galaxies in which the abundance of 5 Å grains is reduced by more than a factor of 10 compared to the solar neighbourhood. The possibility that reduction in the carbon abundance in galaxies (but not the silicon) is the cause of the anomalous colours needs to be explored, especially for the SMC. (c) Several galaxies, for example M31, have 25-60-100 μm colours consistent with cirrus but have very low values of S(25)/S(12), implying excess radiation at 12

215

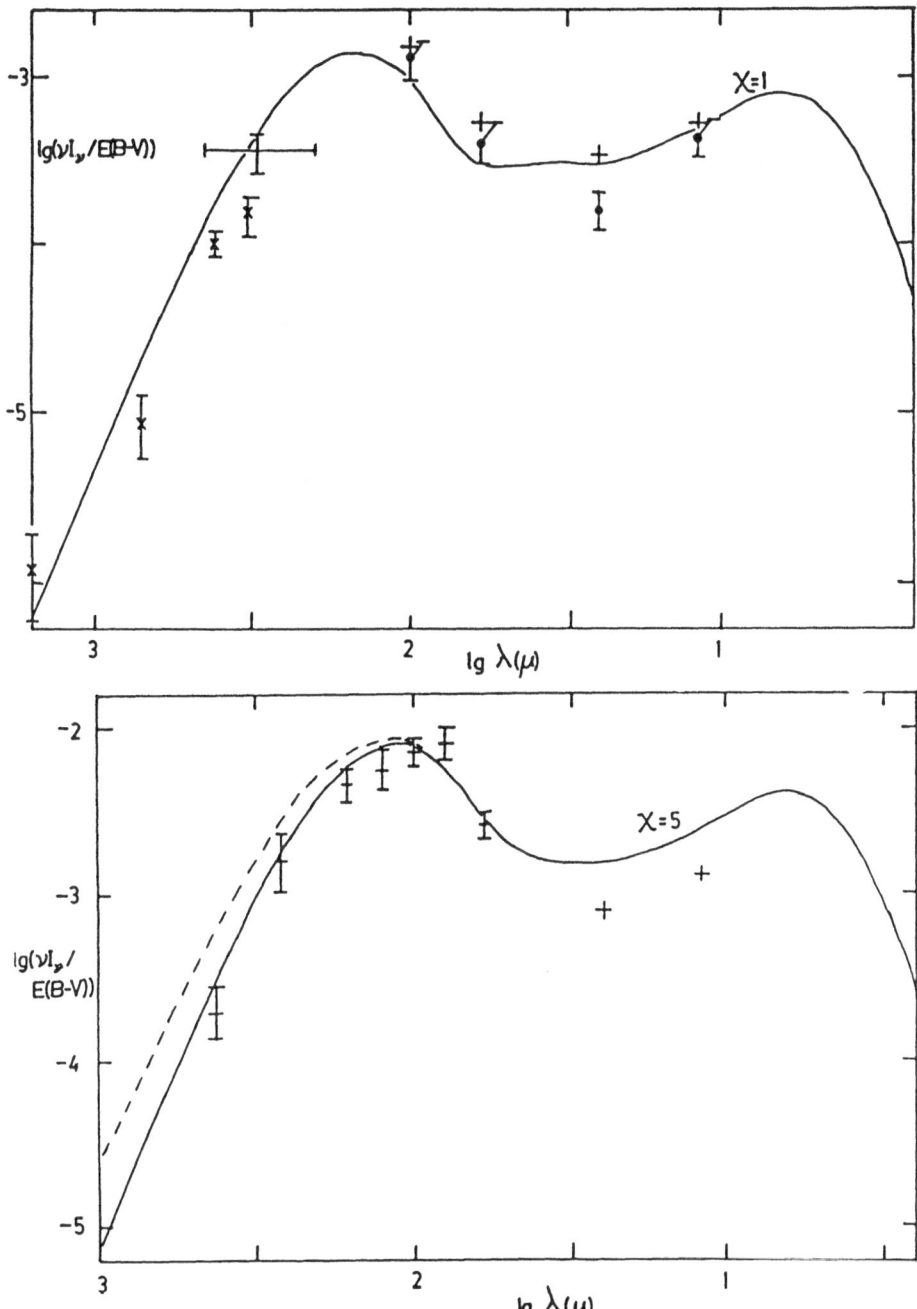

Fig 11: Predicted emissivity, compared with observations (a) towards the Galactic pole (X=1, data from Boulanger & Perault 1989, Halpern et al 1988, Fabbri et al 1988, assumed E(B-V)=0.05), (b) towards the central regions of the Galaxy (X=5, data from Beichman 1987 and refs therein, assumed E(B-V)=6.1). The broken curve shows the effect of assuming the wavelength at which the absoption efficiency of the grains steepens is > 1mm.

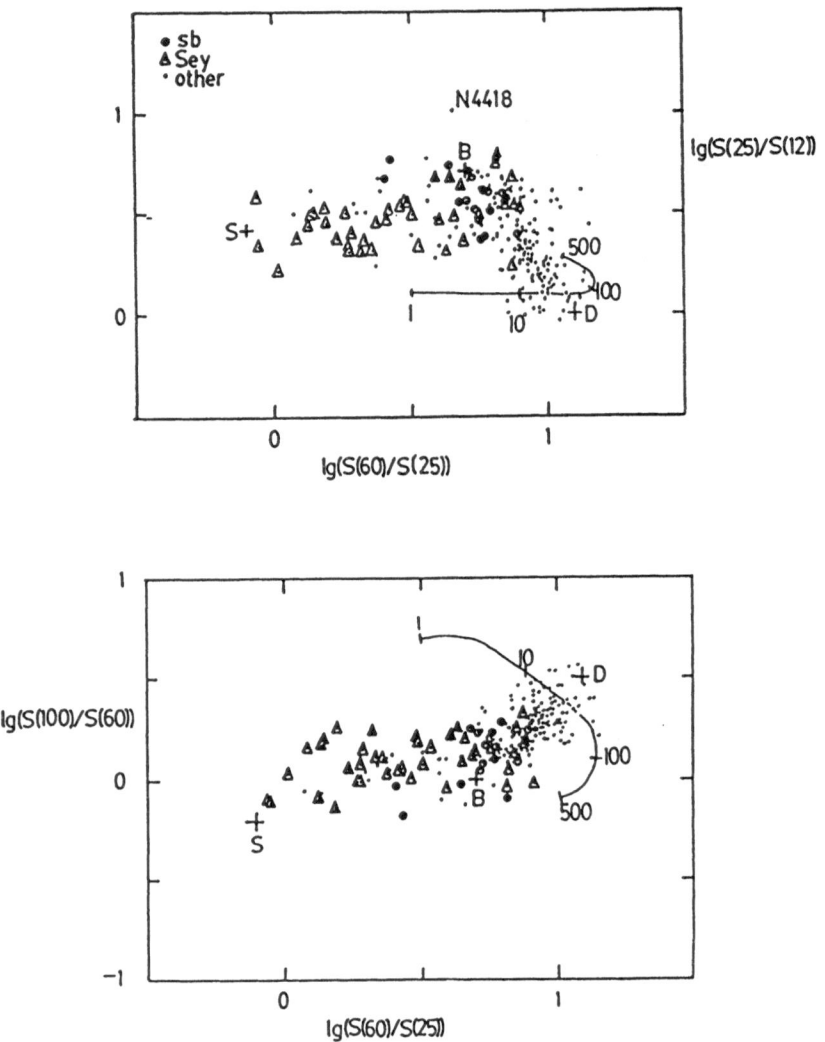

Fig 12: Predicted IRAS colour-colour diagrams for dust model of section 5. (a) and (b) Data for unresolved IRAS galaxies with good quality fluxes in all 4 bands from Rowan-Robinson and Crawford (1989). S, B denote the location of Seyfert and starburst components. The curved lines denote the cirrus models of section 5, labelled with the value of X.

217

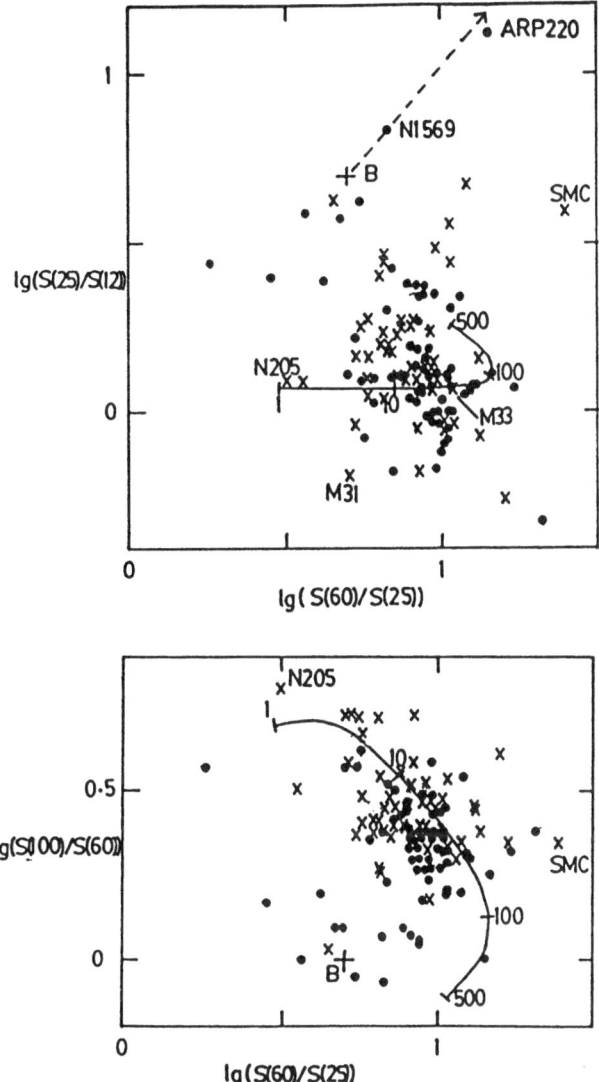

Fig 12 (c) and (d). Data for resolved IRAS galaxies with coadded fluxes in all four bands from Rice et al (1988, crosses) and Young et al (1989, filled circles). Only galaxies with fluxes brighter than 0.4 Jy in all four bands were included. The broken line in Fig 12c shows the effect of reddening ($A_V = 40$) on the starburst component.

μm. Possible explanations of this are a strong contribution from circumstellar dust shells (Soifer et al 1986, Rowan-Robinson and Chester 1987: though for M31 the spatial distribution of the 12 μm radiation does not differ from that at longer wavelengths) or an unusually strong contribution from PAH/very small grains. 8-13 μm spectroscopy of these galaxies would be very valuable.

Discrepant colours occasionally result from poorly determined fluxes, especially at 25 μm where not all IRAS detectors were functioning. For this reason galaxies with fluxes lesss than 0.4 Jy in any band were omitted from Fig 12. However in general experience suggests that IRAS colours are accurate to 0.1 in $\log_{10}$ and that any discrepancy greater than this has a real cause.

## 6. FAR INFRARED AND SUBMILLIMETRE MAPPING OF GALAXIES

Prior to the launch of IRAS rather little information on the spatial extent of far infrared emission in galaxies was available. Some of the earlier work has been reviewed by Telesco (1988). One of the most significant pre-IRAS studies was by Smith (1982), who produced a 170 μ map of the disk of M51, which showed that the bulk of the far infrared emission in M51 is produced by dust associated with the diffuse gas in the disk of the galaxy.

IRAS extended data is still under active study by several groups. Detailed maps have been produced of M31 (Habing et al 1986, Soifer et al 1987 and Walterbos and Schwering 1987), M33 (Rice et al 1989) and of the Magellanic Clouds (Schwering 1988). Rice et al (1988) have published coadded IRAS maps for all galaxies with optical extent greater than 8'. Higher resolution images may be expected for many of these galaxies from the use of maximum entropy and other deconvolution techniques now under active study at IPAC and elsewhere (eg Canterna et al 1989).

Subsequent studies have for the most part concentrated on wavelengths longer than 100 μ. Stark et al (1988) have mapped 4 Virgo spirals at 160 and 350 μ and shown that that there is no evidence for a grain component whose emission peaks beyond 200 μ, a prediction of grain models with emissivity $Q_\nu \propto \nu$ at wavelenths > 100 μ. Eales and Wynn-Williams (1989) have measured 350 and 450 μ fluxes at locations centred on several galaxies with a 100" beam. 160 μm maps have been published of NGC4449 (Thronson et al 1987), NGC4214 (Tnronson et al 1988), NGC 4485 and 4490 (Thronson et al 1989a) and NGC1569 and 3593 (Hunter et al 1989). Thronson et al (1989b) have published maps of IC10 at 95 and 160 μ and given fluxes for several other galaxies. Eckart et al (1989) have mapped Centaurus A at 50 and 100 μ, and have separated the cirrus and starburst components. Smith et al(1989a) have mapped M82 at 450 μm, Smith et al (1989b) have mapped M83 at 100 μm and Engargiola and Harper (1989) have mapped NGC6946 at 100, 160 and 250 μ. Hughes et al (1989) have mapped 8 IRAS galaxies at 800 μm with JCMT and given some 450 and 1100 μm data for some of them. The importance of the longer wavelengths is that the most reliable estimates of dust mass can be obtained at these wavelengths. Fig 13 shows far infrared spectra of selected galaxies from this latter study, with theoretical fits derived from the models described in section 5. Fig 14a-d

219

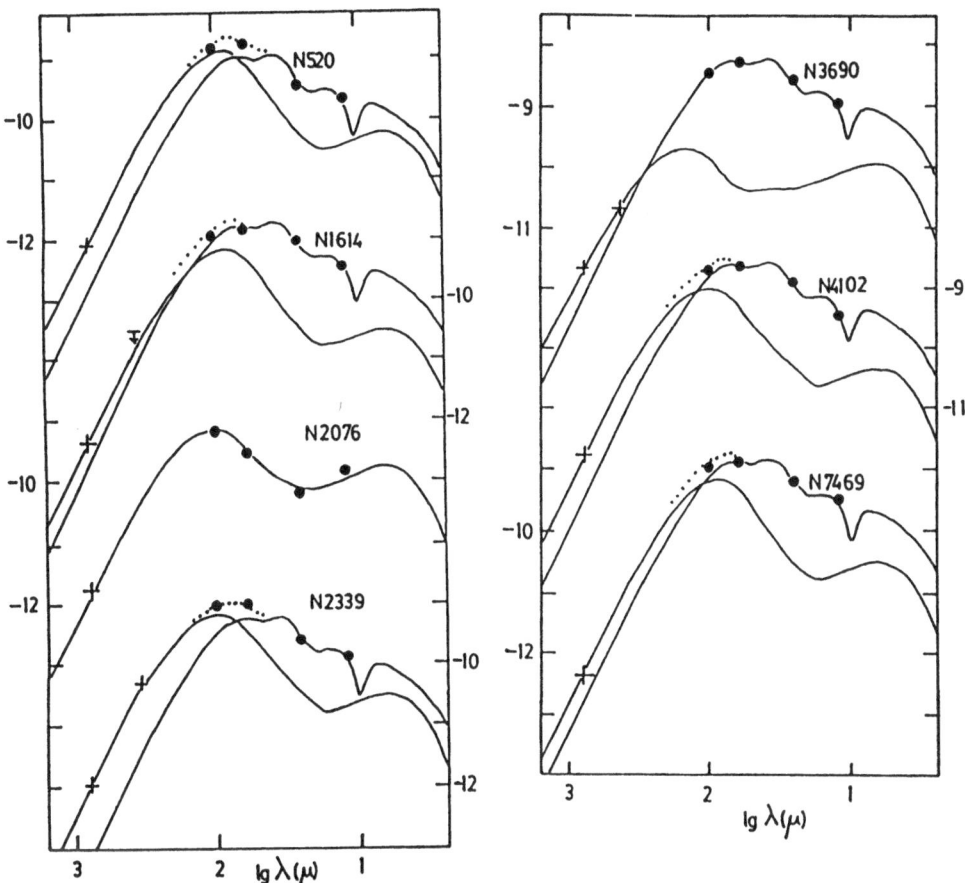

Fig 13: Far infrared and submillimetre spectra of galaxies mapped by Hughes et al (1989) at 800 μm, compared with models of section 5. For NGC 2076 a pure cirrus model, with depletion of the smallest grains, is satisfactory. For the other galaxies both cirrus and starburst components are required (the dotted curve indicates the total predicted flux). Parameters for the models are given in Table 2.

220

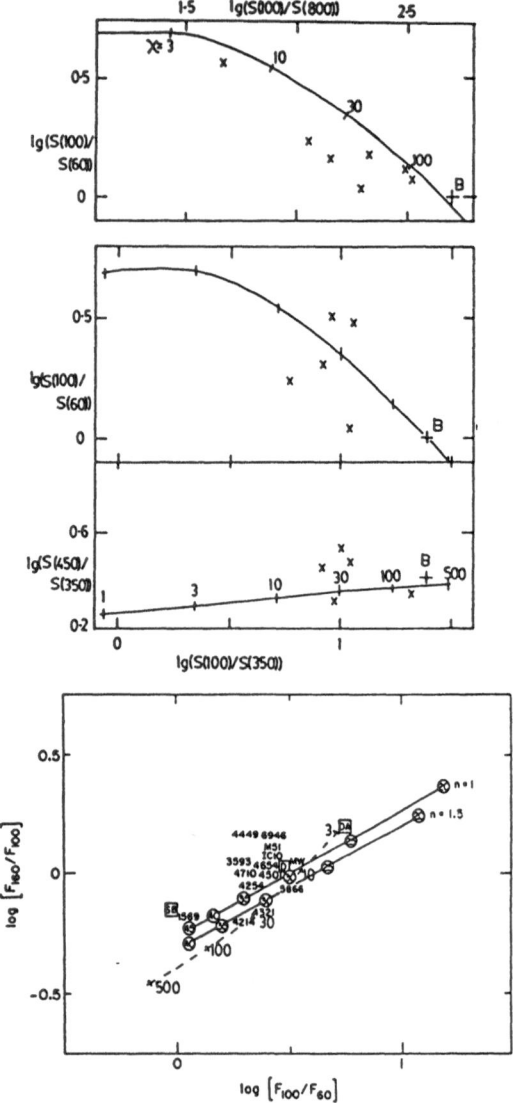

Fig 14: Far infrared colour-colur diagrams, compared with predictions of interstellar grain model of section 5. (a) 100/60 versus 100/800 (data from Hughes et al 1989), (b) 100/60 and (c) 450/350 versus 100/350 (data from Eales et al 1989). (d) 160/100 versus 100/60 (data from Thronson et al 1989b: solid curves, $\nu^n B_\nu (T)$ fits; broken curve, cirrus model of section 5).

show colour-colour diagrams derived from the work of Thronson et al (1989b), Hughes et al (1989) and Eales and Wynn-Williams (1989) compared with the predictions of the models of section 5.

## 7. DETERMINATION OF DUST MASS IN GALAXIES

Hildebrand (1983) gave a prescription for deriving dust masses in galaxies from far infrared data which has been widely used. Young et al (1989) have used the Hildebrand prescription to conclude that the average gas-to-dust ratio in galaxies is 1200. Draine (1989) has given a discussion of the derivation of dust masses which emphasizes some of the difficulties. He emphasizes that here is considerable disagreement about the grain opacity at long wavelengths, though this disagreement is somewhat exaggerated by illustrating the most extreme of the models discussed by Rowan-Robinson (1986) in which $Q_\nu \propto \nu$ all the way to 1 mm. Draine shows that if only IRAS observations are available of galaxies, then the derivation of dust mass is very uncertain, since several rather different models could in principle be fitted to the same observations. However provided a significant proportion of the 12-100 $\mu$m emission from a galaxy is due to cirrus, and fluxes are available in all four IRAS bands, a good separation into cirrus and starburst components can be made, and reasonable estimates of dust mass derived. Observations at long wavelength (>300 $\mu$m) are very valuable in tying down the value of X, the radiation field intensity, and are essential if the 12-100 $\mu$m spectrum is dominated by a starburst.

Table 2 gives dust masses derived from the study of Hughes et al (1989) based on the grain model of Rowan-Robinson (1989) described in section 5 above. Comparison of the dust mass in the cirrus component with the neutral hydrogen masses given by Young et al (1989) shows normal gas-to-dust ratios for this model in most cases. However since in many cases the neutral hydrogen in a galaxy extends well beyond the optical image, whereas the bulk of the infrared emission is generally located within the optical image, there may be a tendency to underestimate the total dust mass from far infrared observations. Dust in the outer parts of a galaxy, illuminated with a starlight intensity much lower than in the central regions, may contribute only a very small fraction of the total infrared flux. Sensitive observations at long wavelengths with large beam-throws will be needed to characterize such dust.

It is important when modelling the far infrared emission from dust in galaxies to take account of the fact that several grain components are present, at different temperatures. Calculations based on the assumption of a single composite grain model and a single temperature are unlikely to yield accurate results. However the cirrus models of Fig 10 can be approximately fitted at long wavelengths with a $\lambda^2 B(T)$ curve, with the values of T as given in Table 3 for different X. The validity of this fit is for $\lambda > 1700/T$ $\mu$m. Also given for these models are the values of log$\{S(100)/S(60)\}$ and log $\{M_d/S(100\mu m)$ $D^2\}$, log $\{M_d/S(800\mu m)$ $D^2\}$. Note that whereas $M_d/S_{100}$ $D^2 \propto X$, $M_d/S_{800}$ $D^2 \propto X^{0.3}$, so much more accurate dust masses can be obtained if long wavelength observations are available.

TABLE 2: PARAMETERS FOR GALAXIES MAPPED BY HUGHES ET AL (1989)

| galaxy | distance (Mpc) (H=50) | cirrus model | | | | starburst model |
| | | X | depletion of 5 Å grains | log $M_d$(C) | log M(HI) | log $M_d$(SB) |
|---|---|---|---|---|---|---|
| NGC520 | 45.4 | 30 | 90% | 7.25 | 10.10 | 5.94 |
| NGC1614 | 92.9 | 30 | 90% | 7.57 | 9.88 | 6.83 |
| NGC2076 | 48.4 | 10 | 50% | 7.73 | | - |
| NGC2339 | 46.7 | 10 | 90% | 7.50 | 10.05 | 5.75 |
| NGC3690 | 62.1 | 1 | - | 8.34 | <9.73 | 6.97 |
| NGC4102 | 19.7 | 10 | 90% | 6.97 | 9.02 | 5.60 |
| NGC7469 | 102.0 | 30 | 90% | 7.63 | 9.90 | 6.80 |

TABLE 3: CIRRUS MODEL PARAMETERS FOR DUST MASS DETERMINATION

| X= | 1 | 3 | 5 | 10 | 20 | 30 | 50 | 100 | 200 | 500 | SB |
|---|---|---|---|---|---|---|---|---|---|---|---|
| log{S(100)/S(60)} | 0.69 | 0.70 | 0.64 | 0.54 | 0.44 | 0.34 | 0.24 | 0.13 | 0.03 | -0.11 | 0.0 |
| $T(\nu^2 B_\nu)$ [a] | 16 | 19 | 21 | 24 | 27 | 29 | 31 | 34 | 37 | 43 | 40 |
| $\log(M_d/S_{100} D^2)$ [b] | 4.08 | 3.42 | 3.16 | 2.83 | 2.60 | 2.37 | 2.17 | 1.93 | 1.67 | 1.44 | 1.33 |
| $\log(M_d/S_{800} D^2)$ [b] | 4.99 | 4.84 | 4.78 | 4.70 | 4.62 | 4.57 | 4.52 | 4.43 | 4.34 | 4.19 | |

a    valid for $\lambda > 1700/T$ μm
b    solar masses/(Jy $Mpc^2$)

## 8. DUST IN ELLIPTICALS AND LENTICULARS

There has been a growing realization that ellipticals and lenticulars have a significant interstellar medium, and that interesting amounts of star formation take place there. There has been a decade of work on HI emission from ellipticals and more recently CO observations in several cases (see eg the reviews by Wardle & Knapp (1986) and Schweizer (1987) ). Although the majority of the galaxies detected by IRAS are spirals, quite a number of ellipticals and lenticulars were detected (eg Jura 1986, Jura et al 1987, Knapp et al 1989). Thronson and Bally (1987) have studied the IRAS colour-colour diagrams for these galaxies and conclude that they occupy the same region of the diagrams as spiral galaxies (and, for that matter, star forming regions in our Galaxy). About 2/3rds of the sample they studied have the colours characteristic of cirrus and 1/3rd those of dusty regions surrounding young stars. The star formation rate they derive (0.1-1 $M_0$ /year) is comparable to the mass-loss rate for evolved stars in these galaxies, but mergers and gas infall may also contribute significantly. Bally and Thronson (1989) studied the IRAS data for a sample of 74 S0 galaxies which had known single-dish radio fluxes. 30% were detected in all 4 IRAS bands and 80% were detected in at least one band. The galaxies divided into those which followed the infrared-radio relation for spirals, for which the radio emission is presumably due to normal star formation, and those with excess radio emission, presumably due to an active nucleus and jets or lobes. A small number showed a slight excess of infrared to radio. Similar conclusions were reached by Walsh et al (1989). Knapp et al (1989) report that 2/3rds of a sample of several hundred SOs are detected by IRAS at 60 and 100 μm.

Thronson et al (1989c) examined the IRAS data for 150 lenticular and elliptical 'shell' galaxies (Malin and Carter 1983), which are believed to be the result of low velocity mergers. Although some of the galaxies showed evidence for enhanced star formation, the majority did not and they concluded that either (1) the merging galaxies are almost always E or S0 with only modest amounts of interstellar gas, or (2) the time-scale for creation and maintenance of the shell is longer than the time-scale for the starburst event, or (3) the formation of a shell structure requires a mass difference between the galaxies of a factor 10-100, so only a small fraction of the i.s.m. is heated or participates in star formation.

Walsh and Knapp (1989) find that the ellipticals detected by IRAS tend preferentially to be those with dust lanes visible in the optical. However the infrared properties are not strongly dependent on the visible dust content. They also find a slightly enhanced 100 μm detection rate for ellipticals with shells, boxy isophotes or inner discs, all of which are evidence of a recent merger, a result which is not necessarily inconsistent with that of Thronson et al (1989b).

It is unfortunate that the infrared sources associated with early type galaxies are almost all rather weak, so that there is little immediate prospect of detection at wavelengths > 300 μm, and hence of accurate dust mass determinations.

REFERENCES

Allamandola, L.J., Tielens, A.G., & Barker, J.R., 1985, Ap.J. 290, L25

Allen, M., & Robinson, G.W., 1975, Ap.J. 195, 81

Andriesse, C.D.,1978, A.A. 66, 169

Bally, J., & Thronson, H.A.,Jr, 1989, A.J. 97, 69

Beichman, C.A., 1987, A.R.A.A. 25

Bernard, J.P, & Desert, X., 1989, in "Interstellar Medium in Galaxies" eds M.Shull & H.Thronson Jr (Univ. of Wyoming)

Boulanger, F., Baud, B., & van Albada, G.D., 1985, A.A. 144, L9

Boulanger, F., Beichman, C., Desert, F.X., Helou, G., Perault, M., & Ryter, C., 1988, Ap.J. 332, 328

Boulanger, F., & Perault, M., 1988, Ap.J. 330, 964

Canterna, R., Hackwell, J.A., & Grasdalen, G.L., 1989, in "Interstellar Medium in Galaxies" eds M.Shull & H.Thronson Jr (Univ. of Wyoming)

Crawford, J., & Rowan-Robinson, M., 1986, MNRAS 221, 923

Cox, P., & Mezger, P.G., 1989, Astron.Astrophys.Review

Desert, F.X., and Dennefeld, M., 1988, A.A. 206, 227

Donn, B., 1968, Ap.J. 152, L129

Draine, B.T., 1988a, 22nd ESLAB Symposium, Infrared Spectroscopy in Astronomy

Draine, B.T., 1988b, in IAU Symposium 135, Interstellar Dust, eds L.J.Allamandola & A.G.G.M.Tielens (Reidel)

Draine, B.T., 1989, this volume

Draine, B.T., & Anderson, N., 1985, Ap.J. 292, 494

Draine, B.T., & Lee, H.M., 1984, Ap.J. 285, 89

Duley, W.M.,1973, Nature Phys.Sci. 244, 57

Duley, W.W., 1987, MNRAS 229, 203

Duley, W.W., & Williams, D.A., 1981, MNRAS 196, 269

Duley, W.W., & Williams, D.A., 1988a, MNRAS 231, 969

Duley, W.W., & Williams, D.A., 1988b, MNRAS 230, 1p

Eckart, A., Cameron, M., Rothermel, H., Wild, W., Zinnecker, H., Olberg, M., Rydbeck, G., & Wiklind, T., 1989, in "Interstellar Medium in Galaxies" eds M.Shull & H.Thronson Jr (Univ. of Wyoming)

Eales, S.A., and Devereux, N., 1989, in "Interstellar Medium in Galaxies" eds M.Shull & H.Thronson Jr (Univ. of Wyoming)

Eales, S.A., Wynn-Williams, G., & Duncan, W.D., 1989, Ap.J. 339, 859

Efstathiou, A., & Rowan-Robinson, M., 1989, MNRAS (submitted)

Engargiola, G., & Harper, D.A., 1989, in "Interstellar Medium in Galaxies" eds M.Shull & H.Thronson Jr (Univ. of Wyoming)

Fabbri, R., Guidi, I., Natale, V., & Ventura, G., 1988, preprint

Gillett, F.C., Forrest, W.J., & Merrill, K.M., 1973. Ap.J. 183, 87

Gosh, S.K., & Drapatz, S., 1989, A.A.

Greenberg, J.M., 1968, in Stars and stellar Systems, Vol 7, ed. Middlehurst et al (Chicago Univ.Press), p.221

Guhathakutra, P., & Draine, B.T., 1989, Ap.J. (in press)

Habing, H.J., et al, 1984, Ap.J. 278, L59

Halpern, M., Benford, R., Meyer, S., Muehlner, D., & Weiss, R., 1988, Ap.J. 332, 596

Helou, G., 1986, Ap.J. 311, L33

Helou, G., 1988, in IAU Symposium 135, Interstellar Dust in Galaxies

Herter, T., Shupe, D.L., & Chernoff, D.F., 1989, Ap.J. (in press)

Hildebrand, R.H., 1983, QJRAS 24, 267

Hong, S.S., & Greenberg, J.M., 1980, AA 88, 194

Hughes, J., Rowan-Robinson, M., Lawrence, A., & Crawford, J., 1989 (in preparation)

Hunter, D.A., Thronson, H.A.Jr, Casey, S., & Harper, D.A., 1989, Ap.J. 341, 697

de Jong, T., et al, 1984, Ap.J. 278, L67

de Jong, T., & Brink, K., 1987, in Star Formation in Galaxies, ed. C.L.Persson, p.323

Jones, A.P., Duley, W.W., & Williams, D.A., 1987, MNRAS 229, 213

Jura, M., 1986, Ap.J. 306, 483

Jura, M., Kim, D.W., Knapp, G.R., & Guhathahurta, P., 1987, Ap.J. 312, L11

Knapp, G.R., Guhathakurta, P., Kim, D.-W., & Jura, M., 1989, Ap.J. Supp. (in press)

Leech,K.J., Lawrence, A., Rowan-Robinson, M., Walker, D., & Penston, M.V., 1988, MNRAS 231, 977

Leech, K.J., Penston, M.V., Terlevich, R., Lawrence, A., Rowan-Robinson, M., & Crawford, J., 1989, MNRAS 240, 349

Leene & Cox, 1987, A.A. 174, L1

Leger, A., & Puget, J.L., 1984, A.A. 128, 212

Leisawitz, D., 1989, in "Interstellar Medium in Galaxies" eds M.Shull & H.Thronson Jr (Univ. of Wyoming)

Malin, D.F., & Carter, D., 1983, Ap.J. 274, 534

Mathis, J.S., Rumpl, W., & Nordsieck, K.H., 1977, Ap.J. 217, 425

Mathis, J.S., Mezger, P.G., & Panagia, N., 1983, A.A. 128, 212

Mathis, J.S., & Whiffen, G., 1989, Ap.J. 341, 808

Pajot, F., Boisse, P., Gispert, R., Lamarre, J.M., Puget, J.-L., & Serra, G., 1986 , A.A. 157, 393

Persson, C., & Helou, G., 1987, Ap.J. 314, 513

Platt, J.R., 1956, Ap.J. 123, 486

Price, S.D., 1981, A.J. 86, 193

Puget, J.L., & Leger, A., 1989, A.R.A.A. 27

Purcell, E.M., 1976, Ap.J. 206, 685

Rice, W., Lonsdale, C.J., Soifer, B.T., Neugebauer, G., Kopan, E.L., Lloyd, L.A., de Jong, T., & Habing, H.J.,1988, Ap.J. Supp. 68, 91

Rice, W., Boulanger, F., Viallefond, F., Soifer, B.T., & Freedman, W.L., 1989, Ap.J. (in press)

Roche, P.F., 1988, in 22nd ESLAB Symposium, Infrared Spectroscopy in Astronomy

Rowan-Robinson, M., 1979, Ap.J. 234, 111

Rowan-Robinson, M., 1982, in Submillimeter Astronomy, ed. P.Phillips & J.Beckman (CUP), p.47

Rowan-Robinson, M., 1986, MNRAS 219, 737

Rowan-Robinson, M., 1987a, in Star Formation in Galaxies, ed. C.Persson, p.133

Rowan-Robinson, M., 1987b, in Starbursts and Galaxy Evolution, eds T.X.Thuan, T.Montmerle & J.T.T.Van (Edition Frontieres) p.235

Rowan-Robinson, M., & Chester, T., 1987, Ap.J. 313, 413

Rowan-Robinson, M., Lock, T.D., Walker, D.W., & Harris, S., 1986, MNRAS 222, 273

Rowan-Robinson, M., & Crawford, J., 1986, in Light on Dark Matter, ed.    F.P.Israel (Reidel) p.421

Rowan-Robinson, M., & Crawford, J., 1989, MNRAS 238, 523

Ryter et al, 1987, A.A. 186, 312

Sakato, A., Wada, S., Tanabe, T., & Onaka, T., 1983, Nature 301, 493

Sakato, A., Wada, S., Tanabe, T., & Onaka, T., 1984, Ap.J. 287, L51

Schweizer, F., 1987, in IAU Symposium 127, Structure and Dynamics of    Elliptical Galaxies (Reidel)

Schwering, P., 1988, Ph.D. thesis, Univ. of Leiden

Sellgren, K., 1984, Ap.J. 277, 623

Soifer, B.T., Houck, J.R., & Neugebauer, G., 1987, A.R.A.A. 25, 187

Soifer, B.T., Rice, W.L., Mould, J.R., Gillett, F.C., Rowan-Robinson, M., &    Habing, H.J.,1986, Ap.J. 304, 651

Smith, J., 1982, Ap.J. Ap.J. 261, 463

Smith, P.A., Brand, P.W.J.L., Puxley, P.J., Mountain, C.M., Gear, W.K., & Nakai,    N., 1989a, in "Interstellar Medium in Galaxies" eds M.Shull & H.Thronson Jr (Univ. of Wyoming)

Smith, B.J., Lester, D.F., & Harvey, P.M., 1989b, in "Interstellar Medium in Galaxies" eds M.Shull & H.Thronson Jr (Univ. of Wyoming)

Stark, A.A., Davidson, J.A., Harper, D.A., Pernic, R., Loewenstain, R., & Casey,    S., 1989, Ap.J. (in press)

Telesco,, 1988, A.R.A.A. 26, 343

Telesco, C.M., Becklin, E.E., & Wynn-Williams, G., 1984, Ap.J. 282, 427

Telesco, C.M., Decher, R., & Joy, M., 1989, Ap.J. (in press)

Thronson, H.A., Jr, & Bally, J., 1987, Ap.J. 319, L63

Thronson, H.A.Jr, Hunter, D.A., Telesco, C.M., Harper, D.A., & Decher, R., 1987, Ap.J. 317, 180

Thronson, H.A.Jr, Hunter, D.A., Telesco, C.M., Greenhouse, M., & Harper, D.A., 1988, Ap.J. 334, 605

Thronson, H.A. Jr, Hunter, D.A., Casey, S., Latter, W.B., & Harper, D.A., 1989a, Ap.J. 339, 803

Thronson, H.A.Jr, Hunter, D.A., Casey, S., & Harper, D.A., 1989b, Ap.J. (in    press)

Thronson, H.A., Jr, Bally, J., & Hacking, P., 1989b, A.J. 97, 363

Tielens, A.G.G.M., & Allamandola, L.J., 1987, in Interstellar Processes, eds D.Hollenbach & H.A.Thronson Jr (Reidel) p.397

Walsh, D.E.P., Knapp, G.R., Wrobel, J.M., & Kim, D.-W., 1989, Ap.J. 337, 209

Walsh, D., & Knapp, J., 1989, in "Interstellar Medium in Galaxies" eds M.Shull & H.Thronson Jr (Univ. of Wyoming)

Walterbos, R.A.M., & Schwering, P.B.W., 1987, A.A.Supp. 180, 27

Wardle, M., & Knapp, G.R., 1986, A.J. 91, 23

Williams, D.A., 1989, preprint

Young, J.S., Xie, S., Kenney, J.D.P., & Rice, W.L., 1989, Ap.J. Supp. (in press)

# ELLIPTICAL GALAXIES WITH DUST LANES

W.W. Zeilinger[1], F. Bertola, G. Galletta
*Department of Astronomy University of Padova, Italy*

ABSTRACT. Optical and far-infrared surveys reveal that more than 40% of the elliptical galaxies contain dust, the most prominent group being dust-lane ellipticals. The observed decoupling of the angular momenta of gas and stars indicates that the dust lane is most probably due to a second event in the history of these galaxies. The dust lane is an important tracer of the potential of the underlying galaxy. IRAS data indicate a predominantly cool dust component which mass is very small compared to the total galaxy mass. There is evidence that the·dust grain properties differ from those generally assumed for spiral galaxies.

# 1    Introduction

Elliptical galaxies were always considered to be *"by definition"* essentially gas and dust free systems. But there is now an increasing evidence for the presence of a significant amount of interstellar medium in elliptical galaxies. The presence of neutral and ionized gas in some ellipticals are examples of this evidence. X-ray observations suggest that several ellipticals are surrounded by extensive halos of hot gas that fuel the central gaseous disks via cooling flows (Biermann, Kronberg, 1983; Nulsen, Stewart, Fabian, 1984; Forman, Jones, Tucker, 1985).

A distinct subgroup of elliptical galaxies is formed by systems whose stellar body is crossed by a dust lane. The lane consists of dust and gas and is the projection of a ring or of a narrow disk seen edge-on. Dust-lane ellipticals, although traditionally classified as S0 galaxies, have a number of features that distinguish them from "normal" $S0_2$ and $S0_3$ galaxies for which a (major-axis) dust lane is a normal property (Sandage, 1961). This subgroup is characterized by the presence of a lane of absorbing matter which crosses the galaxy body at various orientations, the majority being either along the minor or major axis of the light distribution. Bertola, Galletta (1978) pointed out the case of minor-axis dust-lane ellipticals, NGC 5128 (Cen A) being the prototype. Successively in a survey carried out by Hawarden et al. (1981) also galaxies having dust lanes along their major axis or along intermediate axes have been identified and included in the subgroup.

On the basis of the observational evidence (Bertola, 1972; Bertola, Capaccioli, 1975; Illingworth, 1977) it is now generally accepted that ellipticals are not rotationally flat-tened (Binney, 1978). The orientation of the dust lane is an important tracer of the

---

[1]Affiliated to the Astrophysics Division of the Space Science Department of ESA

*E. Bussoletti and A. A. Vittone (eds.), Dusty Objects in the Universe, 227–234.*

potential of the underlying galaxy, because it indicates one of the two permitted planes of the galaxy. In addition, the dust lane represents a good laboratory to study the physical properties of the interstellar medium. Dust as a gas tracer allows estimates on the gas mass and on the frequency distribution of gas in ellipticals.

# 2 Properties of dust-lane ellipticals

In recent years the class of dust-lane ellipticals has been subject of several detailed studies. After the first systematic survey by Hawarden et al. (1981) the number of new candidate objects grew steadily. This was made possible with improved detector technology and more sophisticated data analysis, allowing the detection of weak dust features against the steep intensity gradient of the galaxy (Sparks et al., 1985). More than 200 galaxies have been identified sofar as candidate obects (Zeilinger, 1987). Sadler, Gerhard (1985) estimated that about 40% of the nearby ellipticals contain dust structures (dust lanes, dust spots, etc.).

## 2.1 Morphological properties

An homogenous compilation of *bona fide* dust-lane ellipticals has been presented by Bertola (1987) with thirty minor-axis, nine major-axis and seven skew-axis dust-lane systems. Given the morphology of minor-axis dust lanes galaxies, the identification of these objects is rather unambigous. In fact, there are no cases of S0 galaxies known where the disk contains the minor axis of the bulge.

Major-axis dust-lane ellipticals may be on the contrary easily confused with normal S0 galaxies, because at low values of the disk/bulge ratio, the detection of the stellar disk is very difficult. This problem is also reflected in the morphological classifications of major galaxy catalogues where those galaxies appear among the class of S0s.

The dust lane can be considered as formed by material acquired from outside which is settled, or on the way to settle, in an equilibrium configuration within the galaxy. The orientation of the dust lane with respect to the galaxy ellipsoid will consequently depend on the impact parameters and on the intrinsic shape of the system. The material not yet settled give rise to a warped structure of the dust lane in the outer regions.

## 2.2 Photometric and kinematic properties

In Table 1 a compilation is given on the available photometric and kinematic data concerning the galaxies listed by Bertola (1987). Surface photometry shows that the luminosity profiles closely follow the $r^{1/4}$–law typical of elliptical galaxies, confirming the morphological selection criteria.

Since the most general shape of ellipticals is a triaxial ellipsoid and the infalling mass is supposed to be very small compared to the total galaxy mass, there are two possible configurations of equilibrium: namely, the planes defined either by the major and intermediate axis or by the minor and intermediate axis. Therefore, depending on the infall angle, both minor-axis and major-axis dust lanes should be observable. The orientation of the angular momentum of the gas associated with the dust lane depends also on the infall angle. Since it is in general observed that the stellar angular momentum vector lies along the minor axis of the galaxy, it is expected to observe in the case of minor-axis dust-lane ellipticals orthogonal angular momenta and in the case of major-axis dust-lane ellipticas a roughly equal distribution of galaxies having either co- or counter-rotating gas with respect to the stellar rotation.

As a matter of fact, in all galaxies sofar observed, with exception of the rather faint objects Anon 0632–629, Anon 0641–412 and IC 4320, ionized gas in the position angle of the dust lane is observed. In all the minor-axis dust-lane ellipticals the angular momenta of gas and stars are found be orthogonal as expected. Concerning the major-axis dust-lane ellipticals a first evidence for the expected counter-rotation was found in the E0 galaxy NGC 5898 (Bertola, Bettoni, 1988). For this galaxy it was concluded on basis of geometrical considerations that the projected rotation axis of the stars coincides with the axis of the dust ring and that therefore the two angular momenta are intrinsically anti-parallel. The definite proof was found a in sample of four major-axis dust-lane ellipticals. In two of them (Anon 1029–459 and NGC 3528) counter-rotating gas is found while in the other two (NGC 4370 and NGC 5745) the gas rotates in the same sense as the stars (Bertola, Buson, Zeilinger, 1988).

Further support to the accretion hypothesis may give the presence of the warps. The models of Tubbs (1980) and Simonson (1982) for NGC 5128 explain the warped structure in terms of an ongoing settling process, where the warped parts have not yet reached the equlibrium configuration. In this way the warps are explained as a transient phenomenon that will vanish when all the material has settled in one of the permitted planes. Van Albada, Kotanyi, Schwarzschild (1982) interpreted the warps to be stationary, assuming a slow tumbling motion around the minor axis of the triaxial figure of the elliptical body. This model requires a retrograde tumbling motion with respect to the motions within the warp. However, the majority of the cases of warped minor-axis dust-lane ellipticals show prograde motions along the major axis. The model of van Albada, Kotanyi, Schwarzschild (1982) can be retained only if the stellar streaming is opposite to the tumbling motions in these cases.

Another explanation of the warped structure in some cases is given by Sparke, Casertano (1988) who showed that a self-gravitating thin disk of material which is subject to the potential of a flattened spheroid can have a discrete mode of vertical oscillation. In the case of the warped major-axis dust-lane elliptical Anon 1029–459 a disk mass of $>10^8$ M$_\odot$ is necessary in order to explain the observed warp by a discrete bending mode (Sparke, Casertano, 1989).

# 3  Properties of the dust-lane

In the following chapter an overview of the physical properties of the dust and associated gas is given. Early studies of infrared galaxy spectra revealed a close similarity to those of molecular clouds in our Galaxy (Rieke, Lebovsky, 1979) This suggests that the thermal far-infrared radiation emitted by normal galaxies originates partly from cool interstellar dust grains which are heated by disk stars and partly in molecular clouds which are heated by newly formed stars. Recent works indicate that, contrary to the expectation, dust is also a rather common phenomenon in early-type galaxies. Knapp et al., (1989) found that about 70% of early-type galaxies with $m_B < 14^{mag}$ are detected in the 60 $\mu$m and 100 $\mu m$ IRAS bands. Using the the standard "line-add" routines the IRAS database was analyzed for *bona fide* dust-lane ellipticals and 23 galaxies were identified in the four IRAS bands out of a sample of 46 objects (Bertola, Rifatto, Sulentic, Zeilinger, 1989).

## 3.1 The dust component

It is generally assumed that the dust consists of carbon-silicate grains with a bimodal range of sizes, having diameter of about 0.1 $\mu m$ and 0.01 $\mu m$ respectively. The only dust-lane elliptical where IRAS resolved spatially the far-infrared emission is NGC 5128. It turns out that the emission follows closely the star-forming regions in the dust lane, reproducing also the warps in the infrared. Marston, Dickens (1988) noted an excess of the 12 $\mu m$ emission in NGC 5128, which may be attributed to evolved M stars that are more typical of ellipticals than of disk galaxies (Rowan-Robinson et al., 1986), although the far-infrared colours seem to indicate a disk system. This apparent contradiction could be an evidence for an accretion event between an elliptical and a late-type system. This may be a very important intrinsic property to all ellipticals with dust lanes, but the verification will have to wait for ISO (Infrared Satellite Observatory) in order that these objects can be investigated with a similar resolution as NGC 5128 with IRAS.

Assuming pure black body radiation and that the dust emissivity is proportional to the frequency, lower limits of the dust mass can be calculated. The predominantly cool dust component ($T_{dust} \approx 20$ K) in dust-lane ellipticals results to be very small compared to the total galaxy mass, typically ranging between $10^4 < M_{dust} < 10^5$ $M_\odot$. Similar results are reported using estimates on the mean absorption and geometric properties of the dust lane (Sadler, Gerhard, 1985; Gallagher, 1986).

These estimates, however, are based on the assumption that the dust grain properties are the same throughout the universe. There are evidences that in spiral galaxies the properties of the dust component are similar to those of the galactic dust (see e.g. Elmegreen, 1980; Keel, 1983). But there are indications that these properties may be different for early-type galaxies. In the case of the minor-axis dust-lane elliptical IC 4320 (Warren-Smith, Berry, 1983) the extinction law shows an approximately linear variation with inverse wavelength with an unusually strong slope. The ratio of total-to-selective extinction results to have a value of only $R[A_V/E_{B-V}] = 1.9$, compared with an average value of 3.1 for the Galaxy (Savage, Mathis, 1979). The authors interpreted this reddening law in terms of smaller dust grains with respect to our Galaxy. This may be due to different formation mechanisms or may imply that the dust grains in IC 4320 have been substantially modified in the last $5 \times 10^8$ years, being the result of gradual destruction. Further evidences for systematically smaller dust grain sizes in dust-lane ellipticals were pointed out by Brosch et al. (1985) in the case of NGC 7070A and by Rifatto (1989) for NGC 2534. In this case, the above cited limits of the dust mass should be revised.

## 3.2 The gas component

It is convenient to distinguish two gas components in elliptical galaxies, namely a cool ($<10^2$ K) and a warm ($\approx 10^4$ K) one. It appears that about 15% of all ellipticals contain measureable amounts of neutral gas (see Knapp, 1987 for a review). The $M_{HI}/L$ ratio is found to be a good indicator of the morphological type of the galaxy. The distribution of $M_{HI}/L$ for ellipticals and S0s are very similar, yet different from spirals. In order to explain this behaviour, Wardle, Knapp (1986) suggested an external origin for most of the neutral hydrogen in early-type galaxies. In our sample of dust-lane ellipticals the typical $M_{HI}/L$ ratios agree with those derived by Wardle, Knapp (1986) for early-type galaxies.

The properties of molecules in the dust lane are poorly known. Only NGC 5128 has been observed sofar. The molecular material is found to be closely associated with the dust lane, its mass being about $2 \times 10^8$ $M_\odot$. The absorption features in the $^{12}CO$ and $^{13}CO$ lines indicate that the properties of the molecular clouds are comparable to those of our Galaxy (Eckart et al., 1989; Cameron, 1989). In the $\log(M(H_2)) - \log(L_{IR})$ relation early-type galaxies appear to constitute a low-luminosity continuation of the correlation for spiral galaxies (Wiklind, Henkel, 1989). This may have important implications on the star formation efficiency in early-type galaxies, which could be as high as in spirals. Dust-lane ellipticals, being the result of an accretion event should be expected to be an intermediate case between spirals and "normal" early-type galaxies.

The ionized gas component, in the galaxies we are considering, is associated with the optical dust lane. In NGC 5128 a series of H II regions could be identfied (e.g. Dufour et al., 1979; Möllenhoff, 1981a). Spectroscopic observations (Möllenhoff, 1981b) indicate two types of emission regions in the dust lane. The first group (outer H II regions) consists of normal giant H II regions near the outer rim of the dust lane, O and B stars being the exciting sources. There are also indications for the presence of Wolf-Rayet stars. The temperatures are rather low, about 6500 K. A slight overabundance of oxygen and nitrogen with respect to the Orion nebula is noted. The second component is formed by the inner H II regions, which are hotter than the outer ones and show a similar chemical composition. A strong red stellar continuum is found, with emission lines arising from low ionization states ([N II], [S II]). The rather high [N II]/H$\alpha$ line ratios can be explained by shock heating, yielding a quite high temperature ($\approx 10^5$ K).

The ionized gas in dust-lane ellipticals generally appear to be in ordered rotation with a velocity higher than that of the stars.

It is worth to mention that a similar morphology to dust-lane ellipticals is also found in some ellipticals having gaseous disks but no dust. Both major-axis gaseous disks (e.g. NGC 7097 - Caldwell, Kirshner, Richstone, 1986) and minor-axis disks (e.g. NGC 5077 - Bertola et al., 1989) are found.

Kotanyi, Ekers (1979) pointed out that the dust lanes tend to align perpendicularly to the radio source, when present. This indicates that the radio axis is related to the rotation axis of the gas and that radio emission may be causally related to the presence of dust.

The presence of dust and gas in elliptical galaxies is a rather common phenomenon. The estimate from optical data that about 40% of ellipticals contain dust (e.g. Sadler, Gerhard, 1985; Ebneter et al., 1988) is supported by the far-infrared data with an estimate of frequency of about 50% (Jura et al., 1987; Knapp et al., 1989). Finally, about 60% of all ellipticals contain ionized gas (Phillips et al., 1986).

232

**Table 1:** Data on dust-lane ellipticals.

| object | $m_B$ | $cz$ | $\log(L_{IR}/L_B)$ | $\varepsilon$ | $V_m$ | $\sigma_0$ | $V_m/\sigma_0$ | dust | $L_\bullet,L_{gas}$ | ref. |
|---|---|---|---|---|---|---|---|---|---|---|
| A 0151–498 | 13.12 | 6170 | – | 0.33 | 50 | 257 | 0.19 | minor | ⊥ | 10 |
| NGC 1052 | 11.53 | 1471 | –1.22 | 0.31 | 96 | 245 | 0.39 | minor | ⊥ | 7 |
| NGC 1947 | 11.49 | 1160 | –1.02 | 0.10 | 57 | 150 | 0.38 | minor | ⊥ | 5 |
| A 0609–331 | | 11300 | – | 0.50 | 160 | 265 | 0.60 | minor | ⊥ | 8 |
| A 0632–629 | | 8550 | – | 0.43 | 145 | 215 | 0.67 | minor | ? | 8 |
| A 0641–412 | | 10900 | – | 0.14 | 100 | 160 | 0.63 | minor | ? | 8 |
| A 1029–459 | 12.67 | 2764 | –0.47 | 0.30 | 210 | 260 | 0.81 | major | ↑↓ | 3,4 |
| NGC 3528 | 12.38 | 3635 | – | 0.35 | 120 | 270 | 0.44 | major | ↑↓ | 3,4 |
| NGC 4370 | 10.06 | 750 | –1.68 | 0.47 | 60 | 120 | 0.50 | major | ↑↓ | 3,4 |
| NGC 4374 | 10.08 | 1000 | –1.72 | 0.20 | 68 | 290 | 0.23 | skew | ↑↓ | 5 |
| NGC 4589 | 12.00 | 1824 | –1.00 | 0.19 | 100 | 215 | 0.47 | minor | ⊥ | 9 |
| NGC 5128 | 7.96 | 497 | –0.27 | 0.23 | 80 | 140 | 0.57 | minor | ⊥ | 1,12 |
| NGC 5266 | 12.27 | 2880 | –0.66 | 0.29 | 140 | 200 | 0.70 | minor | ⊥ | 6,8,11 |
| IC 4320 | 13.76 | 6670 | – | 0.05 | 40 | 270 | 0.15 | minor | ? | 5 |
| NGC 5363 | 11.08 | 1121 | –1.02 | 0.44 | 135 | 199 | 0.67 | minor | ⊥ | 10 |
| NGC 5745 | 14.02 | 7020 | –0.01 | 0.47 | 220 | 160 | 1.38 | major | ↑↓ | 3,4 |
| NGC 5898 | 12.60 | 2160 | – | 0.00 | 80 | 230 | 0.35 | E0 | ↑↓ | 2 |
| NGC 7070A | 13.40 | 2382 | – | 0.60 | 30 | 100 | 0.30 | skew | ⊥ | 10 |

col. 1: object identifier
col. 2: B magnitude
col. 3: heliocentric velocity [km s$^{-1}$]
col. 4: far-infrared excess
col. 5: observed ellipticity
col. 6: maximum rotational velocity [km s$^{-1}$]
col. 7: central velocity dispersion [km s$^{-1}$]
col. 8: ratio of maximum rotational velocity and central velocity dispersion
col. 9: orientation of the dust lane
col. 10: distribution of angular momenta of gas and stars
col. 11: References:

1 Bertola, F., Galletta, G., Zeilinger, W.W.: 1985, Ap. J., **292**, L51.
2 Bertola, F., Bettoni, D.: 1988, Ap. J., **329**, 102.
3 Bertola, F., Galletta, G., Kotanyi, C., Zeilinger, W.W.: 1988, M.N.R.A.S., **234**, 733.
4 Bertola, F., Buson, L.M., Zeilinger, W.W.: 1988, Nature, **335**, 705.
5 Bertola, F., Galletta, G., Zeilinger, W.W.: 1989, in preparation.
6 Caldwell, N.: 1984, Ap. J., **278**, 96.
7 Davies, R.L., Illingworth, G.D.: 1986, Ap. J., **302**, 234.
8 Möllenhoff, C., Marenbach, G.: 1986, A. & A., **154**, 219.
9 Möllenhoff, C., Bender, R.: 1989, A. & A., **214**, 61.
10 Sharples, R.M., Carter, D., Hawarden, T.G., Longmore, A.J.: 1983, M.N.R.A.S., **202**, 37
11 Varnas, S., Bertola, F., Galletta, G., Freeman, K., Carter, D.: 1987, Ap. J., **313**, 69.
12 Wilkinson, A., Sharples, R.M., Fosbury, R., Wallace, P.T.: 1986, M.N.R.A.S., **218**, 297.

# References

van Albada, Kotanyi, C.G., Schwarzschild: 1982, M.N.R.A.S., **198**, 303.
Bertola, F.: 1972, in Proc. 15$^{th}$ Meeting of Ital. Astr. Soc., p. 199.
Bertola, F.: 1987, in *"Structure and Dynamics of Elliptical Galaxies"* IAU Symp. 127, p. 135, ed. T. de Zeeuw (Reidel:Dordrecht).

Bertola, F., Bettoni, D.: 1988, Astrophys. J., **329**, 102.

Bertola, F., Buson, L.M., Zeilinger, W.W.: 1988, Nature, **335**, 705.

Bertola, F. Capaccioli, M.: 1975, Astrophys. J., **200**, 439.

Bertola, F. Galletta, G.: 1978, Astrophys. J., **226**, L115.

Bertola, F., Bettoni, D., Danziger, I.J., Sadler, E.M., Sparke, L.S., de Zeeuw, T.: 1989, preprint.

Bertola, F., Rifatto, A., Sulentic, J.W., Zeilinger, W.W.: 1989, in preparation.

Biermann, P., Kronberg, P.P.: 1983, Astrophys. J., **268**, L69.

Binney, J.J.: 1978. M.N.R.A.S., **183**, 501.

Brosch, N., Greenberg, J.M., Grosbøl, P.J.: 1985, Astron. Astrophys., **143**, 399.

Caldwell, N., Kirshner, R.P.,Richstone, D.O.: 1986, Astrophys. J, **305**, 136.

Cameron, M.: 1989, this meeting.

Dufour, R.J., van den Bergh, S., Harvel, C.A., Martins, D.H., Schiffer, F.H., Talbot, R.J., Talent, D.L., Wells, D.C.: 1979, Astron. J., **84**, 284.

Ebneter, K., Djorgovski, S., Davis, M.: 1988, Astron. J., **95**, 422.

Eckart, A., Cameron, M., Rothermel, H., Wild, W., Zinnecker, H., Rydbeck, G., Olberg, M., Wiklind, T.: 1989, preprint.

Elmegreen, D.M.: 1980, Astrophys. J. Suppl., **43**, 37.

Forman, W., Jones, C., Tucker, W.: 1985, Astrophys. J., **293**, 102.

Gallagher, J.S.: 1986, Pub. Astr. Soc. Pac., **98**, 81.

Hawarden, T.G., Elson, R.A.W., Longmore, A.J., Tritton, S.B., Corwin, H.G.Jr.: 1981, M.N.R.A.S., **196**, 747.

Jura, M., Kim, D.-W., Knapp, G.R., Guhathakurta, P.: 1987, Astrophys. J, **312**, L11.

Illingworth, G.: 1977, Astrophys. J., **218**, L43.

Keel, W.C.: 1983. Astron. J., **88**, 1579.

Knapp, G.R., Guhathakurta, P., Kim, D. W., Jura, M.: 1989, Astrophys. J. Suppl. **70**, 329.

Kotanyi, C.G., Ekers, R.D.: 1979, Astron. Astrophys., **73**, L1.

Marston, A.P., Dickens, R.J.: 1988, Astron. Astrophys., **193**, 27.

Möllenhoff, C.: 1981a, Astron. Astrophys., **93**, 248.

Möllenhoff, C.: 1981b, Astron. Astrophys., **99**, 341.

Nulsen, P.E.J., Stewart, G.C., Fabian, A.C.: 1984, M.N.R.A.S., **208**, 185.

Phillips, M.M., Jenkins, C.R., Dopita, M.A., Sadler, E.M., Binette, L.: 1986, Astron. J., **91**, 1062.

Rieke, G.H., Lebovsky, M.J.: 1979, Ann. Rev. Astr. Astrophys., **17**, 477.

Rifatto, A.: 1989, this meeting.

Rowan-Robinson, M., Lock, A., Walker, D.W., Harris, S.: 1986, in "Light on Dark Matter", p. 101, ed. F.P. Israel (Reidel:Dordrecht).

Sadler, E.M., Gerhard, O.: 1985, M.N.R.A.S., **214**, 177.

Sandage, A.: 1961, "The Hubble Atlas of Galaxies" (Carnegie Institution:Washington).

Savage, B.D., Mathis, J.S.: 1979, Ann. Rev. Astr. Astrophys., **17**, 73.

Simonson, G.F.: 1982, Ph. D. thesis Yale University.

Sparke, L.S., Casertano, S.: 1988, M.N.R.A.S., **234**, 873.

Sparke, L.S., Casertano, S.: 1989, in "Astrophysical Discs" (Manchester).

Sparks, W.B., Wall, J.V., Thorne, D.J., Jorden, P.R., van Breda, I.G., Rudd, P.J., Jorgensen, H.E.: 1985, M.N.R.A.S., **217**, 87.

Tubbs, A.D.: 1980, Astrophys. J., **241**, 969.

Wardle, M., Knapp, G.R.: 1986, Astron. J., **91**, 23.

Warren-Smith, R.F., Berry, D.S.: 1983, M.N.R.A.S., **205**, 889.

Wiklind, T., Henkel, C.: 1989, Astron. Astrophys, in press.

Zeilinger, W.W.: 1987, Ph. D. thesis University of Vienna.

# OBSERVATIONAL COSMOLOGY IN THE NEAR INFRA-RED

S.T. Chase, N. Mandolesi
*Istituto TE.S.R.E., Bologna, Italy*

ABSTRACT. Although we have a fairly good understanding of the structure of the Universe at recent epochs, its history and evolution remain obscure. Clearly we still lack key pieces of the cosmic puzzle. The study of background radiations should be an effective means of elucidating the development of the Universe at early epochs. In what follows we give a brief overview of the known cosmological backgrounds and some of the major problems with current scenarios. Finally we focus our attention on the least studied of the cosmological backgrounds, namely the Near Infra-Red Background. For attempts to measure this background the problem of foreground contamination by local sources is severe. Most of the Universe is bright at 2–4 $\mu$m! In particular, emission from dust in the Solar System (ZL & IPD) and the Galaxy ('cirrus') needs to be better understood before the extragalactic background can be estimated with confidence.

# 1  Introduction

Although we have a fairly good understanding of the structure of the Universe at recent epochs, its history and evolution remain obscure. One of the major open problems is the question of galaxy formation and the evolution of large-scale structure. There now exists a large body of data on the distribution of luminous matter on scales out to a few hundred Mpc, and also on the general properties of galaxies in the neighbourhood of the local group. Added to this we have considerable information about objects at high redshift, and a growing understanding of the properties of background radiation fields. Despite this, there remains a wide range of quasi-viable theoretical models for the growth of structure in the Universe. Clearly we still lack key pieces of the cosmic puzzle.

The study of background radiations should be an effective means of elucidating the development of the Universe at early epochs, though observations in the Infra-Red are complicated by the presence of local emission due for example to dust in the solar system and associated with the Galaxy. In what follows we give a brief overview of the known cosmological backgrounds and some of the major problems with current scenarios. Finally we will focus our attention on the least studied of the cosmological backgrounds, namely the Near Infra-Red Background.

*E. Bussoletti and A. A. Vittone (eds.), Dusty Objects in the Universe, 235–242.*
© 1990 *Kluwer Academic Publishers.*

# 2 Background Radiation as a Cosmic Probe

## 2.1 The known Background Radiation fields

A) The Microwave Background. The Microwave Background Radiation (MBR) was discovered 25 years ago, and is the most studied and best characterised of the cosmological radiation fields. In the simplest 'standard' cosmologies it propagates freely to us from the epoch of recombination at $z \simeq 1500$. Its small-scale anisotropy should carry an imprint of the density perturbation spectrum from which galaxies and large-scale structure of the Universe are supposed to have grown by gravitational instability. Current limits on the anisotropy amplitude are about $\Delta T/T = 3\ 10^{-5}$, on angular scales of a few arcminutes to a few degrees, and recent results from the **COBE** satellite show the spectrum to be Planckian to better than 1%, with a temperature of $2.735 \pm 0.060$ K (Mather *et al.* 1990).

B) The X-ray Background. It is not clear how much of the X-ray Background (XRB) is really cosmological in origin, and how much is due to known classes of extragalactic objects such as AGNs (see e.g. Setti 1987). Whether any component is truly diffuse, or if the entire background is due to discrete sources, is even less clear. Models which produce the entire XRB by thermal bremsstrahlung from a hot and uniform Inter-Galactic Medium (IGM) require gas densities equivalent to between 0.2–0.3 of the universal closure density, and gas temperatures of 10–15 keV at the present epoch (e.g. Guilbert & Fabian 1986, Taylor & Wright 1989). The energy requirements for such models are very high, and the necessary gas density is in conflict with limits on the baryonic density derived from primordial nucleosynthesis arguments. Small-scale clumping of the gas relaxes the energy and mean density requirements, but requires an even hotter low density medium to pressure-confine the clumps.

C) The Infra-Red Background. At present very little is known about the cosmological Near Infra-Red Background (NIRB), as even its intensity and spectrum are poorly defined. The detection reported by Matsumoto *et al.* (e.g. Matsumoto *et al.* 1988 Ap.J. **332**, 575) is consistent with a 1500 K black body diluted by a factor of $10^{-11}$, and corresponds to $I_\nu \simeq 1.45\ 10^5$ Jy/sr at $\lambda \simeq 3.8\ \mu$m. The source of the NIRB is of great interest to cosmology as it probably originates at $10 < z < 100$, and possibly at even higher redshift. For this reason it is likely to be a valuable probe for investigating the epoch of galaxy formation. Suggestions for a possible origin include: a) incompletely thermalized energy inputs at epochs prior to recombination, such as accretion by primordial 'mini black holes', or the decay/annihilation of exotic particle species (halflife $> 3\ 10^4$ years); b) decaying particles at $z > 50$, but after recombination; c) pregalactic explosions (e.g. Ostriker & Cowie 1981); d) energy input due to discrete sources at $z \simeq 5$–50, such as supermassive objects (SMOs or VMOs), or Pop III stars at $z > 25$ with $T_{eff} > 3\ 10^4$ K, possibly seen through a low optical depth of dust; e) Primaeval Galaxies (PGs) at $z \geq$ 3–5. A good review of the the possibilities is given by Rowan-Robinson & Carr 1988. To determine which of these processes if any contribute to the NIRB, we require information not only about the intensity and spectrum but also about its structure on small angular scales.

## 2.2 Some problems with current scenarios.

Most current models of galaxy formation start with an initial perturbation spectrum at recombination ( $z \simeq 1500$ ), which is assumed to be distributed as a gaussian random field. Linear growth in the epoch to $z \simeq 2$ requires a perturbation amplitude of $\Delta \rho / \rho \geq 3 \ 10^{-4}$ unless biassing is introduced (e.g. Kaiser 1984), in which case the RMS amplitude can be slightly reduced. The non-observation of temperature anisotropies in the MBR to a level of $\simeq 2 \ 10^{-5}$ on angular scales of several arcminutes to a few degrees is a problem for these models unless some form of obscuration is interposed between the observer and the epoch of recombination. The two obvious possibilities are either a re-ionized cosmic medium or a uniform distribution of dust at high redshift. One can however use the observed spectrum of the MBR to place upper limits on the temperature and optical depth of ionized gas between us and the epoch of recombination at $z \simeq 1500$. Furthermore, most scenarios involving a significant optical depth of dust at high redshift also produce a large amount of thermalized radiation which appears as a submillimetre-wave excess in the MBR spectrum, peaking at around 700 $\mu$m (e.g Hogan & Bond 1988, Bond, Carr & Hogan 1986, Negroponte 1986).

The recently released first results from the **COBE** satellite show no evidence for a submillimetre excess such as is mentioned above. This represents a serious constraint on all models in which an early generation of objects such as Pop III stars produce a significant amount of dust. For cosmological models with an isothermal perturbation spectrum, it has been strongly argued by Bond *et al.* and others that the first generation of objects which form must be associations of massive stars of order 200 $M_\odot$, but these will produce a lot of dust and a thermalized submillimetre excess. **COBE** would appear to exclude such models.

The **COBE** data also place strong limits on re-ionization scenarios. Mather *et al.* argue that the lack of distortion in the MBR spectrum as measured by **COBE** imposes a $3\sigma$ limit of $y_c < 10^{-3}$ on the Comptonization parameter, which is basically the product of the optical depth to Compton scattering and the electron temperature of the IGM. Thus, to allow for an optical depth of unity one must restrict the IGM temperature to less than 500 eV, or 5 $10^6$ K, at any time since recombination. Clearly this is a problem for the hot IGM models which have been suggested to explain the XRB, and in fact Mather *et al.* claim that such models can be ruled out. The **COBE** data restricts the XRB due to thermal bremsstrahlung to about 1/36 of the observed value. It will apparently now be necessary to find another source for the XRB, and also alternative reasons for the absence of MBR anisotropies.

How and when galaxies formed is very much an open question, and at present it is not clear whether Primaeval Galaxies (PGs) have been observed or not. In the simple model of Partridge & Peebles (1967) one expects to observe the redshifted $Ly_\alpha$ emission in the 2–4 $\mu$m band, but several searches have apparently failed to reveal this (e.g. Boughn *et al.* 1986, Collins & Joseph 1988). On the other hand, Cowie (1988) argues that flat-spectrum galaxies seen in recent deep surveys (e.g. Tyson & Seitzer 1988) must contain PGs, in order to match observed number counts if $q_0 = 1/2$. Current limits on the luminosity and areal density of PGs do not seriously constrain models of galaxy formation. The areal density of PGs is estimated at anything up to $10^4$ per square degree. A PG undergoing an initial burst of star formation might be 10 to 20 times as luminous as NGC6240, and the currently popular CDM models favour a relatively late epoch for galaxy formation at redshifts of 2 or 3. Even if galaxies form earlier than this, Silk (1986) expects a change in the star forming initial mass function from mainly low-mass to higher mass O-B stars as the redshift drops below $z \simeq 3$. These arguments suggest that PGs should be most luminous at this epoch, and together with the non-detection of redshifted $Ly_\alpha$ emission, that PGs might be dusty objects. In this case, and for redshifts in the range 2 to 5, PGs would be significantly fainter in the K-band and brighter at 3-5 $\mu$m.

## 2.3 Possible contributions to the total NIRB flux

For observational purposes it is most convenient to discuss the NIRB in terms of the following categories of possible components.

1) A truly diffuse NIRB (similar to the CBR) which is essentially isotropic on all scales, but probably with a low level of anisotropy ($\delta I/I < 10^{-4}$) reflecting the primordial perturbation spectrum. Such a background would arise in cases a) and b) of §2.1.

2) A 'quasi-diffuse' background with a moderate level of anisotropy on scales of a few degrees or less, such as might arise as a result of the pregalactic explosions in case c). This would not reflect the full distribution of the primordial perturbation spectrum, but only that of its low-mass peaks where the first generation of supermassive objects ($M \simeq 10^4 - 10^6 \ M_\odot$) would form in isothermal models.

3) A background which is directly due to discrete sources of some hitherto unobserved class of objects as in case d) & e) above, and which might appear either as point-like or extended sources. In the case of Pop III stars or VMOs the small-scale anisotropies would again reflect the distribution of high-density peaks on small mass scales. The amplitude on scales of a few arcseconds would be quite large, with $\delta I/I$ determined by the variance of the number density and the $\log(n)/\log(S)$ relationship of the sources. For an approximately Poissonian distribution of point sources observed to a fixed flux density cutoff at $S_{min}$ = constant in a beam of angular size $\theta$, the RMS value of $\delta I/I$ would scale as $1/\theta$. For Poisson-distributed clumps of sources one would expect $\delta I/I$ to show more complex structure for angular scales characteristic of the mean clump separation and size. Similar considerations apply for PGs (case e) of §2.1), except that the spatial statistics would derive from the epoch at which the evolution of the perturbation spectrum became significantly non-linear, and the magnitude of $\delta I/I$ as a function of beamsize would flatten out at the characteristic angular scale of the PGs.

4) A 'confusion background' due to the integrated light of galaxies, as calculated for example by Franceschini. In this case the spatial distribution of fluctuations in $\delta I/I$ would reflect the clustering of galaxies, with the amplitude distribution (P(D)) again determined by the $\log(n)/\log(S)$ relationship, and by the threshold flux density $S_{min}$ at which a source is considered to be detected. It has been suggested by Bahcall (1984) and others that 'Brown Dwarves' may comprise the 'missing mass' in galaxies. Brown Dwarves may possibly also contribute slightly to the Infra-Red background light (Karimabadi & Blitz 1984).

It is certain that 4) will contribute to the total NIRB flux and to its 'granularity', though how much is unknown. Clearly, attempts to understand the origin of the NIRB must also examine the 'confusion background' via deep integrations at a range of infra-red wavelengths. The question to be settled by observation is, which of the above categories of sources contribute to the total NIRB flux, and to what extent?

# 3 Observations of the NIRB

## 3.1 Sources of difficulty in the detection of a NIRB

The detection of isotropic background radiation is difficult enough from the instrumental point of view, but is further complicated by the presence of local backgrounds such as the Zodiacal Light (ZL) and the integrated starlight from our Galaxy. Ground-based and balloon-borne observations additionally suffer from highly variable atmospheric emission, particularly due to OH lines. Estimates of the contributions from local backgrounds are available at present, and are likely to be improved upon in the near future. The contribution from extragalactic sources is less well-defined, mainly because of poorly understood evolutionary effects. The Zodiacal Light (ZL) and re-radiated thermal emission from inter-planetary dust is the dominant astronomical source of background emission for all wavelengths shorter than about 30 $\mu$m, and will remain a significant contribution at longer wavelengths. Although most of the emission is rather smooth and uniform across the sky, varying by a factor of 2 or 3 with solar elongation and ecliptic latitude for elongations greater than 90° (Murdock & Price 1985), there are several bands and arcs close to the ecliptic plane. These are associated with asteroidal debris and cometary trails, and have angular extents of the order of a degree and the variation of intensity is a few percent of the total ZL intensity. There is an additional temporal variation of about 10% per annum due mainly to the tilt of about 3° between the symmetry plane of the dust and the ecliptic. The small-scale structure is a potential source of confusion when mapping extended objects, and the ZL emission in general must be accurately subtracted from photometric data, especially on faint objects.

The discovery of galactic 'cirrus' by **IRAS** was unexpected, and the nature of the emitting material is rather enigmatic. The cirrus emission is thought to be a mixture of continuum and band/line components due to dust, but the relative contributions are unknown. The 12 and 25 $\mu$m emission is orders of magnitude higher than one would predict for a thermal continuum extrapolated from the 60 and 100 $\mu$m fluxes, though spatial correlation is strong. Speculation on the source of the emission ranges from the non-equilibrium heating of very small dust grains by UV photons (Desert *et al.* 1986; Cox *et al.* 1986), to band emission by PAHs (Puget *et al.* 1985; Allamandola *et al.* 1985). Due to the aforementioned uncertainties it may prove extremely difficult to subtract the cirrus component from the total NIR flux in order to recover the true cosmological NIRB.

The abovementioned 'contaminants' are all non-isotropic and hence, at least in principle, may be separated from an isotropic background if sufficient sky coverage is available. A more serious problem arises in the case of extragalactic contributions to the total NIR flux, since these components are approximately isotropic on scales greater than a few arcminutes, and the integrated flux at 2–5 $\mu$m is hard to estimate due to evolutionary uncertainties. The 2–5 $\mu$m waveband is particularly suitable for the study of galaxy evolution over cosmological timescales. It is well known that 'normal' galaxies have spectra peaking at one to a few microns wavelength. In high-redshift galaxies one is looking at rest-frame emission at 1 to a few $\mu$m where dust absorption in the emitting system is orders of magnitude less than in the UV or visible bands. There is also a strong *positive* K-correction due to the galaxy spectra rising from the mid-IR to $\lambda$ of about 1 $\mu$m. AGNs and 'starbursts' are two categories of galaxies which emit most of their energy in the FIR. Synchrotron emission dominates in all **IRAS** bands in many objects with AGNs, such as luminous BL Lac objects, OVV quasars and some Seyferts (Soifer *et al.* 1989). The emission from starbursts is dominated by thermal dust emission peaking at around 100 $\mu$m, and their nuclear spectra in the 3–13 $\mu$m range are often dominated by bands attributed to PAHs (Roche 1987). These two phenomena may well be related, for instance they may be simultaneously triggered by interactions or mergers.

## 3.2 Future Observations

The **TRIP** balloon-borne infra-red telescope is designed specifically for measurement of the NIRB and also the ZL & IPD. It carries a $32 \times 1$ array of detectors and seven filters with passbands in the range 2–5 $\mu$m. The primary mirror is 25 cm in diameter, and the pixel field of view is $\simeq 1.4$ arcminutes (Ventura *et al.* 1987, Morroi *et al.* 1989). It is planned to fly this instrument from Trapani, Sicily, in July 1990. By making a series of pointed observations at a range of azimuth and zenith angles during a 10 hour flight, it is hoped that a deconvolution of the isotropic NIR background from local sources (including atmospheric OH emission) will be possible. While this instrument may provide an estimate of the NIRB intensity, it does not have the pointing stability necessary to investigate the small-scale structure. It should also be possible to improve on current models of the ZL & IPD using the data from this experiment.

The **DIRBE** experiment on **COBE** covers the wavelength range 1–300 $\mu$m in 10 filters, and has an angular field of view of 0.7°. There is also a polarization capability which will assist in the separation of local dust emission from the total detected flux levels. As yet no results have been released, but this instrument is well suited to measure the spectrum and intensity of the NIRB, as well as its large-scale isotropy. The extensive sky coverage and modest spectral and angular resolution will permit a deconvolution of the extragalactic NIRB from local backgrounds, including starlight. Estimation of the contribution due to galaxies will be unfortunately be model-dependent, however, since the large beamsize will result in a confusion limit of 3–5 mJy with respect to the integrated flux at 3.6 $\mu$m due to galaxies.

The Infra-Red Space Observatory (**ISO**) will probably make significant contributions to our understanding of the small-scale stucture of the NIRB. It carries an infrared camera, **ISOCAM**, which is extremely well suited to perform very deep integrations by virtue both of its broad-band sensitivity and the pixel sizes available in the $3'$ camera field of view. Source confusion quickly becomes a serious problem as the field of view increases, and this is likely to be a limiting factor in using the long wavelength photometer, **ISOPHOT**. With **ISOCAM**, integration to a sensitivity comparable to the 1 $\sigma$ confusion limit (about 5 $\mu$Jy per pixel in a $6''$ per pixel mode at 3–4 $\mu$m) will require about 4 hours per $3'$ field. This flux level is about 1/20 of the mean NIRB intensity reported by Matsumoto *et al.* Deep integrations of a few hours duration at 5 to 8.5 $\mu$m on 'empty' fields with **ISOCAM** are expected to detect significant numbers of galaxies and AGNs at redshifts of up to about 1. Primaeval galaxies (PGs) and disc-forming systems may be detectable to even higher redshifts. Proposals to perform deep integrations on 'empty fields' at 3–4 $\mu$m and 5–8.5 $\mu$m on a sample of 10 or 15 fields with low cirrus contamination have already been suggested for the **ISOCAM** core programme, and a similar proposal at 12 $\mu$m may also be made. For each field the database will consist of about 700 pixels in three colours, and it should then be possible to extract information about the small-scale structure of the NIRB, number counts of various classes of galaxies out to redshifts of 1 to 3, and possibly also to detect PGs if they are dusty. For redshifts in the range 2 to 5 a dusty PG would have an angular size of 4″ to 6″ and a flux density in the range 30 to 100 $\mu$Jy at 3 to 8 $\mu$m wavelength. Even if only a few percent of PGs are as luminous as this, they will be detectable at 3$\sigma$ in 1 hour out to a redshift of 3 in the **ISOCAM** bands at 3 to 8 $\mu$m. and there is a good chance of seeing 2 or 3 in 10 **ISOCAM** frames. It has also been suggested that the near- and mid-IR colours of galaxies can be used to derive approximate photometric redshifts for large samples of faint objects (Rieke *et al.*). Studies of the evolution of galaxies are necessarily statistical, and may be pursued in two ways. One can try to select a suitable sample of objects at various redshifts, and observe them in some detail in order to discern a trend with epoch in their photometric or spectral properties. Alternatively one can do deep number counts in at least two colours, and then try to draw conclusions about the high-redshift population from the resulting data in conjunction with ground-based data in the optical and near infrared. The two approaches are complimentary, and will probably both be attempted with **ISO**.

ISO can also help to clarify the nature of the galactic 'cirrus' by spectrally resolving the emission detected by the broad-band **IRAS** filters. For this purpose it will probably be best to use CVFs at the shorter wavelengths, with medium resolution (R $\simeq$ 200) spectra covering the range longwards of about 16 $\mu$m. The total observing time for a bright patch of cirrus (flux density 10–40 Jy at 60–100 $\mu$m) will be about 10 hours.

# 4   Conclusions

Our understanding of proceses at early epochs leading to the presently observed structure in the Universe is clearly incomplete. Further study of cosmic backgrounds is a potentially powerful means to improve upon our present ignorance. In particular the NIRB is deserving of deeper investigation, as a better understanding of the contributing factors to the total extragalactic background in the 2–4 $\mu$m region is sure to improve our understanding of galaxy evolution. Observation of small-scale structure in the NIRB would be of great significance, as this would provide constraints on models of galaxy evolution, and other physical processes occuring at high redshift. It will be necessary to observe many fields with angular resolution of a few arcminutes, and integrate to the confusion limit at several wavelengths, in order to achieve this goal. Such observations will best be made with a camera, and some progress may be achieved with ground-based work, but atmospheric emission is a problem. Significant results will probably not be gained until ISO is operational, since **ISOCAM** is ideally suited to this purpose. Unfortunately the problem of foreground contamination by local sources is severe. Most of the Universe is bright at 2–4 $\mu$m! In particular, emission from dust in the Solar System (ZL & IPD) and the Galaxy ('cirrus') needs to be better understood before the extragalactic background can be estimated with confidence.

# References

Allamandola, L.J., Tielens, A.G.G.M., Baker, J.R.: 1985, Ap.J.(Lett.) **290**, L25

Bahcall: 1984, Ap.J. **287**, 926

Boughn, S.P., Saulson, P.R., Uson, J.M.: 1986, Ap.J. **301**, 17

Bond, J.R., Carr, B.J., Hogan, C.J.: 1986, Ap.J. **306**, 428

Collins, C.A., Joseph, R.D.: 1988, M.N.R.A.S. **235**, 209

Cowie, L.L.: 1988, in *"The Post-Recombination Universe"* p. 1, eds. N. Kaiser, A.N. Lasenby; Kluwer Academic Publishers

Cox, P., Krugel, E., Mezger, P.G.: 1985, Astr. Astrophys. **155**, 380

Desert, F.X., Boulanger, F., Shore, S.N.: 1986, Astron. Astrophys **160**, 295

Guilbert, P.W., Fabian, A.C.: 1986, M.N.R.A.S. **220**, 439

Hogan, C.J., Bond, J.R.: 1988, 'Origin and Anisotropy of the Cosmic Submillimetre Background', in *'The Post-Recombination Universe'* p. 141, eds. N. Kaiser, A.N. Lasenby; Kluwer Academic Publishers

Kaiser, N.: 1984, Ap.J. **284**, L9

Karimabadi, Blitz: 1984, Ap.J. **283**, 167

Mather, J.C., *et al.*: 1990, 'A Preliminary Measurement of the Cosmic Microwave Background Spectrum by the Cosmic Background Explorer (COBE) Satellite', COBE preprint No. 90–01

Matsumoto, T., Akiba, M., Murakami, H.: 1988, Ap.J. **332**, 575

Morroi, R., Cortiglioni, S., Greco, C., Mandolesi, N., Morigi, G., Palazzi, E., Rossetti, E., Sanzani, G., Ventura, G., Franceschini, A.: 1989, "Measurement of diffuse near IR background radiation with the TRIP experiment", Proceedings del III Convegno Nazionale di Astronomia Infrarossa, Gallipoli, 1989; in press

Murdock, T.L., Price, S.D.: 1985, Astron. J. **90**, 375

Negroponte, J.: 1986, M.N.R.A.S. **222**, 19

Ostriker, J.P., Cowie, L.L.: 1981, Ap.J. **273**, L127

Partridge, R.B., Peebles, P.J.E.: 1967, AP.J., **147**, 868

Puget, J-L., Leger, A., Boulanger, F.: 1985, Astron. Astrophys. **142**, L19

Setti, G.: 1987, in IAU symposium No. 124, *'Observational Cosmology'*, pp579, eds. A. Hewitt, G. Burbidge, L.Z. Fong; Reidel, Dordrecht

Soifer,B.T., Houck, J.R., Neugebauer, G.: 1989, "The IRAS view of the extragalactic sky", preprint

Rieke, Eisenhardt, Lebovsky: 1987, Ap.J. **316**, 70

Roche, P.F.: 1987, "Infrared Features in Extragalactic Objects" in "Polycyclic Aromatic Hydrocarbons and Astrophysics", eds. Leger et al. Reidel

Rowan-Robinson, M., Carr, B.J.: 1988, in *'The Post-Recombination Universe'* p. 125, eds. N. Kaiser, A.N. Lasenby; Kluwer Academic Publishers

Silk, J., 1986, Ap.J. **297**, 1

Taylor, G.B, Wright, E.L.: 1989, Ap.J. **339**, 619

Tyson, J.A., Seitzer,P.: 1988, Ap.J. **335**, 552

Ventura, G., Brighenti, A., Calzolari, P., Cazzola, G., Cortiglione, S., Giovannini, G., Mandolesi, N., Morigi, G., Negroponte, J., Partridge, R.B., Salinari, P.: 1987, Memorie SAIT **58**, 351

# INTRACLUSTER EXTINCTION OF LIGHT FROM GALAXIES IN VIRGO

**A. Biviano[1], G. Giuricin[1,2], F. Mardirossian[1,2], M. Mezzetti[1,2], Y. Rephaeli[3]**
[1] *Department of Astronomy, University of Trieste, Italy*
[2] *C.I.R.A.C., Trieste, Italy*
[3] *School of Physics and Astronomy, Tel Aviv University, Israel*

ABSTRACT. We have analysed existing data on colours and Tully-Fisher distances of a sample of spiral galaxies in the Virgo cluster. The data seem to indicate a significant correlation between the colour-excesses and distances of these galaxies. This correlation may be due to extinction by intracluster (IC) dust. However, due to uncertainties in the observational data, and the limited size of the sample, such an interpretation may not yet be construed as an unequivocal evidence for dust in the Virgo cluster of galaxies.

## 1 Introduction

Hot gas in cores of clusters of galaxies is metal-enriched (for a review, see Sarazin 1986), and therefore likely to be mostly of galactic origin. Our knowledge of interstellar media motivates the expectation that the IC gas may also contain detectable amount of dust. Dwek, Rephaeli and Mather (1989) reviewed the observational evidence –starting with the work of Zwicky (1952)– for IC dust. An amount of IC dust corresponding to visual extinction at a level of about 0.2 mag is typically deduced.

The possibility of the existence of IC dust is not only interesting, but also important in the study of the properties and evolution of clusters of galaxies. Observational evidence for IC dust is very scant and not very convincing. We have begun a systematic study of all available data on colours and distances of galaxies in a few clusters. The data are quite limited; only in the case of the Virgo cluster the size of the data base renders it marginally statistically meaningful. We describe here the results of a correlation analysis on the distribution of colour excess vs. Tully-Fisher distance data on 36 spirals galaxies within 6° region centred on M87 in the Virgo cluster. We use $(B - V)_T^0$ colours, with distances obtained via the blue Tully-Fisher (TF) relation (Tully and Fisher 1977).

## 2 The Correlation Analysis

To date, the possibility of the existence of IC dust was studied mainly through measurements of the light and counts of distant sources seen in the background of clusters. With

*E. Bussoletti and A. A. Vittone (eds.), Dusty Objects in the Universe,* 243–246.

the increased sensitivity and amount of data acquisition, it is beginning to be possible to test the existence of IC dust through its effect on the colours of the cluster galaxies. Relatively precise colours and distances are required in order for a correlation to be seen in the distribution of the colour vs. IC distance of galaxies in the cluster. Being closest to us and well measured, Virgo is the obvious (and currently the only) cluster for which such a study may be marginally feasible.

We collected the data relevant to our analysis from the following sources: galaxy morphological types were taken from Binggeli et al. (1985), the catalogues of Sandage and Tammann (1981), and de Vaucouleurs et al. (1976) (hereafter reffered as "RC2"). Total corrected blue magnitudes $B_T^0$, and total corrected colour-indices $(B-V)_T^0$ were taken from RC2. Corrected 21 cm line-width parameters, $V_{maz}$ (as defined in Bottinelli et al. 1983), and $W_R^i$ (as defined in Tully and Fouqué 1985), were taken from Hoffman et al. (1989), Pierce and Tully (1988), Bottinelli et al. (1984), Tully and Shaya (1984), and Bottinelli, Gouguenheim and Paturel (1982).

Galaxy distances were estimated by means of the TF relation fitted on the brightest spiral members of the Virgo cluster, located within 6° from M87:

$$M_{B_T^0} = -6.2 - 6.03 \log V_{maz} \tag{1}$$

$$M_{B_T^0} = -2.2 - 6.89 \log W_R^i \tag{2}$$

where $M_{B_T^0}$ is the absolute $B_T^0$ magnitude. We adopted a mean distance modulus of 31.0 magnitudes for the cluster (in agreement with current estimates, see e.g. Pierce and Tully 1988), to obtain the constants in the TF relations.

Colour excesses, $E(B-V)_T^0$, were determined as follows:

$$E(B-V)_T^0 \equiv (B-V)_T^0 - <(B-V)_T^0>_{dV77} \tag{3}$$

where $<(B-V)_T^0>_{dV77}$ are the mean standard corrected colours per morphological type taken from de Vaucouleurs (1977).

The correlation analysis was performed by computing on our data-sample the Kendall non-parametric correlation coefficient (see, e.g., Kendall 1948). The coupled effect of a Colour-Magnitude (CM) relation in spirals (see, e.g., Griersmith 1980) and a Malmquist bias might induce an artificial correlation between colour excesses and distances. In order to remove this spurious effect, we applied either a statistical correction or a correction to the colours.

As Kendall (1948) has shown, given three sets of data it is possible to give an estimate of the true correlation between two of these sets, independent of their correlations with the third. This is done by using the Kendall partial rank correlation coefficient, defined as (see, e.g., Siegel 1956):

$$K_{de.m} = \frac{K_{de} - K_{me} K_{dm}}{\sqrt{(1 - K_{me}^2)(1 - K_{dm}^2)}} \tag{4}$$

where $K_{de.m}$ denotes the Kendall partial correlation coefficient, between the variable $d$ (distance-moduli) and the variable $e$ (colour excesses), independent of the third variable

$m$ (magnitudes). The others are the usual (Kendall) correlation coefficients between the indexed variables.

Alternatively, we corrected the observed $(B-V)_T^0$ colours, using the following relation that we derived from Griersmith (1980):

$$(B - V)_T^c = 0.973(B - V)_T^0 + 0.027(M_{B_T^0} - < M_{B_T^0} >) \qquad (5)$$

A mean absolute magnitude $< M_{B_T^0} >$ was adopted for each of the samples in order to set the zero point of the relation.

# 3   Results

The application of the two different methods of correction yielded very similar results. We found a strong correlation (at a significance level $\geq$ 99%) between the colour-excesses and the distance-moduli of the spirals in the Virgo region, while no significant correlation was found in a comparison sample of field spirals (taken from Haynes and Giovanelli 1984, and Davis and Seaquist 1983). Moreover, a stronger correlation exists between the above quantities in a Virgo subsample composed of those galaxies with the higher quality 21 cm data.

In order to estimate the amount of light extinction associated with the Virgo cluster, we divided the Virgo sample into three subsamples. The first subsample consists of "foreground" galaxies, i.e. all galaxies with a distance $D$ from us: $D < 13.35$ Mpc. The second and third subsamples contain "cluster" galaxies, i.e. all galaxies with 13.35 Mpc $\leq D \leq 18.35$ Mpc, and "background" galaxies, i.e. all galaxies with $D > 18.35$ Mpc (the distance-modulus we adopted for the Virgo cluster corresponds to a distance of 15.85 Mpc). The mean corrected colour excesses, $< E(B-V)_T^c >$, of the three subsamples are: $-0.02 \pm 0.02$, $+0.04 \pm 0.03$ and $+0.10 \pm 0.04$, for 11, 10 and 15 galaxies, respectively. The difference between the background $< E(B - V)_T^c >$ and the foreground $< E(B - V)_T^c >$ is: $+0.12 \pm 0.04$, which corresponds to a blue extinction $A_B = R_B \times 0.12$, where $R_B$ is likely to be in the range between $\sim 2$ (Warren-Smith and Berry 1983, Brosch et al. 1988, Capaccioli et al. 1989, Rifatto 1989) and $\sim 4$ (see, e.g., Johnson 1968).

A naive interpretation of the above result would resort to IC dust in Virgo. Indeed, a continuous ejection of (gas and) dust by the cluster galaxies can replenish grains destroyed by sputtering by the hot IC gas. An extensive and detailed treatment of IC dust within the context of such a model has been given by Dwek, Rephaeli and Mather (1989). A basic result from this latter study is that IC dust should be significantly depleted in the central, but not in the outer region of a cluster. This is interesting and possible; however, we caution that the sample of 36 spiral galaxy is rather limited. Though it is not clear how, uncertainties in the intrinsic colours of spirals and in distances deduced from TF relations may have given rise to the correlation seen in the data. Our results are therefore preliminary, and more high quality data are needed before the method applied here can lead to a more definite result concerning the existence and amount of IC dust in Virgo, and possibly other nearby clusters.

246

# Acknowledgements

This work was partially supported by the *Ministero per l'Università e per la Ricerca scientifica e tecnologica*, and by the *Consiglio Nazionale delle Ricerche (CNR–GNA)*.

# References

Binggeli, B., Sandage, A., Tammann, G.A.: 1985, Astrophys. J., **90**, 1681.

Bottinelli, L., Gouguenheim, L., Paturel, G.: 1982, Astron. Astrophys Suppl., **47**, 171.

Bottinelli, L., Gouguenheim, L., Paturel, G., de Vaucouleurs, G.: 1983, Astron. Astrophys., **118**, 4.

Bottinelli, L., Gouguenheim, L., Paturel, G., de Vaucouleurs, G.: 1984, Astron. Astrophys. Suppl., **56**, 381.

Brosch, N., Mayo Greenberg, J., Grosbol, P.J.: 1988, Astron. Astrophys., **143**, 399.

Capaccioli, M., Cappellaro, E., Della Valle, M., D'Onofrio, M., Rosino, L., Turatto, M.: 1989, preprint.

Davis, L.E., Seaquist, E.R.: 1983, Astrophys. J. Suppl., **53**, 269.

de Vaucouleurs, G.: 1977, *"The evolution of galaxies and stellar populations"*, Eds.: B.M. Tinsley, R.B. Larson, Yale Univ. Obs.

de Vaucouleurs, G., de Vaucouleurs, A., Corwin, H.: 1976, *"2$^{nd}$ reference catalogue of bright galaxies"*, Univ. of Texas Press, Austin - "RC2".

Dwek, E., Rephaeli, Y., Mather, J.: 1989, Astrophys. J., in press.

Griersmith, D.: 1980, Astron. J., **85**, 1295.

Haynes, M.P., Giovanelli, R.: 1984, Astron. J., **89**, 758.

Hoffman, L.G., Lewis, B.M., Helou, G., Salpeter, E.E., Williams, H.: 1989, Astrophys. J. Suppl., **69**, 65.

Johnson, H.L.: 1968, *"Nebulae and interstellar matter"*, eds.: B.M. Middlehurst, L.H.Aller, Univ. Chicago Press.

Kendall, M.G.: 1948, *"Rank correlation methods"*, Griffin, London.

Pierce, M.J., Tully, R.B.: 1988, Astrophys. J., **330**, 579.

Rifatto, A.: 1989, preprint.

Sandage, A., Tammann, G.A.: 1981, *"A Revised shapley-ames catalog of bright galaxies"*, Carnegie Inst. of Washington Pub. 635, Washington, D.C..

Sarazin, C.L.: 1986, Rev. Modern Phys., **58**, 1.

Siegel, S.: 1956, *"Non-parametric statistics: for the behavioral sciences"*, McGraw-Hill, N.Y..

Tully, R.B., Fisher, J.R.: 1977, Astron. Astrophys., **54**, 661.

Tully, R.B., Fouqué, P.: 1985, Astrophys. J. Suppl., **58**, 67.

Tully, R.B., Shaya, E.J.: 1984, Astrophys. J., **281**, 31.

Warren-Smith, R.F., Berry, D.S.: 1983, M.N.R.A.S., **205**, 889.

Zwicky, F.: 1952, P.A.S.P., **64**, 242.

# AXIALLY SYMMETRIC RADIATIVE TRANSFER MODELS OF LOW MASS PROTOSTARS

A. Efstathiou and M. Rowan-Robinson
Astronomy Unit, School of Mathematical Sciences,
Queen Mary and Westfield College, London E1 4NS.

**Abstract.** In this paper we present a modification to the method of Efstathiou and Rowan-Robinson (1989) for the solution of the axially symmetric radiative transfer problem in dust clouds, which is more appropriate for modeling protostellar candidates. In contrast to previous attempts to solve the axisymmetric problem, the method integrates the equation of radiative transfer exactly in the sense that it does not neglect any terms in the equation and treats multiple anisotropic scattering. The results of several computed models are presented showing the dependence of the emergent spectrum on the model parameters and viewing angle. We also present a very good fit to the observed spectrum of IRS5 L1551 and compare it with previous work.

## 1. Introduction

In the currently accepted scenario (Shu, Adams and Lizano 1987) the process of low-mass star formation is broken down into four evolutionary phases. The first phase involves the formation of a molecular cloud core that is becoming increasingly vulnerable to gravitational collapse and which can be approximated to a singular isothermal sphere with the density following $r^{-2}$. The second phase begins with the collapse of the cloud core. Shu (1977) has shown that in the case of pure collapse without rotation an isothermal sphere collapses self-similarly. The infall is initiated from "inside out" by an expansion wave which propagates outward into the ambient static molecular cloud core at the speed of sound, and inside of which the density approaches the free-fall form,

$$\rho \propto r^{-3/2} \tag{1}$$

Terebey, Shu and Cassen (1984; herafter TSC) also included rotation in the collapse as a small perturbational effect and showed that outside the centrifugal radius,

$$R_C \equiv \frac{G^3 M^3 \Omega^2}{16\alpha^8} \tag{2}$$

(where M is the mass of the protostar, $\Omega$ is the angular rotation rate assumed to be constant, and $\alpha$ is the sound speed), equation (1) is still a good approximation. Inside of $R_C$, where the effects of rotation become important, their solution is matched to the

247

*E. Bussoletti and A. A. Vittone (eds.), Dusty Objects in the Universe, 247–254.*
© 1990 *Kluwer Academic Publishers.*

inner region solution of Cassen and Moosman (1981) who applied uniform rotation to the solution of Shu (1977) and found that the trajectories of infalling matter follow,

$$\zeta = \frac{\cos\theta_0 - \cos\theta}{sin^2\theta_0 cos\theta_0} \tag{3}$$

where $\zeta \equiv \frac{Rc}{r}$ and $\theta_0$ is the polar angle of the asymptotically radial streamline.

By applying conservation of mass along a stream tube TSC found that the density will be given by,

$$\rho = Cr^{-3/2}\left(1 + \frac{\cos\theta}{\cos\theta_0}\right)^{-1/2}(1 + 2\zeta P_2(\cos\theta_0))^{-1} \tag{4}$$

where $P_2(\cos\theta)$ is the usual second order Legendre polynomial.

In the third phase of this scenario a strong stellar wind pushes toward the area of least resistance (i.e the rotational poles of the system) and forms two collimated jets. This corresponds to the bipolar outflow phase. Eventually the stellar wind widens and sweeps out material in nearly all the sphere around the protostar, leading to the fourth stage of protostellar evolution, a T-Tauri star with a surrounding nebular disc.

Further progress in the understanding of low mass star formation has been hampered by the problems associated with testing this scenario with detailed axisymmetric radiative transfer calculations. Adams, Lada and Shu (1987) presented models of protostars calculated with the method of Adams and Shu (1986). Apart from neglecting scattering completely they treat the axisymmetric problem approximately by means of an "an equivalent spherical envelope". In this paper we set to solve the axisymmetric problem exactly and present models of deeply embedded objects being in the second or third stage of their evolution in the scenario outlined above.

## 2. Method of solution

The mathematical and computational details of the method of solution of the axially symmetric radiative transfer problem in dust clouds have been discussed in detail in Efstathiou and Rowan-Robinson (1989; herafter ER) where the method was applied to flared discs around a cool star. The method of solution is in fact a generalisation of the method of Rowan-Robinson (1980) for the solution of the spherically symmetric problem. The radiation intensity is split into three components depending on how photons last interacted with dust in the cloud. The first component consists of unattenuated radiation from the central source, the second of thermal emission from grains and the third of scattered light from grains according to a specified phase function. All three components are calculated by ray-tracing along carefully selected directions to ensure proper evaluation of the moments of the intensity and good coverage of all solid angles. After calculation of the intensity and its moments, the temperature distribution in the cloud is corrected using the radiative balance condition,

$$\Delta T = \frac{\int_0^\infty Q_{\nu,abs}(J_\nu - B_\nu[T(r,\Theta)])d\nu}{\int_0^\infty Q_{\nu,abs}\frac{\partial B_\nu}{\partial T}d\nu} \tag{5}$$

where $Q_{\nu,abs}$ is the grain absorption efficiency, $J_\nu$ is the average intensity and $B_\nu$ is the Planck function.

The flux constancy condition can no longer be coupled with the radiative balance condition for the temperature correction procedure, as in a spherically symmetric calculation (Rowan-Robinson 1980), because it is now a global one,

$$r^2 \int_0^{\pi/2} H \sin\Theta d\Theta = const. \tag{6}$$

In equation (6) $H = \int_0^\infty H_\nu d\nu$ is the flux integrated over all frequencies and $\Theta$ is the azimuthal polar coordinate.

The program developed to implement the above method allows a lot of flexibility in the specification of the geometry and the density distribution. ER used a power-law parameterisation of the form,

$$\rho(r,\theta) \propto r^{-2}\Theta^\gamma \tag{7}$$

where $\gamma$ was treated as a free parameter. With a density distribution of this form they obtained a very good fit to the spectrum of VY CMa, a famous late M supergiant which has long been thought to be surrounded by a circumstellar disc (Herbig 1970, Rowan-Robinson and Harris 1982).

For the purpose of modeling protostellar sources the above model was extended to a full axisymmetric calculation covering the whole of the $r-\Theta$ plane. Apart from the modification in the geometry we also need an analytic specification of the density distribution that is appropriate for these systems. For this purpose we have used the axisymmetric density profile of TSC which is obtained by solving equation (3) for $\theta_0$ and substituting into (4).

## 3. Model parameters

The most crucial parameter in these models is the centrifugal radius $R_C$ which determines the extent and degree of flattening of the inner part of the cloud. The constant $C$ in equation (4) is incorporated into the expression for $\tau_{uv}$ which for a given viewing angle $\theta_v$ (measured from the equatorial plane) will be given by,

$$\tau_{uv}(\theta_v) = Q_{\nu,ext} \int_{r_1}^{r_2} \rho(r, \pi/2 - \theta_v) dr \tag{8}$$

The solution predicts the formation of a ring at $r = R_C$ which is a feature of many calculations of protostellar collapse (Bodenheimer and Black 1978, Boss 1980). We have made no attempt to eliminate the resulting density enhancement around $R_C$ as it is confined to a very small area and does not affect our radiative transfer calculation as long as $R_C$ does not coincide with one of the grid points, in which case the radiative balance condition breaks down. The effect of this centrifugal ring is to give a deeply absorbed near and mid-infrared spectrum when the source is viewed directly along the equatorial plane.

Each model is characterised by four parameters: the effective temperature of the protostar $T_s$ (which is assumed to radiate as a blackbody), the ratio of the inner and outer cloud radii $r_1/r_2$, the ratio of the centrifugal radius $R_C$ to the outer radius $R_C/r_2$ and the optical depth of the cloud for a line of sight in the equatorial plane $\tau_{uv}(0)$. The optical depth is of course varying with viewing angle. The grain properties assumed in these models are those of the composite grain model of Rowan-Robinson (1982).

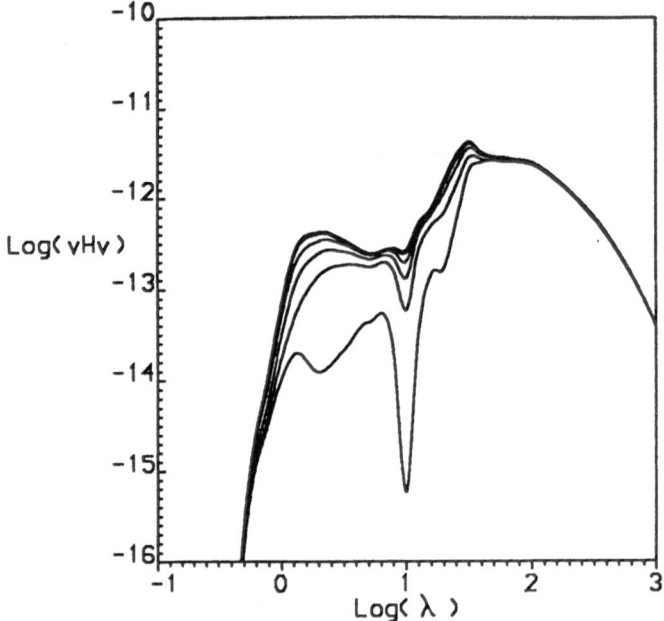

**Figure 1.** *Variation of the spectrum of a model with $R_C = 10^3 \times r_1$, $r_1/r_2 = 10^{-5}$, $T_s = 4,000K$ and $\tau_{uv}(0) = 300$ with viewing angle. In order of increasing short wavelength $Log\ \nu H_\nu$, $\theta_v = 0$ (edge-on), $\pi/16, \pi/8, 3\pi/16, \pi/4,$ and $\pi/2$.*

Before going on to compare theoretically predicted spectra with observations it is interesting to look at the effect of different parameters on the shape of the emergent spectrum.

## 4. Dependence of emergent spectra on model parameters.

We restrict our attention to models of deeply embedded objects in the second or third stage of protostellar evolution. Such objects are expected to have effective temperatures in the range $3,000 - 6,000K$. We find that the emergent spectrum is not very sensitive to changes of $T_s$ in this range, and so for the models presented here we have assumed an average value of $T_s = 4,000K$. For the grain melting temperature we have assumed the "canonical" value of $T_1 = 1,000K$. The protostar is therefore sitting in a dust free cavity of radius $\sim 20R_*$ (where $R_*$ is the protostar radius). The dust cloud is assumed to span five orders of magnitude (i.e. $r_1/r_2 = 10^{-5}$). The spectra are computed at a distance of $\sim 2 \times 10^8 R_*$ and at viewing angles $0, \pi/16, \pi/8, 3\pi/16, \pi/4$ and $\pi/2$. It is generally found that the spectrum varies very little for viewing angles greater than $\pi/4$ so we are covering mainly the 0 to $\pi/4$ range where the orientation becomes of crucial importance.

Fig. 1 shows the variation of the spectrum of a deeply embedded source ($\tau_{uv}(0) = 300$) and a very extended disc ($R_C = 10^3 r_1$). The corresponding optical depths for the viewing angles for which the spectrum is computed are 300, 110, 86, 74, 69 and 65. With increasing $\theta_v$ the

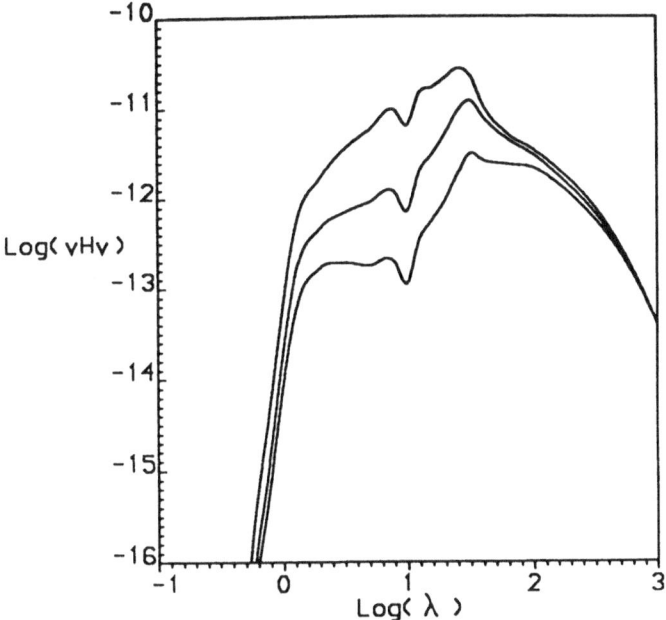

**Figure 2.** *Variation of the average spectrum ($\theta_v = \pi/4$) with $R_C$. In order of increasing mid-infrared $Log \nu H_\nu$, $R_C = 10^3 \times r_1$, $10^2 \times r_1$ and $10 \times r_1$. All models assume $r_1/r_2 = 10^{-5}$, $T_s = 4,000K$ and an optical depth $\tau_{uv}(\pi/4) = 90$.*

optical depth to the protostar decreases and therefore the short wavelength flux originating from the inner part of the cloud increases. Note that the edge-on view is severely absorbed as we are looking through the centrifugal barrier. The far infrared flux remains unaffected for wavelengths longer than about $60\mu m$. The near and mid infrared spectra, however, are very much dependent on the viewing angle (for $\theta_v < \pi/4$) which should therefore be taken into account when attempting detailed modeling of protostellar sources. Estimates of the orientation of objects in the bipolar ouflow phase will prove very useful in constraining further the model parameters and determining whether our models of cloud collapse are satisfactory.

The effect of changing the centrifugal radius is shown in Fig. 2. A lower value of $R_C$ may represent either a lower rotation rate $\Omega$, an earlier phase in the collapse and evolution of the system so that there has not been enough time for the disc to grow, or a higher sound speed $\alpha$ (c.f. equation (2)). The result is that the isodensity contours become less flattened and there is therefore less variation in the spectra from edge-on to face-on views. $R_C$ also has profound effects on the shape of the spectral energy distribution irrespective of the viewing angle. The effect of increasing $R_C$ is to change the slope of the near and mid-infrared spectrum as a result of the change in the slope of the density profile. This is a very interesting and useful result that can be used as a diagnostic of the extent of the disc.

In Fig. 3 we compare spectra of models of decreasing equatorial optical depth while

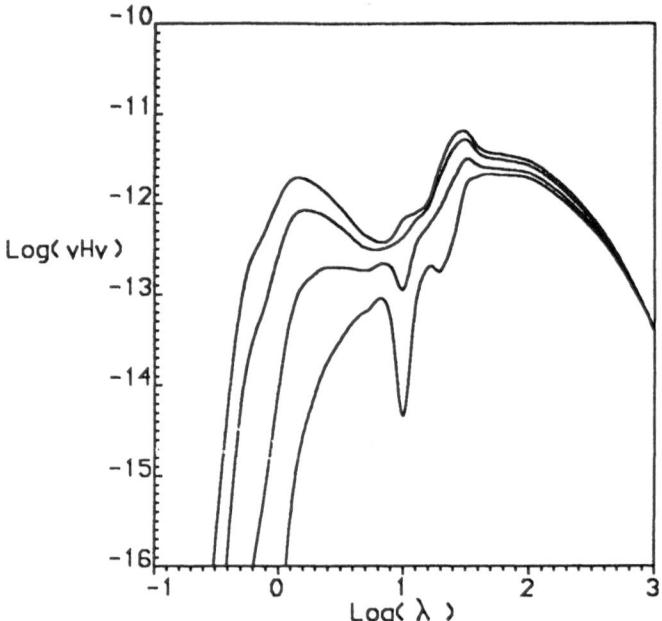

**Figure 3.** *Variation of the average spectrum ($\theta_v = \pi/4$) with $\tau_{uv}$. In order of increasing short wavelength flux, $\tau_{uv}(\pi/4) = 180, 90, 45, 30$. All models assume $r_1/r_2 = 10^{-5}$, $R_C = 10^3 r_1$ and $T_s = 4,000K$.*

keeping the centrifugal radius constant ($R_C = 10^3 r_1$). For the viewing angle we assume the average value of $\pi/4$. This sequence of models may be seen to represent the spectral evolution of a source in later stages when the envelope has been diluted whereas accretion still takes place through the disc (which is still optically thick). Pursuing this idea even further it is interesting to see that as the optical depth decreases the spectrum assumes a double peak form (with the disc emission peaking in the far infrared and the attenuated starlight in short wavelengths) which is characteristic of objects such as VSSG23 (Chini 1981, Wilking and Lada 1983).

## 5 Comparison with the observations

Models computed with the method described here have been succesful in fitting the observed spectra of several protostellar candidates (Efstathiou 1989, also Efstathiou and Rowan-Robinson 1989a). We present here a model fit to the spectrum of IRS5 L1551, the prototypical bipolar outflow source in the Taurus molecular cloud. Fig. 4 shows a fit to the observed spectrum of IRS5 L1551 with a model having the following parameters ($T_s = 6,600K$, $r_1/r_2 = 2 \times 10^{-5}$ and $R_C/r_2 = 2 \times 10^{-2}$). The optical depth along the required viewing angle $\theta_v = \pi/8$ is $\tau_{uv} = 140$. The data are taken from Cohen and Schwartz (1983), Cohen *et al.* (1984), Davidson and Jaffe (1984) and IRAS data. The model predicts a high value of $R_C$ and therefore a highly flattened density distribution which is consistent with

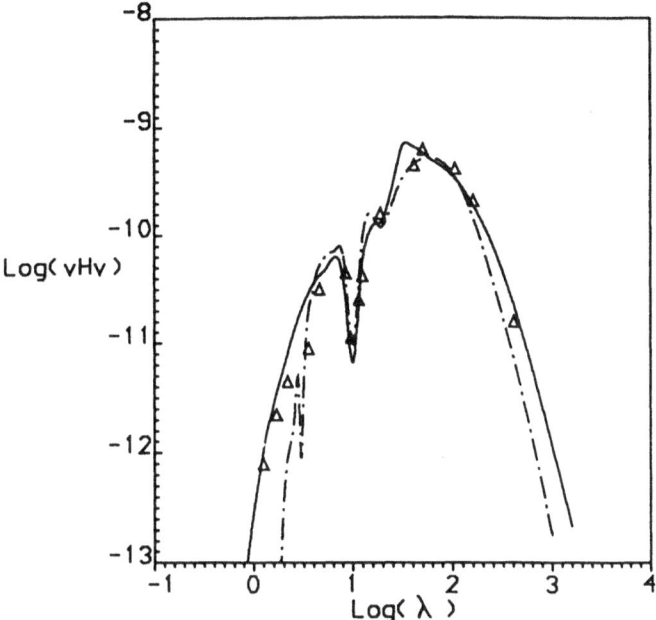

**Figure 4.** *Model fit to IRS5 L1551 (solid line). The model assumes $T_s = 6,600K$, $r_1/r_2 = 2 \times 10^{-5}$, $R_C/r_2 = 2 \times 10^{-2}$ and an optical depth of 140 along a viewing angle of $\pi/8$ (see text for data references). Also shown for comparison (dot-dashed line) is the model of Adams, Lada and Shu (1987).*

the observed morphology and what one would expect by theory for an object in the bipolar outflow phase. On the basis of the ratio of radial and tangential velocities of Herbig-Haro objects associated with IRS5 L1551 the viewing angle is estimated to be about 20° (Strom, Grasdalen and Strom 1974, Cudworth and Herbig 1979). This value agrees rather well with that predicted by our model. Also shown in Fig. 4 is the model fit of Adams, Lada and Shu (1987). Although their model shows a good agreement with the observations in the mid and far infrared, in shorter wavelengths their fit is very poor due to the neglect of scattering and the approximate treatment of the geometry.

**Conclusions.**

For the first time we have carried out detailed radiative transfer calculations for the dust clouds surrounding young stellar objects taking into account the axisymmetric geometry. We have shown the effect of different model parameters on the predicted spectral energy distributions and their dependence on the viewing angle. The centrifugal radius $R_C \sim M^3\Omega^2/\alpha^8$ is the most important parameter that determines the shape of the emergent spectral energy distribution.

We have also presented a model that gives a very good overall fit to the observed spectrum of IRS5 L1551, the prototypical bipolar outflow source. Detailed modeling of a large sample of protostellar sources is promising to be very fruitful in further understanding of the evolution of young stellar objects and pointing out any theoretical inconsistencies. Any further observational constraints such as rotation rates, gas temperatures and orientations will prove invaluable.

### Acknowledgements

We would like to thank J.Hughes for useful discussions and suggestions. A.E. was supported in part by an ORS award and studentships awarded by Drapers' Co./Queen Mary College and the University of London.

## References

Adams, F.C., and Shu, F.H., 1986, *Astrophys.J.*, 308, 836.
Adams, F.C., Lada, C.J., and Shu, F.H., 1987, *Astrophys.J.*, 312, 788.
Bodenheimer, P., and Black, D.C., 1978, in *'Protostars and Planets I'*, ed.T.Gehrels,
Univ. of Arisona Press, Tucson.
Boss, A.P., 1980, *Astrophys.J.*, 237, 563.
Cassen, P., and Moosman, A., 1981, *Icarus*, 48, 353.
Chini, R., 1981, *Astr.Astrophys.*, 99, 346.
Cohen, M., and Schwartz, R.D., 1983, *Astrophys.J.*, 265, 877.
Cohen, M., Harvey, P.M., Schwartz, R.D., and Wilking, B.A., 1984, *Astrophys.J.*, 278, 671.
Cudworth, K.M., and Herbig, G.H., 1979, *Atron.J.*, 84, 548.
Davidson, K., and Jaffe, D.T., 1984, *Astrophys.J.Lett.*, 255, L103.
Efstathiou, A., 1989, *Ph.D. thesis*, University of London, in preparation.
Efstathiou, A., and Rowan-Robinson, M., 1989, *Mon.Not.R.astr.Soc.*, submited.
Efstathiou, A., and Rowan-Robinson, M., 1989a, in preparation.
Herbig, G.H., 1970, *Astrophys.J.*, 162, 557.
Rowan-Robinson, M., 1980, *Astrophys.J.Suppl.*, 234, 111.
Rowan-Robinson, M., 1982, *Mon.Not.R.astr.Soc.*, 201, 289.
Rowan-Robinson , M., and Harris, S., 1982, *Mon.Not.R.astr.Soc.*, 200, 197.
Shu, F.H., 1977, *Astrophys.J.*, 214, 488.
Shu, F.H., Adams, F.C., and Lizano, S., 1987, *Ann.Rev.Astr.Astrophys.*, 25, 23.
Strom, S.E., Grasdalen, G.L., and Strom, K.M., 1974, *Astrophys.J.*, 191, 111.
Terebey, S., Shu, F.H., and Cassen, P. 1984, *Astrophys.J.*, 286, 529.
Wilking, B.A., and Lada, C.J., 1983, *Astrophys.J.*, 274, 698.

# THE INFRARED SPACE OBSERVATORY (ISO)

M.F. KESSLER[1] and L. METCALFE[1]
[1] *Astrophysics Division,*
*Space Science Department of ESA,*
*Postbus 299, 2200 AG Noordwijk,*
*The Netherlands.*

ABSTRACT. The Infrared Space Observatory (ISO) will provide astronomers with unique facilities for making imaging, spectroscopic, photometric and polarimetric observations at infrared wavelengths ($2.5 - 200\,\mu$m) for a period of at least 18 months. Two-thirds of ISO's observing time will be available to the astronomical community via the traditional route of submission of proposals. Its 60-cm diameter cryogenically-cooled telescope will be equipped with a suite of four complementary and versatile instruments, currently being developed by consortia of scientific institutes. ISO is a fully approved and funded project of the European Space Agency (ESA) with a foreseen launch date of May 1993.

## 1. Introduction

The study of dusty objects and dust itself is the natural province of infrared astronomy. It is at infrared wavelengths that the bulk of their thermal energy is emitted. Also, many spectral features giving clues as to the chemical composition of the dust occur in this region. Yet, despite its rich scientific promise, the $3-200\,\mu$m spectral range has remained relatively under-explored due to difficulties imposed by the terrestrial atmosphere on observations at these wavelengths. IRAS, during its brief 10-month orbital mission, returned such a rich harvest of data that, 6 years later, fresh science is still being produced. Compared to IRAS, the Infrared Space Observatory (ISO) will have a longer lifetime, more sophisticated instrumentation, greater sensitivity, wider wavelength range and better angular resolution. ISO will give astronomers the capability of routinely making high sensitivity observations at these wavelengths, but for detailed study of individual objects rather than as a survey mission.

An ISO Science Team provides advice to ESA on all scientific aspects of ISO during the lifetime of the project. This team consists of the Principal Investigators of the four instruments (Catherine Cesarsky, Saclay; Peter Clegg, London; Thijs de Graauw, Groningen; Dietrich Lemke, Heidelberg), five Mission Scientists (Thérèse Encrenaz, Paris; Harm Habing, Leiden; Martin Harwit, Washington; Alan Moorwood, ESO; Jean-Loup Puget, Paris), the ESA Payload Manager (Michel Anderegg, ESTEC) and is chaired by the ESA Project Sci-

*E. Bussoletti and A. A. Vittone (eds.), Dusty Objects in the Universe, 255–266.*
© 1990 *Kluwer Academic Publishers.*

entist (Martin Kessler, ESTEC). The rôles of the Mission Scientists are to provide scientific input to the project and to represent the interests of the general astronomical community.

ISO will be a true observatory, open to the European and American (via NASA) astronomical communities, and will offer a wide range of instrumentation capable of tackling a large variety of astrophysical problems. This paper is designed to give potential observers an overview of the scientific capabilities of the mission and details of the how and when observing time will be allocated.

## 2.  Observing Time and Operations

The observing time on ISO will be allocated on a "per object" basis, as was the case for EXOSAT, rather than on a "per shift" or "per night" basis as is done with IUE and most ground-based telescopes. Nearly two-thirds of the observing time will be available to the scientific community via the traditional route of proposal submission, review and selection. In addition to this *Open Time*, there will also be *Guaranteed Time* for the groups who provide the instruments, for the Mission Scientists and for the Observatory Team, who will be responsible for all scientific operations. The division of time between these two categories will vary as the mission progresses. After launch, it is anticipated that there will be a period of up to 8 weeks during which the operational orbit will be attained, the spacecraft sub-systems switched on and checked out and the scientific performance of the instruments established. Following this, there will be a 1-month period, consisting of 50% *open time* and 50% *guaranteed time*, during which astronomical observations, designated by the *Observing Time Allocation Committee* as being of the highest priority, will be carried out. For the rest of the mission (at least 15 months), 65% of the time will be *open time*.

The first *Call for Observing Proposals* will be issued 18 months before launch. It will contain details of expected instrument performances and will solicit proposals for observations to be carried out in the period from 3 to 10 months after launch. Due to the large number of observations expected to be proposed for ISO, the proposal-handling system will be automated as much as is possible. Thus, proposals must be submitted electronically. ESA intends to issue a software package –to run on the proposers' own computer– to help the community prepare these electronic proposals and to check them before transmission to the Observatory, either via network file transfer or on floppy disks. In order that the best use can be made of ISO's limited lifetime, there will be a review of the implementation of the observing programme about 5 months after launch; if actual instrument performances differ from those predicted, the *Observing Time Allocation Committee* will recommend suitable adjustments to the programme. During the in-orbit phase of the mission, it is planned to issue two further *Calls for Proposals* so as to give the community the opportunity to react to initial results.

The in-orbit operations of the spacecraft and instruments will be carried out by a team of scientists and engineers located at the ISO Control Centre in Villafranca, near Madrid, Spain. This site is currently used by the IUE Observatory. During scientific use, the satellite will always be in real-time contact with the ground segment; however, ISO will be operated according to a detailed, pre-planned schedule in order to maximise the overall efficiency of the mission. The required astronomical flexibility will be provided by the concepts of "branched" and "linked" observations. In the former, a default and one (possibly more) optional set of operations will be defined and prepared in advance of the observation. Then, based on an immediate examination of the incoming data , a decision can be made, in real-time, to branch from one option to another. For safety reasons, there will be restrictions on the range of choices available in a branch. For linked observations, a "test" exposure will be made on an orbit; the results examined over the course of a day or so; and a decision made, off-line, as to the final parameters to be used for the "main" observation to be scheduled several orbits later.

Examination of the scientific data will be carried out both on- and off-line. In close to real-time, a "quick-look" output, adequate for an initial estimate of the success or failure of an observation will be available to the Resident Astronomer and Guest Observer (if present). This assessment will be used to make the branching decisions. Within a few days of an observation being completed, observers will be supplied with a standard set of products from which they may make their astronomical analyses. The distributed products will contain: reformatted raw data; processed data (with instrument-specific artifacts removed); some extracted scientific results to give a rough indication of the main result of the observation; calibration data; and various auxiliary data (e.g. any relevant satellite information). The baseline distribution method is a tape in FITS format but other possibilities are being considered.

## 3. Spacecraft and Mission

The ISO satellite (5.3 m high, 2.3 m wide and 2400 kg at launch) is dominated by its so-called "Payload Module", the upper cylindrical part seen in figure 1. This module, shown in section in figure 2, is essentially a large cryostat, which provides the cooling power necessary to keep the telescope and the scientific instruments at temperatures between 1.8 and 4K for an in-orbit lifetime of at least 18 months. Inside the vacuum vessel is a toroidal tank filled with about 2250 litres of superfluid helium. Some of the infrared detectors are directly coupled to this helium tank and are at a temperature of around 2 K. Apart from these, all other units are cooled using the cold boil-off gas from the liquid helium. This gas is first routed through the optical support structure to cool the telescope and the scientific instruments. It is then passed along the optical baffles and radiation shields before being vented to space. For the last 72 hours before launch, when access to the satellite is impossible as the launcher is being prepared, ISO's cooling needs are met by evaporating normal liquid helium from a small 60 litre tank using electrical heating and by routing this gas along the normal ISO cooling loops.

Suspended in the middle of the main helium tank is the telescope, which has an aplanatic Ritchey-Chrétien configuration with an effective aperture of 600 mm and an overall f/ratio of 15. A weight-relieved fused-silica (Herasil I) primary mirror and a solid fused-silica secondary mirror have been selected as the telescope optics. The optical quality of these mirrors is adequate for diffraction-limited performance at a wavelength of 5 $\mu$m. (Note, however, that the ISO system is only capable of diffraction-limited performance at wavelengths above $\sim$ 10 $\mu$m due to the pointing performance). Stringent control of stray light, particularly from bright infrared sources outside the telescope's field of view, is necessary in order to ensure that the system sensitivity is not degraded. This control is accomplished by

258

(i) a light-tight shield around the instruments (ii) equipping the mirrors with baffles (iii) by the main optical baffle, which protects the telescope from off-axis stray light and thermal self-emission from the sunshade (iv) the sunshade, which prevents earth and sun radiation from entering the telescope and (v) imposition of viewing constraints.

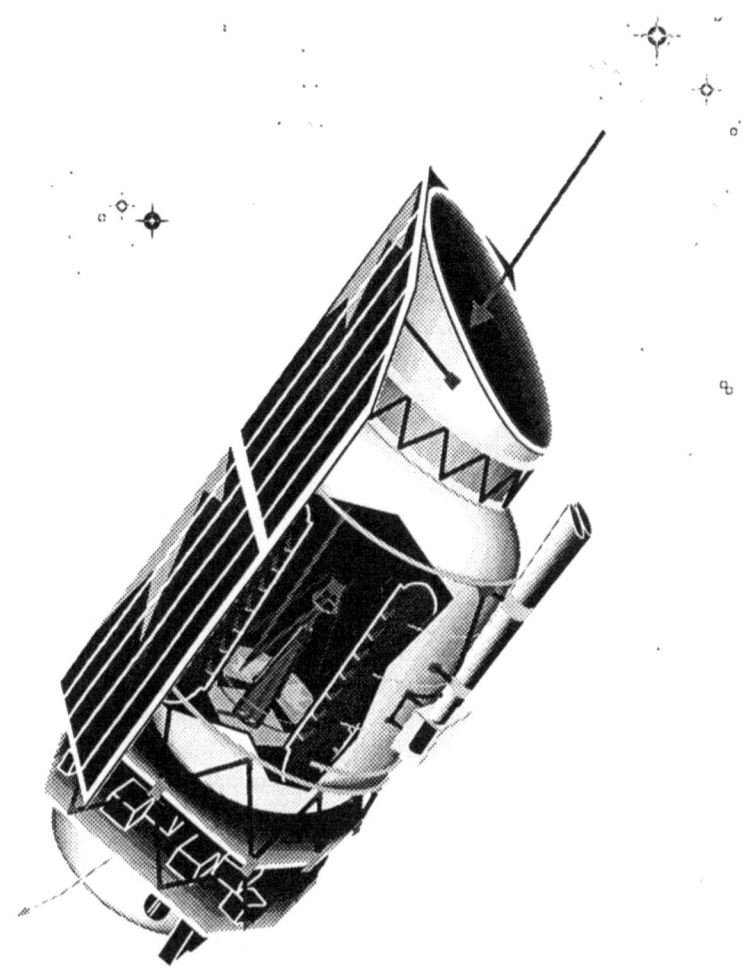

*Fig. 1. Computer Graphic of ISO in Orbit*

The scientific instruments are mounted on the opposite side of the optical support structure to the primary mirror, each one occupying an 80 degree segment of the cylindrical volume available. The 20 arc minute total unvignetted field of view of the telescope is split up between the four instruments by a pyramidal mirror. Thus, each instrument simultaneously receives a 3 arc minute unvignetted field centred on an axis at an angle of 8.5 arc minute to the telescope optical axis. To view the same target with different instruments, the satellite has to be repointed.

Mounted on one side of the outer vacuum vessel is a sunshield, which prevents the sun from shining directly on the cryostat and also carries the solar cells. On the other side are the two star trackers.

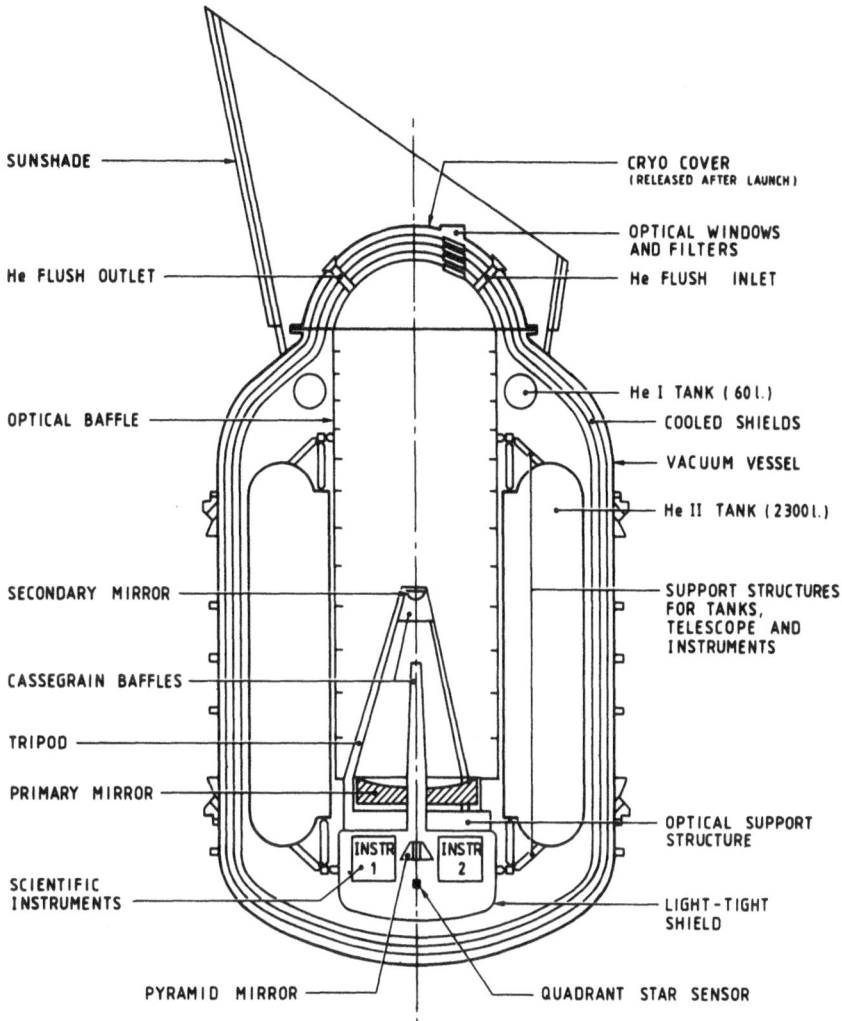

Fig. 2. Schematic of the Payload Module

The "classical" spacecraft functions are provided by a so-called "Service Module", which can be seen underneath the payload module in figure 1. These functions include the structure and the load path to the launcher, the solar array mounted on the sunshield, and sub-systems for thermal control, data handling, power conditioning, telemetry and telecommand (using two antennas with a nominal down-link bit rate of 33 kbps of which about 24 kbps are dedicated to the scientific instruments), and attitude and orbit control. The last provides the three-axis stabilisation and the raster pointing facilities needed for the mission. The requirement on relative pointing error (equivalent to jitter) is 2.7 arc secs ($2\sigma$, half cone, over a period of 30secs. The requirement on the absolute pointing error (the angular separation between the commanded direction and the instantaneous actual direction) is 11.7 arc secs ($2\sigma$, half cone). Sensors used are: sun acquisition sensors to provide coarse information on the position of the sun; fine sun sensors for accurate sun position information; earth limb sensors, covering both "forbidden" and "warning" zones for the spacecraft attitude with respect to the Earth; star trackers as the prime pointing sensors; a quadrant star sensor on the optical axis of the telescope to calibrate the mis-alignment between the star trackers and the telescope axis; and four rate-integrating gyros in an all-skewed configuration. The main actuators are four reaction wheels, also in an all-skewed configuration. These wheels can be unloaded by use of the hydrazine reaction control subsystem, which is also used for orbit acquisition and maintenance.

After launch by an Ariane-4 vehicle into a transfer orbit, ISO's hydrazine reaction control system will be used to attain the operational orbit, which has a 24-hour period, a perigee height of 1000 km and an apogee height of 70000 km. The inclination to the equator will be between 5 degrees and 20 degrees (to be finalised later). With the one ground station that ESA intends to fund, ISO will be able to make astronomical observations during the best 14 hours of each orbit. However, the scientific return of the mission could be greatly increased by use of a second ground station, which would (i) permit ISO to be operated at maximum sensitivity for up to another 3 hours per day, (i.e. for the entire time that it spends outside the Earth's radiation belts) and (ii) provide additional time during which some of the instruments could be used and spacecraft maintenance be carried out. An international collaboration is being sought by ESA to provide this second ground station.

## 4. Scientific Instruments

The ISO scientific payload consists of an imaging photopolarimeter (ISOPHOT), a camera (ISOCAM), a short wavelength spectrometer (SWS) and a long wavelength spectrometer (LWS). Each instrument is being built by a consortium of scientific institutes using national non-ESA funding under the authority of the Principal Investigator (PI); however, they have been designed as a package to offer complementary facilities to the observers. The four instruments view different adjacent patches of the sky, but, in principle, only one will be operational at a time. However, when the camera is not the prime instrument, it can be operated in a so-called parallel mode, either to gather additional astronomical data or to assist another instrument in acquiring and tracking its target. In order to maximise the scientific return of the mission, the ISOPHOT instrument will be operated during as many satellite slews as possible so as to make a partial sky survey at a wavelength of 200 $\mu$m, a region not explored by IRAS.

The scientific capabilities of the total payload for spectroscopy, photometry, polarimetry and imaging are shown in figure 3. In summary, ISO will be capable of **imaging** in broad and narrow spectral bands at a variety of spatial resolutions across its entire wavelength range of 2.5–200 $\mu$m. **Photometry** and **polarimetry** will be possible in broad and narrow spectral bands across the complete wavelength range with multiple apertures available out to a wavelength of 110 $\mu$m. For **spectroscopy**, a variety of resolving powers, ranging from 50 to $10^4$, will be available at wavelengths from 2.5–180 $\mu$m.

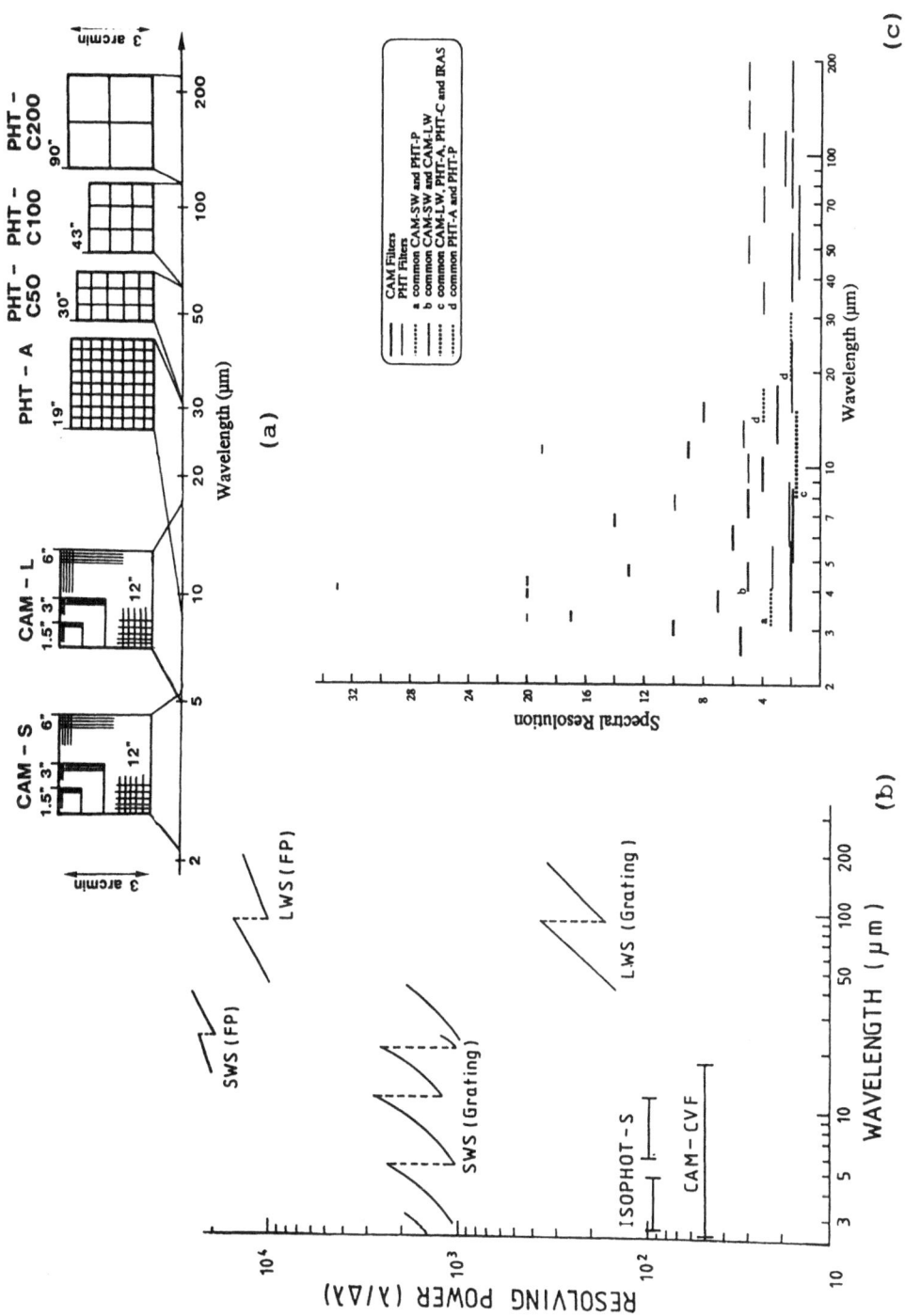

Fig. 3. Scientific Capabilities of the ISO payload.
(a) Imaging. (b) Spectroscopic. (c) Photometric and polarimetric.

262

Investigators:

- PI: D. Lemke, MPI Astronomie,
  Heidelberg, D,
- CoI's from D, DK, E, GB, IRL, USA

Wavelength Range:
- 2.5 - 200 microns,
Spectral Resolving Power:
- broad- and narrow-band filters,
- near-ir grating spectrophotometer with R approximately 90,

Spatial Resolution:
- variable from diffraction-
  limited to wide beam,

Outline Description:
- Four sub-systems: Photopolarimeter (P),
  Camera (C); Imaging array (A);
  Spectrophotometer (S).

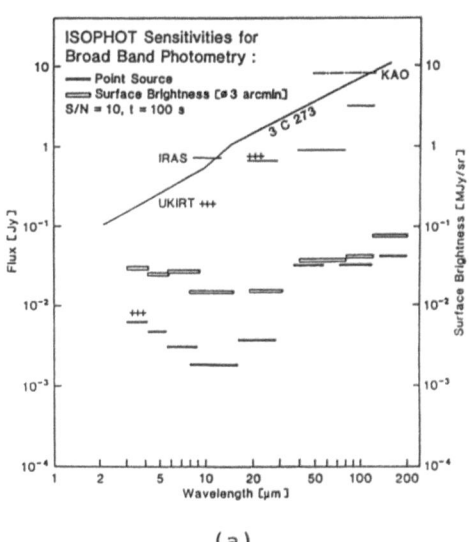

(a)

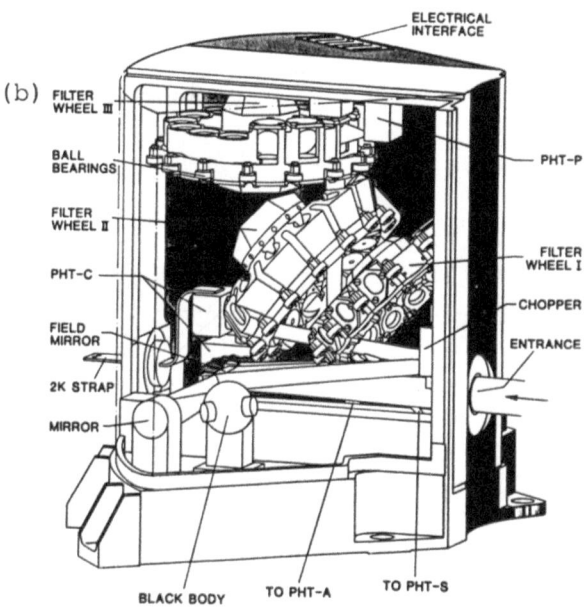

Fig.4. ISOPHOT. (a) ISOPHOT sensitivities for broad-band photometry showing limiting detectable flux scaled to an integration time of 100s for a signal-to-noise ratio of 10. Limiting surface brightness values refer to the full 3' ISOPHOT f.o.v. (b) Schematic of ISOPHOT.

Investigators:

- PI: C.J. Cesarsky, Saclay, F,
- CoI's from F, GB, I, S, USA,

Wavelength Range:
- 2.5 - 17 microns,

Spectral Resolving Power:
- broad- and narrow-band filters,
- circular variable filters,

Spatial Resolution:
- pixel fields of view of 1.5, 3,
  6, and 12 arcseconds,

Outline Description:
- two channels, each with a 32x32
  element detector array.

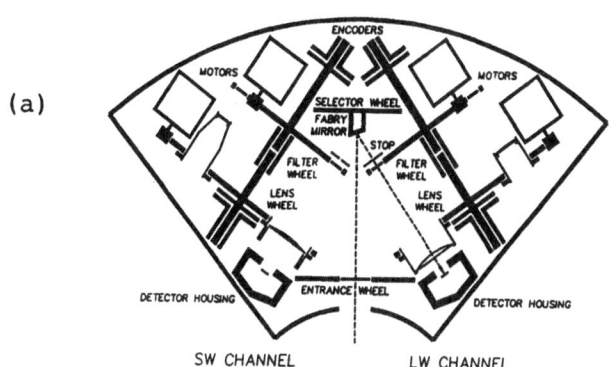

(a)

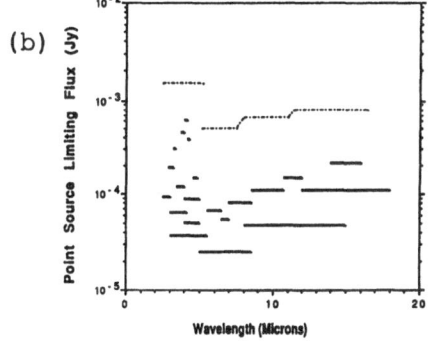
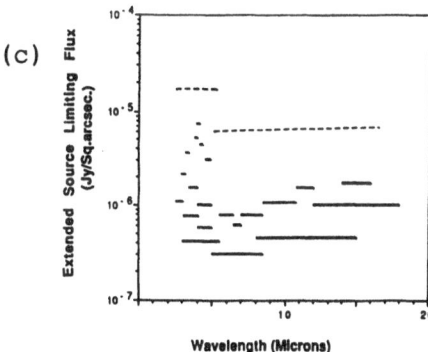

Fig.5. The ISO camera (ISOCAM). (a) Schematic of ISOCAM. (b) and (c) ISOCAM limiting flux for point sources and extended sources, respectively. The sensitivities are scaled to a total observation time of 3600s for a 3 sigma detection using the 6" per pixel f.o.v.

Investigators:
- PI: Th. de Graauw, Space Research,
  Groningen, NL,
- CoI's from D, NL, USA

Wavelength Range:
- 2.5 - 45 microns,
Spectral Resolving Power:
- 1000 across the entire wavelength
  range and 20000 from 15-35 microns,

Spatial Resolution:
- 10 x 20 and 20 x 30 arcseconds,

Outline Description:
- Two gratings and two Fabry-Perot
  Interferometers.

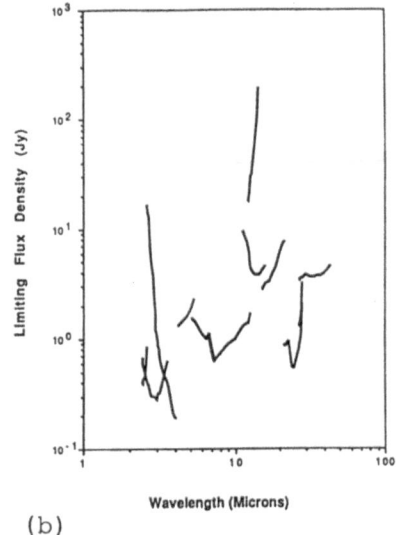

(b)

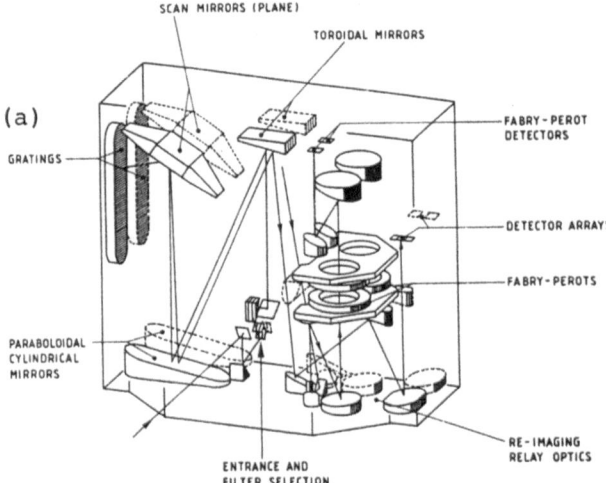

Fig.6. The SWS. (a) Schematic of SWS. (b) Limiting spectral flux density detectable with SWS in an observation time of 15 minutes per point, at a signal-to-noise ratio of 20, in the grating mode.

265

Investigators:
- PI: P.E. Clegg, Queen Mary Westfield,
  London, GB,
- CoI's from F, GB, I, USA,

Wavelength Range:
- 45 - 180 microns,
Spectral Resolving Power:
- 200 and 10000 across the entire
  wavelength range,

Spatial Resolution:
- 1.65 arcminutes,

Outline Description:
- Grating and two Fabry-Perot
  Interferometers.

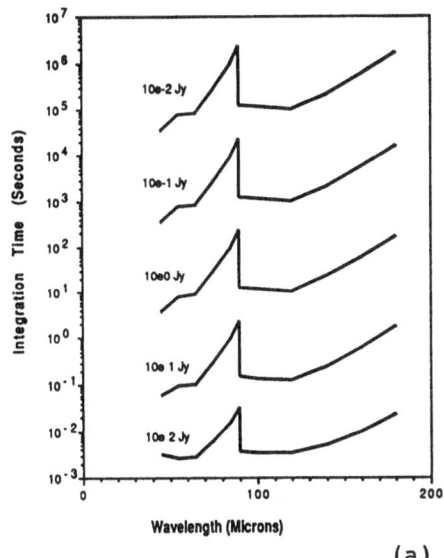

(a)

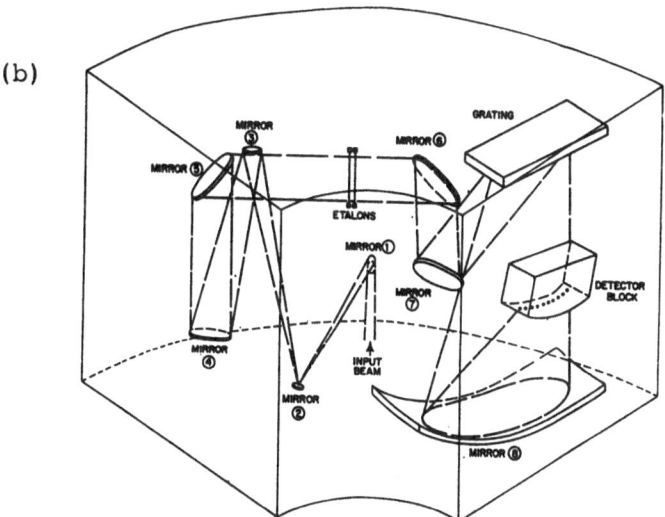

(b)

Fig.7. The Long Wavelength Spectrometer (LWS).
(a) LWS Sensitivity. Integration time required, in grating
mode, to achieve a signal-to-noise ratio of 10, for the
spectral flux densities indicated on the curves.
(b) Schematic of LWS.

A schematic drawing of each instrument together with an outline description and an estimate of its sensitivity are given in figures 4–7. Further details of the individual instruments can be found in Lemke (1989) for ISOPHOT, Cesarsky et al. (1989) for ISOCAM, de Graauw et al. (1989) for SWS and Emery et al. (1985) for LWS.

## 5. Status

ISO results from a mission proposal submitted to ESA in 1979. After a feasibility study (phase A) in 1981-2, it was selected in March 1983 to be the next new start in the ESA Scientific Programme. The definition study (phase B) for the ISO spacecraft was successfully completed early in 1988. The detailed design, development, integration and test phase (phase C/D) was started on 15 March 1988 by an industrial consortium led by Aerospatiale (F). Currently the first model of the Payload Module is being integrated and the first cool-down is scheduled for the end of 1989. The foreseen launch date is May 1993.

The scientific instruments were selected in mid-1985. Their development is well advanced with the first models, the alignment and mass/thermal dummies, having been delivered to ESA. The qualification models, which are very similar to the eventual flight units, are under construction for delivery in Summer 1990.

## 6. Conclusion

ISO is a fully-approved and funded mission, which will offer astronomers unique and unprecedented observing opportunities at infrared wavelengths from 2.5–200 μm for a period of at least 18 months. Two thirds of the observatory's time will be available to the general astronomical community. Both the spacecraft and its selected complement of instruments are in their main development phase and the scheduled launch date is May 1993.

## 7. References

Emery, R.J., et al, "The Long Wavelength Spectrometer (LWS) for ISO",
    Proc. SPIE 589, 194–200 (1985).
Cesarsky, C., Sibille, F. and Vigroux, L., "ISOCAM, a Camera for the Infrared Space
    Observatory", Proc. SPIE 1130, 202–213 (1989).
de Graauw, Th. et al, "The ISO Short Wavelength Spectrometer", Proc. 22nd ESLAB
    Symposium on Infrared Spectroscopy in Astronomy, ESA SP-290, 549–551 (1989).
Lemke, D., Burgdorf, M., Hajduk, Ch., and Wolf, J., "Detectors and Arrays of ISO's
    Photopolarimeter", Proc. SPIE 1130, 194–200 (1989).

# MODULATION AT 10μm WITH ARRAY DETECTORS ?

S. REMY, P.O. LAGAGE
Service d'Astrophysique
CEN Saclay
91191 Gif-Sur-Yvette
France

ABSTRACT. The main difficulty to detect faint astronomical sources through the 10μm atmospheric window is to get rid of the huge photon background generated by the telescope and the atmosphere. This difficulty has been cleared up for observations with photometers using the chopping technique. Now that array detectors are available, other solutions to this problem can be investigated. We present here theoretical investigations on various observing techniques (chopping, nodding, staring), to help in finding the best ways to observe at 10μm with array detectors.

## 1. Are 10μm Ground-Based Observations Still Of Interest ?

Given that a 10μm survey of the whole sky has already been completed by the IRAS satellite (Neugebauer et al., 1984) and that the ISO satellite, scheduled for 1993, will be dedicated to pointed observations (see M. Kessler, these proceedings), the utility of ground-based observations has to be questioned before discussing the observing techniques.

The enormous advantage of observations from a satellite results from the absence of atmosphere which, combined with the cooling of the telescope, permits to have a very low photon background and thus a better sensitivity. For example, the expected sensitivity to extended objects of the ISOCAM camera, one of the four ISO instruments, is several orders of magnitude better than the sensibility expected on ground; even if this advantage is reduced for point sources, it remains at the level of an order of magnitude (Cesarsky et al., 1989). Furthermore, the whole spectral range is accessible, when ground-based observations are limited to a few atmospheric windows. Nevertheless, space observations suffer from one weakness, when compared with ground-based observations: the relatively poor angular resolution. Indeed, the telescopes embarked are much smaller than the large telescopes (an order of magnitude); it follows that the diffraction is larger and the angular resolution poorer.

Thus, subarcsec observations of objects relatively intense are typically in the area to be covered from ground. As examples of programmes particularly exciting, one can mention:

i) the mapping of the 10μm emission of the ultraluminous galaxies discovered by IRAS (for example, Sanders et al., 1988 and references therein) in order to disentangle the star-burst component to the AGN like component,

ii) the mapping of βPic in order to see what happens near the star, where the infrared emission is conjectured to be very low, perhaps because grains have coalesced into larger bodies (Telesco et al., 1988 and references therein).

*E. Bussoletti and A. A. Vittone (eds.), Dusty Objects in the Universe*, 267–271.
© 1990 *Kluwer Academic Publishers*.

## 2. On The Difficulties To Observe At 10μm From Ground

It is well known since a long time that the 10μm observations from ground are difficult; the reason being the large background emitted by the environment (atmosphere and telescope). Indeed, from Mister Planck we know that the emissivity of a blackbody at a temperature T peaks at a wavelength given by formula $\lambda_m = T / 2898$, where $\lambda_m$ is expressed in micron.

Our environment being at 300 K, the peak is right in the 10μm region. Thus, we have to deal with a large amount of undesirable photons. To be quantitative, about $10^{10}$ photons arrive every second in a beam of one arcsec. The $1\sigma$ photon noise associated is $10^5 \times (t_{int})^{1/2}$, where $t_{int}$ is the integration time. Thus, in 1 second of integration, a source detected at $10\sigma$ is already $10^4$ fainter than the background. For a few hours of integration, we can achieve the level of $10^6$. Thus, we really have to search for very faint sources embedded in a large background.

The problem is not new and has already been resolved for the observations with photometers. The solution is obviously modulation, that means take one observation on source, another one off source and make the difference. The frequency at which such a modulation has to be done has been the object of extensive studies (for example, Papoular, 1983 and references therein); the sky temporal fluctuations (see Fig. 1) make the frequency to be higher than about 30Hz. As it is impossible to move the telescope so rapidly, the use of secondary moving mirrors has been developed for infrared observations.

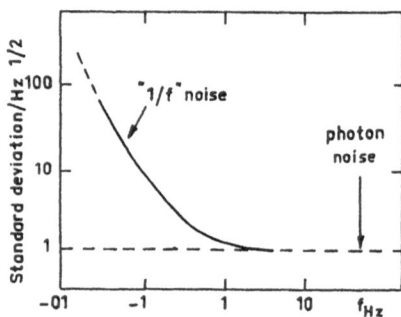

Fig. 1. Typical spectral power densities of background fluctuations

## 3. Observations With Array Detectors

Now that array detectors are available, the problem of the best way of getting rid of the background has to be tackled again. Some answers already exist in the literature but they are somewhat different; for example, Gezari et al. (1989) keep on chopping, when Arens et al. (1988) have tried the staring mode (no modulation).

### 3.1. CHOPPING

Of course the technique developed for photometers is still valid, as an array can be considered as a collection of photometers.

But secondary chopping mirrors are not available on all the telescopes. Furthermore, the question of the need of moving secondary mirrors has to be answered for the future large telescopes; and, in case of positive answer, the frequency at which the modulation has to be done is an important input.

## 3.2. STARING

At first glance, it could be thought that the modulation is no more necessary with array detectors. Indeed, some pixels of the array can be used to monitor the sky evolution; in this case, we have in the same time the sky and the source plus sky, so that we are no longer faced with the problem of temporal fluctuations of the atmosphere. (Of course, this is only possible for sources less extended that the field of view of the array; but as ground-based observations are above all interesting for the detection of objects with weak angular extension, we will only consider this case in the following).

But we meet another problem: the flat field problem. Indeed, as the detectors are not totally identical, we have to correct from this unhomogeneity before making the difference between the pixels on source and off source. This corrective factor is obtained by exposing the array to an uniform background. For the sake of simplicity, let us consider only 2 pixels, pixel 1 and pixel 2, and let us note by $G_1$ and $G_2$ their respective response function; in our notation, B is the background and s the source to be detected. As s is small compared to B, we can make a limited development of $G_1$ so that

$$G_1(B+s) = G_1(B) + dG_1/dB \text{ x s.}$$

Before subtracting the 2 pixels we have to make a flat field correction. If $COR_1$ and $COR_2$ are the flat field correcting factor, the result of the treatment is:

$$COR_1 \text{ x } dG/dB \text{ x } s + (COR_1 \text{ x } G_1(B) - COR_2 \text{ x } G_2(B)).$$

As we can see in the staring mode, we can achieve the detection of an astrophysical source only up to the flat field precision. (Note that the factor in front of s can be calibrated from a reference source).

The flat field precision is limited by several factors, such as the unhomogeneity of the background or the detectors stability and so on... A flat field precision of 1% is easily reached; obtaining $10^{-3}$ is already more difficult; the $10^{-7}$ required in § 2 seems really out of possibility.

Thus, beyond a limit determined by the flat field precision, the staring mode is no more valid and some modulation is indispensable.

## 3.3. MODULATION

The modulation makes a pixel to be successively on source and off source, so that the flat field problem disappears. But, as we will see, observations with arrays still permit to alleviate the problem of the sky temporal fluctuations. The recipe is the following. Let us mark with an indice a the set of images on source and with an indice b the set of images on sky. Then, with our notation, we measure first $G_1(B_a+s)$ and $G_2(B_a)$ and then $G_1(B_b)$ and $G_2(B_b)$, with $B_b = B_a + \Delta B$, where $\Delta B$ is the noise excess defined in §2. The technique consists in making the difference between a and b, then correcting from the flat field (but this time on dG/dB) and to finish, substracting pixel 1 to pixel 2. We end up with the quantity:

$$COR_1 \times dG_1/dB \times s + (COR_2 \times dG_2/dB - COR_1 \times dG_1/dB) \times \Delta B$$

Compared with the formula obtained for photometers $(s + \Delta B)$ we can see that, using the recipe recommended before, we are immunize against the sky fluctuation by a factor equal to the flat field precision. Consequently, we can modulate at frequencies much lower that the usual frequency of photometers.

Note also that the spatial amplitude of the modulation can be such that the source always stays in the field of the array, so that we do not loose effective integration time by modulating.

## 4. Detector Role

Till now we have considered only perfect detectors in the sense that their noise was less important that the photon noise. If it is not the case, the detector and its electronic "1/f" excess noise could play the same role as the sky "1/f" excess. In this case, the detector is the driving force of the method of observation, so that laboratory work is needed to establish the best observing techniques.

## 5. Conclusions

We have shown that, because of the limited flat field precision, some modulation was necessary during 10μm observations from ground even when using array detectors. But, if the detector noise is not dominant, the frequency at which such a modulation have to be done should be much lower than for photometers by using the following method to treat the data:
- difference between the two sets of images (on and off source),
- flat field,
- difference between pixels.

Of course, these theoretical considerations have to be confronted with the sky. And if we want to know whether the modulation frequency can be low enough to modulate only with the telescope, we need to know the sky fluctuations, the flat field precision and so on during in situ measurements.

We plan to do soon such measurements at the CFH telescope by the means of the 10μm camera built by the Service d'Astrophysique and the Observatoire de Lyon for INSU (Institut National des Sciences de l'Univers). The main originality of this camera is that it is equipped with a detector array specially optimized for ground-based observations; these detectors have a large storage capacity, which can absorb the huge photon background in the broad band N filter (Lagage et al., 1989). The other main characteristics are two pixel fields of view (0.5 arsec and 0.8 arcsec), a set of broadband filters covering the L, M, N atmospheric windows, as well as a CVF with a resolution of about 50.

Note that the results discussed here on the basis of ground-based observations are also valid for space observations. In this case, the background is the zodiacal light. Of course, this background is much weaker than the one received on ground (five orders of magnitude, in the case of ISO), so that even for hours of integration the signal to background ratio will never overcome $10^{-4}$. In these conditions, if a very good flat field precision can be achieved the modulation is not needed. Otherwise, some modulation is needed. The temporal fluctuations of the zodiacal light should certainly be very slow, so could be the modulation frequency, at least from the point of view of the sky. But the detector stability will certainly be the dominant factor, especially when we know that the detectors will undergone the cosmic radiation. Extensive laboratory work is needed to assess the question.

## Acknowledgment

It is a pleasure to express our great thanks to R. Papoular for valuable discussions on various aspects of this paper.

## References

Arens, J.F. et al.: 1988, Proceedings of the Int. Workshop "Ground-Based Astronomical Observations with Infrared Array Detectors", University of Hawaii Hilo, 24-26 March 1987, C.G. Wynn-Williams and E.E. Becklin editors.

Cesarsky, C., Sibille, F. and Vigroux, L.: 1989, Proceedings of the International Conf. on "New Technologies for Astronomy", Paris, 25-26 April 1989, J.P. Swings editor, published by SPIE, Vol. 1130, 202.

Gezari, D. et al.: 1989, Proceedings of the 3rd Ames Detector Workshop, in press.

Lagage,P.O.: 1989, Proceedings of the International Conf. on "New Technologies for Astronomy", Paris, 25-26 April 1989, J.P. Swings editor, published by SPIE, Vol. 1130, 169.

Neugebauer, G. et al.: 1984, Astrophys. J., 278, L1-L6.

Papoular, R.: 1983, Astron. Astrophys., 117, 46

Sanders, D.B. et al.: 1988, Astrophys. J., 325, 74.

Telesco, C.M. et al.: 1988, Nature, 335, 51.

# NEAR-INFRARED SPECTRA OF THE EXTENDED YOUNG OBJECTS GL 2591 AND NGC 7538/IRS9

M. Fernandez[1], C. Eiroa[1], K.W. Hodapp[2]
[1]*Observatorio Astronómico de Madrid-IGN, Spain*
[2]*Institute for Astronomy, University of Hawaii, Honolulu*

The 3.08 $\mu$m ice absorption feature is usually found in the line of sight of very young objects deeply embedded in dense molecular clouds. Attempts to fit the feature by means of different kinds of water ices, however, have been unsuccessful, and additional absorptions at both short and long wavelength wings are in fact observed. Some kinds of hydrocarbons seem to be a plausible explanation for the long wavelength wing. The extra absorption at wavelengths shorter than 3.08 $\mu$m has been attributed to ammonia ices and, as an alternative explanation, to scattering by large water ice particles. Many protostars are associated with infrared reflection nebulae. This could be considered an indirect support for the scattering origin of the short wavelength wing of the ice feature, although it does not exclude the ammonia ice explanation at all.

In this context we have observed GL 2591 and NGC 7538/IRS9, two well-known propostars associated with infrared reflection nebulae. In GL 2591 we observed at the position of the central source and at two locations in the associated nebula, whereas the observations in NGC 7538/IRS9 were carried out at the position of the propostar and at a position of the nebular ridge. The data were collected on Mauna Kea Observatory using the NASA 3.0 m telescope, equipped with the common-user spectrometer CGAS. The observed spectral range was 2.4–3.8 $\mu$m.

A summary of our results is following: **GL 2591**. The feature is detected at the three observed positions. There are changes in the wavelength of the maximum absorption with the position, likely indicating the presence of ices at different temperatures. The ice feature at the nebular positions shows an extra absorption with respect to the feature at the central position, which could indicate the influence of scattering. Important is the detection of the 3.3 $\mu$m unidentified emission feature superimposed on the ice absorption. To our knowledge, this is the first detection of that feature in GL 2591. **NGC 7538/IRS9**: we observed a deep ice feature at the position of the propostar, which is essentially similar to previous published ice spectra. The position observed in the reflection nebula shows significant wings at both short and long wavelengths. Scattering might be playing an important role in this nebular spectrum.

A detailed analysis and interpretation of these results will be published elsewhere.

*E. Bussoletti and A. A. Vittone (eds.), Dusty Objects in the Universe*, 273.
© 1990 *Kluwer Academic Publishers.*

# THE 3.3 μm FEATURE, H₂, AND IONIZED GAS IN THE ORION BAR

K. Sellgren[1], A.T. Tokunaga[1], Y. Nakada[2]
[1]*Institute for Astronomy, University of Hawaii, U.S.A.*
[2]*Department of Astronomy, University of Tokyo, Japan*

We present observations of the spatial distribution of the 3.3 μm feature, $H_2$ emission, and ionized gas in the Orion Bar. The Orion Bar is an ionization front $2'$ SE of $\theta^1$C Ori, the star which ionizes the Orion Nebula. We also present results on the 3.3 μm feature width in the Orion Bar. Our results are presented in more detail in Sellgren, Tokunaga, and Nakada (1990).

We have used the IRTF to measure the spatial distribution of the 3.3 μm feature, Br $\alpha$, P $\alpha$, and the Q-branch of $H_2$, along a line perpendicular to the ionization front, and passing through Orion Position 4. The ionized gas traced by Br $\alpha$ and P $\alpha$ peaks $10''$ NW of Position 4, at the ionization front, while the $H_2$ brightness peaks $15''$ SE of Position 4. The 3.3 μm brightness drops dramatically inside the H II region. The 3.3 μm brightness also decreases at the $H_2$ peak. The 3.3 μm spatial distribution appears to be due to destruction of the emitting material within the H II region and extinction of the exciting radiation between the edge of the H II region and the $H_2$ peak, resulting in a maximum between the ionization front and the $H_2$ peak.

We have also used the IRTF's Cooled Grating Array Spectrometer at high spectral resolution $(\lambda/\Delta\lambda = 1400)$, to observe the shape of the 3.3 μm feature in Orion in regions of varying UV intensity: at Position 4, inside the H II region, and at the $H_2$ peak. Previous measurements of the 3.3 μm feature profile (Nagata et al., 1988; Tokunaga et al., 1988; Geballe et al., 1989) find that the central wavelength is constant but that the wavelength of the half power point on the blue side of the 3.3 μm feature is variable. We observe that the blue half power point in Orion is constant at all positions, indicating a constant feature width. Geballe et al. have suggested that variations in 3.3 μm feature width are related to the intensity of the UV field. We suggest instead that the 3.3 μm feature is initially formed with a composition showing a narrow feature, and then is rapidly converted to a composition showing a broad feature.

## References

Geballe, T. R., Tielens, A.G.G.M., Allamandola, L.J., Moorhouse, A., Brand, P. W. J. L.: 1989, Astrophys. J., **196**, 179.

*E. Bussoletti and A. A. Vittone (eds.), Dusty Objects in the Universe, 274–275.*
© 1990 *Kluwer Academic Publishers.*

Nagata, T., Tokunaga, A.T., Sellgren, K., Smith, R. G., Onaka, T., Nakada, Y., Sakata, A.: 1988, Astrophys. J., **326**, 157.

Sellgren, K., Tokunaga, A.T., Nakada, Y.: 1990, Astrophys. J., in press.

Tokunaga, A.T., Nagata, T., Sellgren, K., Smith, R. G., Onaka, T., Nakada, Y., Sakata, A., Wada, S.: 1988, Astrophys. J., **328**, 709.

# THE EXTENDED 3.3 μm EMISSION FEATURE IN THE ORION'S CLOUD

N. Sales[1], M. Giard[2], E. Caux[1], J.M. Lamarre[2], F. Pajot[3], G. Serra[1]
[1]CESR-CNRS/UPS, Toulouse, France
[2]LPSP/IAS-CNRS, Verrières-le-Buisson, Orsay, France

AROME, a balloon borne experiment has been built in order to detect the 3.3 μm emission feature in extended sources (description of AROME is given in M. Giard et al., these proceedings). Particular areas outside the galactic disk have been mapped. The figure presents Orion survey map of 3.3 μm emission in AROME narrow band, and points out an extended emission around Ori A and Ori B. The AROME field of view (0.5°) allows to measure the extended emission around the excited regions.

The 3.3 μm feature's flux integrated over a $5' \times 5'$ area (Sellgren 1981) is eight times lower than the flux measured in the beam of AROME. This shows that the 3.3 μm emission feature is spread out in the interstellar medium and not only located close to the exciting stars. To compare our results to those of Sellgren, we have computed the feature to continuum ratio (R), admitting that the feature bandwidth is smaller than 0.05 μm (see table). One notes that R increrases when we move away from the exciting stars.

| Name ⟶ | Ori A | Ori B |
|---|---|---|
| [1]Feature flux $(10^{-12}$ W.m$^{-2})$ <br> - in 0.5° (AROME) <br> - in 5'x5' (Sellgren 1981) | 48.3 ± 9 <br> 6 ± 2 | 19.2 ± 3 <br> - |
| [2]Feature to continuum ratio (R) <br> - in 0.5° <br> - in 5'x5' | 5.8 <br> 1.5 | 4.3 <br> - |

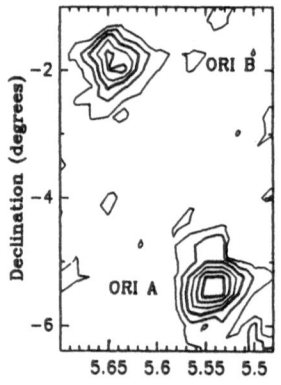

[1]Feature flux $= F(3.3) = (\lambda \cdot I\lambda(NB) - \lambda \cdot I\lambda(WB)) \cdot \Delta\lambda/\lambda$, where $I\lambda(NB)$ and $I\lambda(WB)$ are the spectral densities of the fluxes measured respectively in the narrow and wide bands of AROME, $\lambda = 3.3$ μm and $\Delta\lambda = 0.17$ μm.

[2]Ratio of the flux densities for a $\delta\lambda = 0.05$ μm bandwidth (same units as Sellgren). $R = (F(3.3)/\lambda \cdot I\lambda(WB)) \cdot \lambda/\delta\lambda$.

E. Bussoletti and A. A. Vittone (eds.), Dusty Objects in the Universe, 276.

# EXTRAGALACTIC EXTINCTION:
## THE OPTICAL DATA FOR THE EARLY-TYPE GALAXY NGC 2534.

Agatino RIFATTO
Dipartimento di Astronomia
Università di Padova
Vicolo dell'Osservatorio, 5
I–35122 PADOVA

The dust lane of the early–type galaxy NGC 2534 is used in order to obtain extinction data for the extragalactic dust. In fact, since there are many morphological and dynamical evidences that dust lanes are acquired from outside (e.g.: Bertola F., Galletta G. & Zeilinger W.W., 1985, Ap.J.,**292**, L51; Bertola F. & Bettoni D., 1988, Ap.J., **329**, 102; Bertola F., Galletta G., Kotanyi C. & Zeilinger W.W., 1988, M.N.R.A.S., **234**, 733; Bertola F., Buson L.M. & Zeilinger W.W., 1988, Nature, **335**, 705), these structures become the best laboratory to test the physical properties of the extragalactic dust.

The results show an extinction parameter $R_V$=1.4 which is much smaller than the classical galactic value $R_V$=3.1 adopted in our Galaxy (Savage B.D. & Mathis J.S., 1979, Ann. Rev. Astron. Astr., **17**, 73), implying that the dust sizes are smaller than in our own Galaxy, producing a larger reddening. Fig. 1 shows the observed optical extinction data compared with the extinction law computed using the CCM relation (Cardelli J.A., Clayton G.C. & Mathis J.S., 1988, Ap.J., **329**, L33) for $R_V$=1.4, while all our results are listed in Tab. 1.

The fact that the extragalactic extinction parameter $R_V$ is smaller than in our Galaxy also influences the calibration of the absolute magnitude for the $SN_e$ and so it has deep implications concerning the extragalactic scale distances as emphasized by Tammann (1988, ESO preprint n° 617), Joever (1982, Astrofizika, **18**, 574) and Capaccioli et al. (1989, Ap.J., submitted) who suggest a value for the extinction parameter $R_B=A_B/E_{B-V}$ as small as $\sim 2$ rather than the classical value $R_B \simeq 4$. Then this could be another piece of evidence that the extragalactic properties of the dust are completely different than in our own Galaxy.

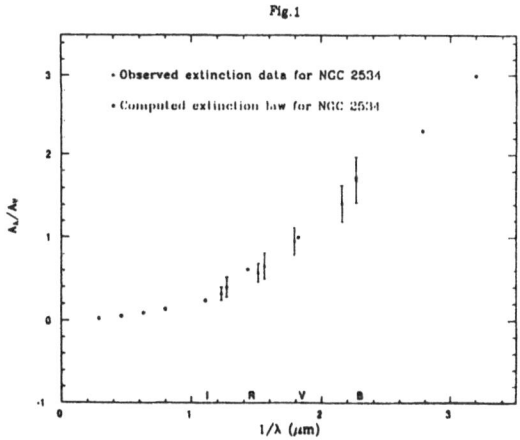

Fig.1

Tab. 1

| NGC 2534 | | | |
|---|---|---|---|
| $\lambda$ (Å) | $1/\lambda$ ($\mu m$) | $A_\lambda$ (mag.) | $A_\lambda/A_V^*$ |
| 4400 | 2.27 | 1.51±.09 | 1.70±.27 |
| 4630 | 2.16 | 1.26±.07 | 1.42±.22 |
| 5500 | 1.82 | .89±.09 | 1.00 |
| 5590 | 1.79 | .85±.06 | .95±.16 |
| 6400 | 1.56 | .58±.08 | .65±.16 |
| 6625 | 1.51 | .51±.05 | .57±.11 |
| 7900 | 1.27 | .36±.07 | .40±.09 |
| 8150 | 1.23 | .29±.04 | .33±.08 |

* V=5500 Å.

*E. Bussoletti and A. A. Vittone (eds.), Dusty Objects in the Universe*, 277.

# POSSIBLE EXISTENCE OF DUST IN NEARBY GALAXY GROUP

M. Girardi[1], G. Giuricin[2,3], F. Mardirossian[2,3], M. Mezzetti[2,3]
[1]*International School for Advanced Studies, Trieste, Italy*
[2]*Department of Astronomy, University of Trieste, Trieste, Italy*
[3]*C.I.R.A.C., Trieste, Italy*

ABSTRACT. The asymmetric distribution of galaxy redshifts in some nearby-group samples is explained as due to the existence of intragroup dust. The existence of a correlation between color excess of galaxies and radial velocity supports the hypothesis made.

An excess of higher redshift galaxies, relative to the brightest member in loose groups, was discovered by Arp (1970, and later confirmed by several authors). We find this effect to be present also in nearby small groups identified by Geller and Huchra (1983) in the Center for Astrophysics Survey.

We propose to explain this effect as due to the presence of diffuse, intragroup dust, together with an unrelaxed evolutionary stage of nearby galaxy groups. The possible existence of intergalactic dust has been discussed by many authors (see, e.g., Shaver, 1987). The not yet virialized status of galaxy groups, belonging to the Local Supercluster, has been discussed by Giuricin et al. (1988), who found that most of them are still in the collapse phase following maximum expansion.

In a collapsing group with intergalactic dust, galaxies on the nearer side of the system are redshifted with respect to the centre of it and less obscured than those, blueshifted, which are on the farther side. For these latter galaxies, in a magnitude limited catalogue, the apparent cut-off magnitude corresponds to a higher intrinsic luminosity, so that their observed number is smaller than that of the redshifted members.

To model a group, we have assumed spherical symmetry, a King density profile for galaxies and dust (with different core radii), a perturbed Hubble flow to account for infall, and a Schechter-type luminosity function. The center of mass of the system is assumed to coincide with the brightest galaxy in the group. The asymmetry $N^+/N^-$ in the redshift distribution (defined as the radio between the number of galaxies with positive velocity excess relatively to the first-rank member, and the number of those with negative velocity excess) produced by the model depends on the amount of dust, and can be about 2-3 for groups at 10 Mpc ($H_0 = 100$ km/s Mpc), and higher for more distant groups, in agreement with the observed values for nearby GH groups.

278

*E. Bussoletti and A. A. Vittone (eds.), Dusty Objects in the Universe, 278–279.*
© 1990 *Kluwer Academic Publishers.*

We have inspected also group samples drawn from the catalogues made by Vennik (1984) and Tully (1987), which are both not complete in magnitude. For the latter sample, when only galaxies with known magnitude are considered, a significant asymmetry is also present, with $N^+ = 117$ and $N^- = 71$. Owing to the incompleteness in magnitude of Tully groups, this asymmetry has to be considered as a lower limit of the true value.

Byrd and Valtonen (1985) suggested that a redshift asymmetry could be produced by interlopers following Hubble flow. However, the asymmetry $N^+/N^-$ which can be produced in this way is at maximum about 2, in the case of full contamination, and rapidly decreases when the fraction of interlopers decreases (if 50% of group members are interlopers, the asymmetry is reduced to 1.2).

If the observed asymmetry is produced by dust, one should except to detect also differential extinction. Galaxies on the nearer side of the group, redshifted with respect to its centre, should present a color excess smaller than that of galaxies on the farther, blueshifted, side. Such a correlation, between color excess of galaxies and the velocity relative to the brightest member, is found to be highly significant for both $B - V$ and $U - B$ colors. Similar results have been obtained for nearby groups selected by Vennik (1984) and by Tully (1987), particularly for those having longer crossing times. The existence of this correlation is also at variance with the explanation proposed by Byrd and Valtonen (1985).

The mean optical depth produced by dust in groups turns out to be on the order of 0.5, the mean density of dust inside groups is on the order of $10^{-30}$ g cm$^{-3}$, with a corresponding mean number density of gas particles on the order of $10^{-4}$ cm$^{-3}$. These values can be reasonably obtained if one considers the ejection of gas, with about solar abundances, by galaxies (first suggested by Larson and Dinerstein, 1975), and if a fraction of metals condensed in grains similar to that of the interstellar space is assumed.

If the presence of dust is not typical of the Local Supercluster, but has a cosmological relevance, the deduced dust density, taking into account the filling factor of groups in the large scale structure of the universe, corresponds to $\Omega_{DUST} \approx 10^{-5} - 10^{-4}$. If one assumes that dust were already present at high redshift, our results are in agreement with the computations of Ostriker and Heisler (1984) and Heisler and Ostriker (1988), who ascribed to dust the cut-off in the number density of quasars at $z \approx 3$. In our case, however, the obscuring objects are not single galaxies, but galaxy groups.

# References

Arp, H.: 1970, Nature, **225**, 103.

Byrd, G.G., Valtonen, M.J.: 1985, Astrophys. J., **289**, 535.

Geller, M.J., Huchra, J.P.: 1983, Astrophys. J. Suppl., **51**, 61.

Giuricin, G., Gondolo, P., Mardirossian, F., Mezzetti, M., Ramella, M.: 1988, Astron. Astrophys., **199**, 85.

Heisler, J., Ostriker, J.: 1988, Astrophys. J., **332**, 543.

Larson, R.B., Dinerstein, H.L.: 1975, P.A.S.P., **87**, 911.

Ostriker, J.P., Heisler, J.: 1984, Astrophys. J., **278**, 1.

Shaver, P.A.: 1987, in *"High Redshift and Primordial Galaxies"*, Third IAP Astrophysics Meet., Paris.

Tully, R.B.: 1987, Astrophys. J., **321**, 280.

Vennik, J.: 1984, Tartu Astrofüüs. Obs. Teated no. **73**, 3.

# Subject index